SPECIAL FUNCTIONS AND THEIR APPLICATIONS

N. N. LEBEDEV

Physico-Technical Institute
Academy of Sciences, U.S.S.R.

Revised English Edition
Translated and Edited by

Richard A. Silverman

DOVER PUBLICATIONS, INC.

New York

Published in Canada by General Publishing Company, Ltd., 30 Lesmill Road, Don Mills, Toronto, Ontario.
Published in the United Kingdom by Constable and Company, Ltd., 10 Orange Street, London WC 2.

This Dover edition, first published in 1972, is an unabridged and corrected republication of the work originally published by Prentice-Hall, Inc., in 1965.

International Standard Book Number: 0-486-60624-4
Library of Congress Catalog Card Number: 72-86228

Manufactured in the United States of America
Dover Publications, Inc.
180 Varick Street
New York, N.Y. 10014

AUTHOR'S PREFACE

This book deals with a branch of mathematics of utmost importance to scientists and engineers concerned with actual mathematical calculations. Here the reader will find a systematic treatment of the basic theory of the more important special functions, as well as applications of this theory to specific problems of physics and engineering. In the choice of topics, I have been guided by the goal of giving a sufficiently detailed exposition of those problems which are of greatest practical interest. This has naturally led to a certain curtailment of the purely theoretical part of the book. In this regard, it should be noted that various useful properties of the special functions which do not appear in the text proper, will be found in the problems at the end of the appropriate chapters.

The book presupposes that the reader is familiar with the elements of the theory of functions of a complex variable, without which one cannot go very far in the study of special functions. However, in order to make the book more accessible to non-mathematicians, I have made a serious attempt to keep to a minimum the required background in complex variable theory. In particular, this has compelled me to depart from the order of presentation found in other treatments of the subject, where the special functions are first defined by certain convenient representations in terms of contour integrals.

The usual elementary course in complex variable theory is adequate for an understanding of most of the material presented here. It is also desirable, but not necessary, to know something about the analytic theory of linear differential equations. I occasionally draw upon other branches of mathematics and physics, but only in connection with certain specific examples, so that lack of familiarity with the relevant information is no obstacle to reading the book.

It is assumed that the reader already appreciates, from his own experience, the need for using special functions. Therefore, I have not made a special point of motivating the introduction of various functions. By the same token, I have always sought the simplest way of defining the special functions and deriving their properties, without concern for historical or other considerations.

The arrangement of the material in the separate chapters is dictated by the desire to make the different parts of the book independent of each other, at least to a certain extent, so that one can study the simplest classes of functions without becoming involved with functions of a more general type. For example, I have separated the theory of the Legendre polynomials and Bessel functions of integral order from the general theory of spherical harmonics and cylinder functions, and I have also constructed the theory of spherical harmonics without recourse to the properties of the hypergeometric function.

The applications of the theory were selected with the aim of illustrating the different ways in which special functions are used in problems of physics and engineering. No attempt has been made to give a detailed treatment of the corresponding branches of mathematical physics. In this regard, most space has been devoted to the application of cylinder functions, and particularly, of spherical harmonics.

In preparing the present second edition of the book I have revised an earlier edition in various ways: Chapter 4 now contains a new version of the theorem on expansions in series of Hermite polynomials, which extends the previous theorem to a larger class of functions. I have also increased the number of examples illustrating the technique of expanding functions in series of Hermite and Laguerre polynomials. In Chapter 5 there is a new section dealing with the theory of Airy functions, which are often encountered in mathematical physics and play an important role in deriving asymptotic representations of various special functions. Chapter 9, devoted to the theory of the hypergeometric function, has been completely revised, and I hope that in its present form, this chapter will be useful to theoretical physicists and others concerned with the application of the hypergeometric function, thereby partially filling a gap in the literature on the subject. I have added many new problems, which serve both as exercise material and as a source of supplementary information not to be found in the text itself. At the same time, I have removed a few problems of no particular interest. Finally, the references have been brought up-to-date.

I would like to take this opportunity to thank I. P. Skalskaya for help in preparing the present edition of my book.

N.N.L.

TRANSLATOR'S PREFACE

For the most part, this edition adheres closely to the revised Russian edition (Moscow, 1963). However, as always with the volumes of this series, I have not hesitated to introduce whatever improvements occurred to me in the course of working through the book. In the present case, two departures from the original text merit special mention:

1. The Bibliography and the references cited in the footnotes have been slanted towards books available in English or the West European languages.

2. Chapters 6 and 8 have been equipped with problems, most of them taken from the excellent collection by Lebedev, Skalskaya and Uflyand (Moscow, 1955).

Finally, it was deemed impractical to build in sufficiently detailed references to numerical tables of the special functions. Here all roads eventually lead to a consultation of the exhaustive bibliography compiled by Fletcher, Miller, Rosenhead and Comrie, or its Russian counterpart by Lebedev and Fedorova.

R.A.S.

CONTENTS

1

THE GAMMA FUNCTION

1.1. Definition of the Gamma Function

One of the simplest and most important special functions is the *gamma function*, knowledge of whose properties is a prerequisite for the study of many other special functions, notably the cylinder functions and the hypergeometric function. Since the gamma function is usually studied in courses on complex variable theory, and even in advanced calculus,[1] the treatment given here will be deliberately brief.

The gamma function is defined by the formula

$$\Gamma(z) = \int_0^\infty e^{-t} t^{z-1}\, dt, \qquad \text{Re}\, z > 0, \tag{1.1.1}$$

whenever the complex variable z has a positive real part Re z. We can write (1.1.1) as a sum of two integrals, i.e.,

$$\Gamma(z) = \int_0^1 e^{-t} t^{z-1}\, dt + \int_1^\infty e^{-t} t^{z-1}\, dt, \tag{1.1.2}$$

where it can easily be shown[2] that the first integral defines a function $P(z)$

---

[1] See D. V. Widder, *Advanced Calculus*, second edition, Prentice-Hall, Inc., Englewood Cliffs, N.J. (1961), Chap. 11.

[2] See E. C. Titchmarsh, *The Theory of Functions*, second edition, Oxford University Press, London (1939), p. 100, noting that the integrand $e^{-t}t^{z-1}$ is analytic in z and continuous in z and t for Re $z > 0$, $0 < t < \infty$, while the first integral is uniformly convergent for Re $z \geqslant \delta > 0$ and the second integral is uniformly convergent for Re $z \leqslant A < \infty$, since then

$$\left|\int_0^1 e^{-t} t^{z-1}\, dt\right| \leqslant \int_0^1 e^{-t} t^{\delta-1}\, dt < \infty, \qquad \left|\int_1^\infty e^{-t} t^{z-1}\, dt\right| \leqslant \int_1^\infty e^{-t} t^{A-1}\, dt < \infty.$$

which is analytic in the half-plane $\operatorname{Re} z > 0$, while the second integral defines an entire function. It follows that the function $\Gamma(z) = P(z) + Q(z)$ is analytic in the half-plane $\operatorname{Re} z > 0$.

The values of $\Gamma(z)$ in the rest of the complex plane can be found by analytic continuation of the function defined by (1.1.1). First we replace the exponential in the integral for $P(z)$ by its power series expansion, and then we integrate term by term, obtaining

$$P(z) = \int_0^1 t^{z-1}\,dt \sum_{k=0}^{\infty} \frac{(-1)^k}{k!} t^k = \sum_{k=0}^{\infty} \frac{(-1)^k}{k!} \int_0^1 t^{k+z-1}\,dt$$
$$= \sum_{k=0}^{\infty} \frac{(-1)^k}{k!} \frac{1}{z+k}, \tag{1.1.3}$$

where it is permissible to reverse the order of integration and summation since[3]

$$\int_0^1 |t^{z-1}|\,dt \sum_{k=0}^{\infty} \left|\frac{(-1)^k}{k!} t^k\right| = \int_0^1 t^{x-1}\,dt \sum_{k=0}^{\infty} \frac{t^k}{k!} = \int_0^1 e^t t^{x-1}\,dt < \infty$$

(the last integral converges for $x = \operatorname{Re} z > 0$). The terms of the series (1.1.3) are analytic functions of z, if $z \neq 0, -1, -2, \ldots$ Moreover, in the region[4]

$$|z + k| \geqslant \delta > 0, \qquad k = 0, 1, 2, \ldots,$$

(1.1.3) is majorized by the convergent series

$$\sum_{k=0}^{\infty} \frac{1}{k!\delta},$$

and hence is uniformly convergent in this region. Using Weierstrass' theorem[5] and the arbitrariness of δ, we conclude that the sum of the series (1.1.3) is a meromorphic function with simple poles at the points $z = 0, -1, -2, \ldots$ For $\operatorname{Re} z > 0$ this function coincides with the integral $P(z)$, and hence is the analytic continuation of $P(z)$.

The function $\Gamma(z)$ differs from $P(z)$ by the term $Q(z)$, which, as just shown, is an entire function. Therefore $\Gamma(z)$ is a meromorphic function of the complex variable z, with simple poles at the points $z = 0, -1, -2, \ldots$ An

---

[3] E. C. Titchmarsh, *op. cit.*, p. 45.

[4] By a *region* we mean an open connected point set (of two or more dimensions) together with some, all, or possibly none of its boundary points. In the latter case, we often speak of an *open region* or *domain*, in the former case, of a *closed region* or *closed domain.*

[5] See A. I. Markushevich, *Theory of Functions of a Complex Variable, Vol. I* (translated by R. A. Silverman), Prentice-Hall, Inc., Englewood Cliffs, N.J. (1965), Theorem 15.6, p. 326.

analytic expression for $\Gamma(z)$, suitable for defining $\Gamma(z)$ in the whole complex plane, is given by

$$\Gamma(z) = \sum_{k=0}^{\infty} \frac{(-1)^k}{k!} \frac{1}{z+k} + \int_1^\infty e^{-t} t^{z-1}\, dt, \qquad z \neq 0, -1, -2, \ldots \quad (1.1.4)$$

It follows from (1.1.4) that $\Gamma(z)$ has the representation

$$\Gamma(z) = \frac{(-1)^n}{n!} \frac{1}{z+n} + \Omega(z+n) \qquad (1.1.5)$$

in a neighborhood of the pole $z = -n$ ($n = 0, 1, 2, \ldots$), with regular part $\Omega(z + n)$.

1.2. Some Relations Satisfied by the Gamma Function

We now prove three basic relations satisfied by the gamma function:

$$\Gamma(z+1) = z\Gamma(z), \qquad (1.2.1)$$

$$\Gamma(z)\Gamma(1-z) = \frac{\pi}{\sin \pi z}, \qquad (1.2.2)$$

$$2^{2z-1}\Gamma(z)\Gamma(z+\tfrac{1}{2}) = \sqrt{\pi}\,\Gamma(2z). \qquad (1.2.3)$$

These formulas play an important role in various transformations and calculations involving $\Gamma(z)$.

To prove (1.2.1), we assume that $\operatorname{Re} z > 0$ and use the integral representation (1.1.1). An integration by parts gives

$$\Gamma(z+1) = \int_0^\infty e^{-t} t^z\, dt = -e^{-t} t^z \Big|_0^\infty + z \int_0^\infty e^{-t} t^{z-1}\, dt = z\Gamma(z)$$

The validity of this result for arbitrary complex $z \neq 0, -1, -2, \ldots$ is an immediate consequence of the principle of analytic continuation,[6] since both sides of the formula are analytic everywhere except at the points $z = 0, -1, -2, \ldots$

To derive (1.2.2), we temporarily assume that $0 < \operatorname{Re} z < 1$ and again use (1.1.1), obtaining

$$\Gamma(z)\Gamma(1-z) = \int_0^\infty \int_0^\infty e^{-(s+t)} s^{-z} t^{z-1}\, ds\, dt.$$

[6] According to this principle, which we will use repeatedly, if $f(z)$ and $\varphi(z)$ are analytic in a domain D and if $f(z) = \varphi(z)$ for all z in a smaller domain D^* contained in D, then $f(z) = \varphi(z)$ for all z in D. The same is true if $f(z) = \varphi(z)$ for all z in any set of points of D with a limit point in D, say, a line segment. See A. I. Markushevich, *op. cit.*, Theorem 17.1, p. 369.

Introducing the new variables

$$u = s + t, \qquad v = \frac{t}{s},$$

we find that[7]

$$\Gamma(z)\Gamma(1 - z) = \int_0^\infty \int_0^\infty e^{-u} v^{z-1} \frac{du\, dv}{1 + v} = \int_0^\infty \frac{v^{z-1}}{1 + v}\, dv = \frac{\pi}{\sin \pi z}.$$

Using the principle of analytic continuation, we see that this formula remains valid everywhere in the complex plane except at the points $z = 0, \pm 1, \pm 2, \ldots$

To prove (1.2.3), known as the *duplication formula*, we assume that $\operatorname{Re} z > 0$ and then use (1.1.1) again, obtaining

$$\begin{aligned} 2^{2z-1}\Gamma(z)\Gamma(z + \tfrac{1}{2}) &= \int_0^\infty \int_0^\infty e^{-(s+t)}(2\sqrt{st})^{2z-1} t^{-1/2}\, ds\, dt \\ &= 4 \int_0^\infty \int_0^\infty e^{-(\alpha^2+\beta^2)}(2\alpha\beta)^{2z-1} \alpha\, d\alpha\, d\beta, \end{aligned}$$

where we have introduced new variables $\alpha = \sqrt{s}$, $\beta = \sqrt{t}$. To this formula we add the similar formula obtained by permuting α and β. This gives the more symmetric representation

$$\begin{aligned} 2^{2z-1}\Gamma(z)\Gamma(z + \tfrac{1}{2}) &= 2 \int_0^\infty \int_0^\infty e^{-(\alpha^2+\beta^2)}(2\alpha\beta)^{2z-1}(\alpha + \beta)\, d\alpha\, d\beta \\ &= 4 \iint_\sigma e^{-(\alpha^2+\beta^2)}(2\alpha\beta)^{2z-1}(\alpha + \beta)\, d\alpha\, d\beta, \end{aligned}$$

where the last integral is over the sector $\sigma: 0 \leqslant \alpha < \infty, 0 \leqslant \beta \leqslant \alpha$. Introducing new variables

$$u = \alpha^2 + \beta^2, \qquad v = 2\alpha\beta,$$

we find that

$$\begin{aligned} 2^{2z-1}\Gamma(z)\Gamma(z + \tfrac{1}{2}) &= \int_0^\infty v^{2z-1}\, dv \int_0^\infty \frac{e^{-u}}{\sqrt{u - v}}\, du \\ &= 2\int_0^\infty e^{-v} v^{2z-1}\, dv \int_0^\infty e^{-w^2}\, dw = \sqrt{\pi}\,\Gamma(2z). \end{aligned}$$

As before, this result can be extended to arbitrary complex values $z \neq 0, -\frac{1}{2}, -1, -\frac{3}{2}, \ldots$, by using the principle of analytic continuation.

We now use formula (1.2.1) to calculate $\Gamma(z)$ for some special values of the variable z. Applying (1.2.1) and noting that $\Gamma(1) = 1$, we find by mathematical induction that

$$\Gamma(n + 1) = n!, \qquad n = 0, 1, 2, \ldots \tag{1.2.4}$$

---

[7] For the evaluation of the integral in the last step, see E. C. Titchmarsh, *op. cit.*, p. 105.

Moreover, setting $z = \frac{1}{2}$ in (1.1.1), we obtain

$$\Gamma(\tfrac{1}{2}) = \int_0^\infty e^{-t} t^{-1/2}\, dt = 2 \int_0^\infty e^{-u^2}\, du = \sqrt{\pi}, \tag{1.2.5}$$

and then (1.2.1) implies

$$\Gamma(n + \tfrac{1}{2}) = \frac{1 \cdot 3 \cdot 5 \cdots (2n - 1)}{2^n} \sqrt{\pi}, \qquad n = 1, 2, \ldots \tag{1.2.6}$$

Finally we use (1.2.2) to prove that the function $\Gamma(z)$ has no zeros in the complex plane. First we note that the points $z = n$ $(n = 0, \pm 1, \pm 2, \ldots)$ cannot be zeros of $\Gamma(z)$, since $\Gamma(n) = (n - 1)!$ if $n = 1, 2, \ldots,$ while $\Gamma(n) = \infty$ if $n = 0, -1, -2, \ldots$ The fact that no other value of z can be a zero of $\Gamma(z)$ is an immediate consequence of (1.2.2), since if a nonintegral value of z were a zero of $\Gamma(z)$ it would have to be a pole of $\Gamma(1 - z)$, which is impossible. It follows at once that $[\Gamma(z)]^{-1}$ *is an entire function.*

1.3. The Logarithmic Derivative of the Gamma Function

The theory of the gamma function is intimately related to the theory of another special function, i.e., the logarithmic derivative of $\Gamma(z)$:

$$\psi(z) = \frac{\Gamma'(z)}{\Gamma(z)}. \tag{1.3.1}$$

Since $\Gamma(z)$ is a meromorphic function with no zeros, $\psi(z)$ can have no singular points other than the poles $z = -n$ $(n = 0, 1, 2, \ldots)$ of $\Gamma(z)$. It follows from (1.1.5) that $\psi(z)$ has the representation[8]

$$\psi(z) = -\frac{1}{z + n} + \Omega(z + n) \tag{1.3.2}$$

in a neighborhood of the point $z = -n$, and hence $\psi(z)$, like $\Gamma(z)$, is a meromorphic function with simple poles at the points $z = 0, -1, -2, \ldots$

The function $\psi(z)$ satisfies relations obtained from formulas (1.2.1–3)[9] by taking logarithmic derivatives. In this way, we find that

$$\psi(z + 1) = \frac{1}{z} + \psi(z), \tag{1.3.3}$$

$$\psi(1 - z) - \psi(z) = \pi \cot \pi z, \tag{1.3.4}$$

$$\psi(z) + \psi(z + \tfrac{1}{2}) + 2 \log 2 = 2\psi(2z). \tag{1.3.5}$$

---

[8] Of course, the regular part $\Omega(z + n)$ in (1.3.2) is not the same as in (1.1.5).

[9] By (1.2.1–3) we mean formulas (1.2.1) through (1.2.3). Similarly, (1.2.1, 4, 6) means formulas (1.2.1), (1.2.4) and (1.2.6), etc.

These formulas can be used to calculate $\psi(z)$ for special values of z. For example, writing

$$\psi(1) = \Gamma'(1) = -\gamma, \tag{1.3.6}$$

where $\gamma = 0.57721566\ldots$ is *Euler's constant*, and using (1.3.3), we obtain

$$\psi(n + 1) = -\gamma + \sum_{k=1}^{n} \frac{1}{k}, \qquad n = 1, 2, \ldots \tag{1.3.7}$$

Moreover, substituting $z = \frac{1}{2}$ into (1.3.5), we find that

$$\psi(\tfrac{1}{2}) = -\gamma - 2 \log 2, \tag{1.3.8}$$

and then (1.3.3) gives

$$\psi(n + \tfrac{1}{2}) = -\gamma - 2 \log 2 + 2 \sum_{k=1}^{n} \frac{1}{2k - 1}, \qquad n = 1, 2, \ldots \tag{1.3.9}$$

The function $\psi(z)$ has simple representations in the form of definite integrals involving the variable z as a parameter. To derive these representations, we first note that (1.1.1) implies[10]

$$\Gamma'(z) = \int_0^\infty e^{-t} t^{z-1} \log t \, dt, \qquad \operatorname{Re} z > 0. \tag{1.3.10}$$

If we replace the logarithm in the integrand by its expression in terms of the *Frullani integral*[11]

$$\log t = \int_0^\infty \frac{e^{-x} - e^{-xt}}{x} \, dx, \qquad \operatorname{Re} t > 0, \tag{1.3.11}$$

we find that[12]

$$\begin{aligned} \Gamma'(z) &= \int_0^\infty \frac{dx}{x} \int_0^\infty (e^{-x} - e^{-xt}) e^{-t} t^{z-1} \, dt \\ &= \int_0^\infty \frac{dx}{x} \left[e^{-x}\Gamma(z) - \int_0^\infty e^{-t(x+1)} t^{z-1} \, dt \right]. \end{aligned}$$

Introducing the new variable of integration $u = t(x + 1)$, we find that the integral in brackets equals $(x + 1)^{-z}\Gamma(z)$. This leads to the following integral representation of $\psi(z)$:

$$\psi(z) = \int_0^\infty \left[e^{-x} - \frac{1}{(x + 1)^z} \right] \frac{dx}{x}, \qquad \operatorname{Re} z > 0. \tag{1.3.12}$$

---

[10] To justify differentiating behind the integral sign, see E. C. Titchmarsh, *op. cit.*, pp. 99–100.

[11] See H. Jeffreys and B. S. Jeffreys, *Methods of Mathematical Physics*, third edition, Cambridge University Press, London (1956), p. 406, and D. V. Widder, *op. cit.*, p. 357.

[12] Here, as elsewhere in this chapter, we omit detailed justification of the reversal of order of integration. An appropriate argument can always be supplied, usually by proving the absolute convergence of the double integral and then using *Fubini's theorem.* See H. Kestelman, *Modern Theories of Integration*, second revised edition, Dover Publications, Inc., New York (1960), Chap. 8, esp. Theorems 279 and 280.

To obtain another integral representation of $\psi(z)$, we write (1.3.12) in the form

$$\psi(z) = \lim_{\delta\to 0}\int_\delta^\infty \left[e^{-x} - \frac{1}{(x+1)^z}\right]\frac{dx}{x} = \lim_{\delta\to 0}\left[\int_\delta^\infty \frac{e^{-x}}{x}\,dx - \int_\delta^\infty \frac{dx}{(x+1)^z x}\right],$$

and change the variable of integration in the second integral, by setting $x + 1 = e^t$. This gives

$$\psi(z) = \lim_{\delta\to 0}\left[\int_\delta^\infty \frac{e^{-t}}{t}\,dt - \int_{\log(1+\delta)}^\infty \frac{e^{-tz}}{1-e^{-t}}\,dt\right]$$

$$= \lim_{\delta\to 0}\left[\int_{\log(1+\delta)}^\infty \left(\frac{e^{-t}}{t} - \frac{e^{-tz}}{1-e^{-t}}\right)dt - \int_{\log(1+\delta)}^{\delta} \frac{e^{-t}}{t}\,dt\right],$$

and therefore, since the second integral approaches zero as $\delta \to 0$,

$$\psi(z) = \int_0^\infty \left(\frac{e^{-t}}{t} - \frac{e^{-tz}}{1-e^{-t}}\right)dt, \qquad \operatorname{Re} z > 0. \tag{1.3.13}$$

Setting $z = 1$ and subtracting the result from (1.3.13), we find that

$$\psi(z) = -\gamma + \int_0^\infty \frac{e^{-t} - e^{-tz}}{1-e^{-t}}\,dt, \qquad \operatorname{Re} z > 0, \tag{1.3.14}$$

or

$$\psi(z) = -\gamma + \int_0^1 \frac{1-x^{z-1}}{1-x}\,dx, \qquad \operatorname{Re} z > 0, \tag{1.3.15}$$

where we have introduced the variable of integration $x = e^{-t}$.

From formula (1.3.15) we can deduce an important representation of $\psi(z)$ as an analytic expression valid for all $z \neq 0, -1, -2, \ldots$, i.e., in the whole domain of definition of $\psi(z)$. To obtain this representation, we substitute the power series expansion

$$(1-x)^{-1} = 1 + x + x^2 + \cdots + x^n + \cdots, \qquad 0 \leqslant x < 1$$

into (1.3.15) and integrate term by term (this operation is easily justified). The result is

$$\psi(z) = -\gamma + \sum_{n=0}^\infty \left(\frac{1}{n+1} - \frac{1}{n+z}\right). \tag{1.3.16}$$

The series (1.3.16), whose terms are analytic functions for $z \neq 0, -1, -2, \ldots$, is uniformly convergent in the region defined by the inequalities

$$|z+n| \geqslant \delta > 0, \quad n = 0, 1, 2, \ldots \quad \text{and} \quad |z| < a,$$

since

$$\left|\frac{1}{n+1} - \frac{1}{n+z}\right| < \frac{a+1}{(n+1)(n-a)}$$

for $n \geqslant N > a$, and the series

$$\sum_{n=N}^{\infty} \frac{a+1}{(n+1)(n-a)}$$

converges. Therefore, since δ is arbitrarily small and a arbitrarily large, both sides of (1.3.16) are analytic functions except at the poles $z = 0, -1, -2, \ldots$, and hence, according to the principle of analytic continuation, the original restriction $\operatorname{Re} z > 0$ used to prove this formula can be dropped. If we replace z by $z + 1$ in (1.3.16), integrate the resulting series between the limits 0 and z, and then take exponentials of both sides, we find the following infinite product representation of the gamma function:

$$\frac{1}{\Gamma(z+1)} = e^{\gamma z} \prod_{n=1}^{\infty} e^{-z/n}\left(1 + \frac{z}{n}\right). \tag{1.3.17}$$

This formula can be made the starting point for the theory of the gamma function, instead of the integral representation (1.1.1).

Finally we derive some formulas for Euler's constant γ. Setting $z = 1$ in (1.3.12–13), we obtain

$$\gamma = -\psi(1) = \int_0^\infty \left(\frac{1}{1+x} - e^{-x}\right)\frac{dx}{x} = \int_0^\infty \left(\frac{1}{1-e^{-t}} - \frac{1}{t}\right)e^{-t}\,dt. \tag{1.3.18}$$

Moreover, (1.3.10) implies

$$\gamma = -\int_0^\infty e^{-t} \log t\, dt, \tag{1.3.19}$$

which, when integrated by parts, gives

$$\gamma = \int_0^1 \log t\, d(e^{-t} - 1) + \int_1^\infty \log t\, d(e^{-t}) = \int_0^1 \frac{1 - e^{-t}}{t}\,dt - \int_1^\infty \frac{e^{-t}}{t}\,dt.$$

Replacing t by $1/t$ in the last integral on the right, we find that

$$\gamma = \int_0^1 \frac{1 - e^{-t} - e^{-1/t}}{t}\,dt. \tag{1.3.20}$$

1.4. Asymptotic Representation of the Gamma Function for Large $|z|$

To describe the behavior of a given function $f(z)$ as $|z| \to \infty$ within a sector $\alpha \leqslant \arg z \leqslant \beta$, it is in many cases sufficient to derive an expression of the form

$$f(z) = \varphi(z)[1 + r(z)], \tag{1.4.1}$$

where $\varphi(z)$ is a function of a simpler structure than $f(z)$, and $r(z)$ converges uniformly to zero as $|z| \to \infty$ within the given sector. Formulas of this type are called *asymptotic representations* of $f(z)$ for large $|z|$. It follows from

(1.4.1) that the ratio $f(z)/\varphi(z)$ converges to unity as $|z| \to \infty$, i.e., the two functions $f(z)$ and $\varphi(z)$ are "asymptotically equal," a fact we indicate by writing

$$f(z) \approx \varphi(z), \qquad |z| \to \infty, \quad \alpha \leqslant \arg z \leqslant \beta. \tag{1.4.2}$$

An estimate of $|r(z)|$ gives the size of the error committed when $f(z)$ is replaced by $\varphi(z)$ for large but finite $|z|$.

We now look for a description of the behavior of the function $f(z)$ as $|z| \to \infty$ which is more exact than that given by (1.4.1). Suppose we succeed in deriving the formula

$$f(z) = \varphi(z)\left[\sum_{n=0}^{N} a_n z^{-n} + r_N(z)\right], \qquad a_0 = 1, \quad N = 1, 2, \ldots, \tag{1.4.3}$$

where $z^N r_N(z)$ converges uniformly to zero as $|z| \to \infty$, $\alpha \leqslant \arg z \leqslant \beta$. [Note that (1.4.3) reduces to (1.4.1) for $N = 0$.] Then we write

$$f(z) \approx \varphi(z) \sum_{n=0}^{\infty} a_n z^{-n}, \qquad |z| \to \infty, \quad \alpha \leqslant \arg z \leqslant \beta, \tag{1.4.4}$$

and the right-hand side is called an *asymptotic series* or *asymptotic expansion* of $f(z)$ for large $|z|$. It should be noted that this definition does not stipulate that the given series converge in the ordinary sense, and on the contrary, the series will usually diverge. Nevertheless, asymptotic series are very useful, since, by taking a finite number of terms, we can obtain an arbitrarily good approximation to the function $f(z)$ for sufficiently large $|z|$. In this book, the reader will find many examples of asymptotic representations and asymptotic series (see Secs. 1.4, 2.2, 3.2, 4.6, 4.14, 4.22, 5.11, etc.). For the general theory of asymptotic series, we refer to the references cited in the Bibliography on p. 300.

To obtain an asymptotic representation of the gamma function $\Gamma(z)$, it is convenient to first derive an asymptotic representation of $\log \Gamma(z)$. To this end, let $\operatorname{Re} z > 0$, and consider the integral representation (1.3.13), with z replaced by $z + 1$, i.e.,

$$\begin{aligned}\psi(z+1) = \frac{\Gamma'(z+1)}{\Gamma(z+1)} &= \int_0^\infty \left(\frac{e^{-t}}{t} - \frac{e^{-tz}}{e^t - 1}\right) dt \\ &= \int_0^\infty \frac{e^{-t} - e^{-tz}}{t}\, dt + \frac{1}{2}\int_0^\infty e^{-tz}\, dt - \int_0^\infty \left(\frac{1}{2} - \frac{1}{t} + \frac{1}{e^t - 1}\right) e^{-tz}\, dt,\end{aligned}$$

or

$$\frac{\Gamma'(z+1)}{\Gamma(z+1)} = \log z + \frac{1}{2z} - \int_0^\infty \left(\frac{1}{2} - \frac{1}{t} + \frac{1}{e^t - 1}\right) e^{-tz}\, dt,$$

where we have used (1.3.11). Integrating the last equation between the limits 1 and z, and bearing in mind that

$$\log \Gamma(z+1) = \log \Gamma(z) + \log z,$$

we find that[13]

$$\log \Gamma(z) = \left(z - \frac{1}{2}\right) \log z - z + 1 + \int_0^\infty \left(\frac{1}{2} - \frac{1}{t} + \frac{1}{e^t - 1}\right) \frac{e^{-tz} - e^{-t}}{t}\, dt, \tag{1.4.5}$$

where $\operatorname{Re} z > 0$. It should be noted that the function

$$f(t) = \left(\frac{1}{2} - \frac{1}{t} + \frac{1}{e^t - 1}\right) \frac{1}{t}, \tag{1.4.6}$$

appearing in the integrand in (1.4.5), is continuous for $t \geqslant 0$, with $f(0) = \frac{1}{12}$, as can easily be verified by expanding $f(t)$ in a power series in a neighborhood of the point $t = 0$.

To simplify (1.4.5), we evaluate the integral

$$\mathscr{I} = \int_0^\infty f(t) e^{-t}\, dt. \tag{1.4.7}$$

This can be done by using the following trick: If

$$\mathscr{J} = \int_0^\infty f(t) e^{-t/2}\, dt, \tag{1.4.8}$$

then

$$\mathscr{J} - \mathscr{I} = \int_0^\infty e^{-t/2}\left[f(t) - \frac{1}{2} f\left(\frac{t}{2}\right)\right] dt = \int_0^\infty \left(\frac{e^{-t/2}}{t} - \frac{1}{e^t - 1}\right) \frac{dt}{t}.$$

It follows that

$$\begin{aligned} \mathscr{J} = (\mathscr{J} - \mathscr{I}) + \mathscr{I} &= \int_0^\infty \left(\frac{e^{-t/2} - e^{-t}}{t} - \frac{e^{-t}}{2}\right) \frac{dt}{t} \\ &= \int_0^\infty \left[-\frac{d}{dt}\left(\frac{e^{-t/2} - e^{-t}}{t}\right) + \frac{e^{-t} - e^{-t/2}}{2t}\right] dt \\ &= -\left.\frac{e^{-t/2} - e^{-t}}{t}\right|_0^\infty + \frac{1}{2}\int_0^\infty \frac{e^{-t} - e^{-t/2}}{t}\, dt = \frac{1}{2} + \frac{1}{2}\log\frac{1}{2}. \end{aligned} \tag{1.4.9}$$

On the other hand, substituting $z = \frac{1}{2}$ into (1.4.5), we find that

$$\mathscr{J} - \mathscr{I} = \tfrac{1}{2}\log \pi - \tfrac{1}{2}, \tag{1.4.10}$$

and hence

$$\mathscr{I} = 1 - \tfrac{1}{2}\log 2\pi. \tag{1.4.11}$$

Using this result, we can write (1.4.5) in the form

$$\log \Gamma(z) = (z - \tfrac{1}{2}) \log z - z + \tfrac{1}{2}\log 2\pi + \omega(z), \tag{1.4.12}$$

[13] The choice of the path of integration is unimportant. To justify integration behind the integral sign, we use an absolute convergence argument (cf. footnote 12, p. 6).

where

$$\omega(z) = \int_0^\infty f(t)e^{-tz}\,dt, \qquad \operatorname{Re} z > 0. \tag{1.4.13}$$

Since $f(t)$ decreases monotonically as t increases,[14] the integral (1.4.13) also converges for $\operatorname{Re} z = 0$, $\operatorname{Im} z \neq 0$.[15]

Using (1.4.12) and (1.4.13), we can easily derive an asymptotic representation of $\Gamma(z)$. First let $|\arg z| \leqslant \pi/2$, and integrate (1.4.13) by parts, obtaining

$$\omega(z) = \frac{1}{z}\left[f(0) + \int_0^\infty f'(t)e^{-tz}\,dt\right]. \tag{1.4.14}$$

Since $f'(t) \leqslant 0$, $|f'(t)| = -f'(t)$, we have

$$|\omega(z)| \leqslant \frac{1}{|z|}\left[f(0) - \int_0^\infty f'(t)\,dt\right] = \frac{2f(0)}{|z|},$$

i.e.,

$$|\omega(z)| \leqslant \frac{1}{6|z|}, \qquad |\arg z| \leqslant \frac{\pi}{2}. \tag{1.4.15}$$

Then, taking exponentials of both sides of (1.4.12), we find that

$$\Gamma(z) = e^{(z-\frac{1}{2})\log z - z + \frac{1}{2}\log 2\pi}\,[1 + r(z)], \qquad |\arg z| \leqslant \frac{\pi}{2}, \tag{1.4.16}$$

where

$$r(z) = e^{\omega(z)} - 1.$$

According to (1.4.15),

$$|r(z)| \leqslant \frac{C}{|z|}, \tag{1.4.17}$$

where C is an absolute constant (we assume that z is bounded away from zero, i.e., $|z| \geqslant a > 0$). Thus $r(z)$ is of order $|z|^{-1}$ as $|z| \to \infty$, a fact indicated by writing[16]

$$r(z) = O(|z|^{-1}), \tag{1.4.18}$$

and hence (1.4.16) is an asymptotic representation of $\Gamma(z)$ in the indicated sector.

To derive an asymptotic representation of $\Gamma(z)$ which is valid in other

---

[14] This follows at once from the expansion

$$f(t) = 2\sum_{k=1}^\infty \frac{1}{t^2 + 4\pi^2 k^2}.$$

See K. Knopp, *Theory and Applications of Infinite Series* (translated by R. C. H. Young), Blackie and Son, Ltd., London (1963), p. 378.

[15] E. C. Titchmarsh, *op. cit.*, p. 21.

[16] We say that $f(z)$ *is of order* $\varphi(z)$ as $z \to z_0$, and write $f(z) = O(\varphi(z))$ as $z \to z_0$ if the inequality $|f(z)| \leqslant A|\varphi(z)|$ holds in a neighborhood of z_0, where A is some constant. If z_0 is not explicitly mentioned, then $z_0 = \infty$.

sectors of the complex plane, we proceed as follows: Let δ be an arbitrarily small fixed positive number, and let

$$\frac{\pi}{2} \leqslant \arg z \leqslant \pi - \delta. \tag{1.4.19}$$

Since $\arg(-z) = \arg z - \pi$, this implies

$$-\frac{\pi}{2} \leqslant \arg(-z) \leqslant -\delta.$$

It follows from (1.2.1–2) that

$$\Gamma(z) = \frac{\pi}{-z\Gamma(-z)\sin \pi z}, \tag{1.4.20}$$

where, according to (1.4.16) and (1.4.18),

$$\Gamma(-z) = e^{-(z+\frac{1}{2})(\log z - \pi i) + z + \frac{1}{2}\log 2\pi}[1 + O(|z|^{-1})]. \tag{1.4.21}$$

On the other hand, in the sector (1.4.19),

$$\begin{aligned} \sin \pi z &= \frac{e^{\pi i z} - e^{-\pi i z}}{2i} = -\frac{e^{-\pi i z}}{2i}(1 - e^{2\pi i z}) \\ &= -\frac{e^{-\pi i z}}{2i}\left(1 - \frac{1}{z} z e^{2\pi i z}\right) = -\frac{e^{-\pi i z}}{2i}[1 + O(|z|^{-1})], \end{aligned} \tag{1.4.22}$$

since $ze^{2\pi i z}$ is bounded in this sector. Substituting (1.4.21–22) into (1.4.20), we again arrive at formula (1.4.16). A similar result is obtained for the sector

$$-(\pi - \delta) \leqslant \arg z \leqslant -\frac{\pi}{2}.$$

Finally, therefore, in any sector

$$|\arg z| \leqslant \pi - \delta,$$

we have the asymptotic representation

$$\Gamma(z) = e^{(z-\frac{1}{2})\log z - z + \frac{1}{2}\log 2\pi}[1 + O(|z|^{-1})]. \tag{1.4.23}$$

Considerations resembling those just given, but much more complicated,[17] lead to the more exact formula

$$\Gamma(z) = e^{(z-\frac{1}{2})\log z - z + \frac{1}{2}\log 2\pi}\left[1 + \frac{1}{12z} + \frac{1}{288z^2} - \frac{139}{51840z^3} + O(|z|^{-4})\right]. \tag{1.4.24}$$

If $z = x$ is a positive real number, then (1.4.16) becomes *Stirling's formula*

$$\Gamma(x) = \sqrt{2\pi}\, x^{x-\frac{1}{2}} e^{-x}[1 + r(x)], \tag{1.4.25}$$

---

[17] See G. N. Watson, *An expansion related to Stirling's formula, derived by the method of steepest descents*, Quart. J. Pure and Appl. Math., **48**, 1 (1920).

where for $r(x)$ we have a sharper estimate than that given by (1.4.17). In fact, if $z = x > 0$, then

$$|\omega(x)| \leqslant f(0) \int_0^\infty e^{-xt}\, dt = \frac{1}{12x}, \tag{1.4.26}$$

so that

$$|r(x)| \leqslant e^{1/12x} - 1. \tag{1.4.27}$$

Finally, we note that (1.2.4) and (1.4.25) imply the following asymptotic representation of the factorial:

$$n! \approx \sqrt{2\pi}\, n^{n+1/2} e^{-n}, \qquad n \to \infty. \tag{1.4.28}$$

1.5. Definite Integrals Related to the Gamma Function

The class of integrals which can be expressed in terms of the gamma function is very large. Here we consider only a few examples, mainly with the intent of deriving some formulas that will be needed later.

Our first result is the formula

$$\int_0^\infty e^{-pt} t^{z-1}\, dt = \frac{\Gamma(z)}{p^z}, \qquad \operatorname{Re} p > 0, \quad \operatorname{Re} z > 0, \tag{1.5.1}$$

which is easily proved for positive real p by making the change of variables $s = pt$, and then using the integral representation (1.1.1). The extension of (1.5.1) to arbitrary complex p with $\operatorname{Re} p > 0$ is accomplished by using the principle of analytic continuation.

Next consider the integral

$$B(x, y) = \int_0^1 t^{x-1}(1 - t)^{y-1}\, dt, \qquad \operatorname{Re} x > 0, \quad \operatorname{Re} y > 0, \tag{1.5.2}$$

known as the *beta function*. It is easy to see that (1.5.2) represents an analytic function in each of the complex variables x and y. If we introduce the new variable of integration $u = t/(1 - t)$, then (1.5.2) becomes

$$B(x, y) = \int_0^\infty \frac{u^{x-1}}{(1 + u)^{x+y}}\, du, \qquad \operatorname{Re} x > 0, \quad \operatorname{Re} y > 0. \tag{1.5.3}$$

Setting $p = 1 + u$, $z = x + y$ in (1.5.1), we find that

$$\frac{1}{(1 + u)^{x+y}} = \frac{1}{\Gamma(x + y)} \int_0^\infty e^{-(1+u)t} t^{x+y-1}\, dt, \tag{1.5.4}$$

and substituting the result into (1.5.3), we obtain

$$\begin{aligned} B(x, y) &= \frac{1}{\Gamma(x + y)} \int_0^\infty e^{-t} t^{x+y-1}\, dt \int_0^\infty e^{-ut} u^{x-1}\, du \\ &= \frac{\Gamma(x)}{\Gamma(x + y)} \int_0^\infty e^{-t} t^{y-1}\, dt = \frac{\Gamma(x)\Gamma(y)}{\Gamma(x + y)}. \end{aligned} \tag{1.5.5}$$

Thus we have derived the formula

$$B(x, y) = \frac{\Gamma(x)\Gamma(y)}{\Gamma(x + y)}, \tag{1.5.6}$$

relating the beta function to the gamma function, which can be used to derive all the properties of the beta function.

PROBLEMS

1. Prove that

$$|\Gamma(iy)|^2 = \frac{\pi}{y \sinh \pi y}, \qquad |\Gamma(\tfrac{1}{2} + iy)|^2 = \frac{\pi}{\cosh \pi y}$$

for real y.

2. Using (1.5.6), verify the identity

$$\int_0^\infty \frac{\cosh 2yt}{(\cosh t)^{2x}}\, dt = 2^{2x-2} \frac{\Gamma(x + y)\Gamma(x - y)}{\Gamma(2x)}, \qquad \operatorname{Re} x > 0, \qquad \operatorname{Re} x > |\operatorname{Re} y|.$$

3. Prove that

$$\int_0^{\pi/2} \cos^\nu \theta \, d\theta = \int_0^{\pi/2} \sin^\nu \theta \, d\theta = \frac{\sqrt{\pi}}{2} \frac{\Gamma\left(\frac{\nu + 1}{2}\right)}{\Gamma\left(\frac{\nu}{2} + 1\right)}, \qquad \operatorname{Re} \nu > -1,$$

$$\int_0^{\pi/2} \cos^\mu \theta \sin^\nu \theta \, d\theta = \frac{1}{2} \frac{\Gamma\left(\frac{\mu + 1}{2}\right)\Gamma\left(\frac{\nu + 1}{2}\right)}{\Gamma\left(\frac{\mu + \nu}{2} + 1\right)}, \qquad \operatorname{Re} \mu > -1, \qquad \operatorname{Re} \nu > -1.$$

4. Verify the formula

$$\Gamma(3z) = \frac{3^{3z - \frac{1}{2}}}{2\pi} \Gamma(z)\Gamma(z + \tfrac{1}{3})\Gamma(z + \tfrac{2}{3}). \tag{i}$$

5. Derive the formula

$$3\psi(3z) = \psi(z) + \psi(z + \tfrac{1}{3}) + \psi(z + \tfrac{2}{3}) + 3 \log 3.$$

Hint. Calculate the logarithmic derivatives of both sides of (i).

6. Derive the following integral representation of the square of the gamma function, where $K_0(t)$ is Macdonald's function (defined in Sec. 5.7):

$$\Gamma^2(z) = 2^{2-2z} \int_0^\infty t^{2z-1} K_0(t)\, dt, \qquad \operatorname{Re} z > 0.$$

Hint. Use formulas (5.10.23), (1.5.1) and the integral in Problem 2.

7. Derive the asymptotic formulas

$$\Gamma(z + \alpha) = e^{(z+\alpha-\frac{1}{2})\log z - z + \frac{1}{2}\log 2\pi}[1 + O(|z|^{-1})],$$

$$\frac{\Gamma(z + \alpha)}{\Gamma(z + \beta)} = z^{\alpha-\beta}\left[1 + \frac{(\alpha - \beta)(\alpha + \beta - 1)}{2z} + O(|z|^{-2})\right],$$

where α and β are arbitrary constants, and $|\arg z| \leqslant \pi - \delta$.

Hint. Use the results of Sec. 1.4.

8. Derive the asymptotic formula

$$|\Gamma(x + iy)| = \sqrt{2\pi}\, e^{-\frac{1}{2}\pi|y|}|y|^{x-\frac{1}{2}}[1 + r(x, y)],$$

where as $|t| \to \infty$, $r(x, y) \to 0$ uniformly in the strip $|x| \leqslant \alpha$ (α is a constant).

9. Show that the integral representation

$$\frac{1}{\Gamma(z)} = \frac{1}{2\pi i}\int_C e^t t^{-z}\, dt$$

holds for arbitrary complex z, where $t^{-z} = e^{-z \log t}$, $|\arg t| < \pi$, and C is the contour shown in Figure 13, p. 115.

10. The *incomplete gamma function* $\gamma(z, \alpha)$ and its *complement* $\Gamma(z, \alpha)$ are defined by the formulas

$$\gamma(z, \alpha) = \int_0^\alpha e^{-t}t^{z-1}\, dt, \qquad \operatorname{Re} z > 0, \quad |\arg \alpha| < \pi,$$

$$\Gamma(z, \alpha) = \int_\alpha^\infty e^{-t}t^{z-1}\, dt, \qquad |\arg \alpha| < \pi,$$

so that

$$\gamma(z, \alpha) + \Gamma(z, \alpha) = \Gamma(z).$$

Prove that for fixed α, $\Gamma(z, \alpha)$ is an entire function of z, while $\gamma(z, \alpha)$ is a meromorphic function of z, with poles at the points $z = 0, -1, -2, \ldots$

11. Derive the formulas

$$\gamma(z + 1, \alpha) = z\gamma(z, \alpha) - e^{-\alpha}\alpha^z,$$

$$\Gamma(z + 1, \alpha) = z\Gamma(z, \alpha) + e^{-\alpha}\alpha^z.$$

12. Derive the following representation of $\gamma(z, \alpha)$:

$$\gamma(z, \alpha) = \sum_{k=0}^{\infty} \frac{(-1)^k \alpha^{k+z}}{k!(k + z)}, \qquad z \neq 0, -1, -2, \ldots$$

2

THE PROBABILITY INTEGRAL AND RELATED FUNCTIONS

2.1. The Probability Integral and Its Basic Properties

By the *probability integral* is meant the function defined for any complex z by the integral

$$\Phi(z) = \frac{2}{\sqrt{\pi}} \int_0^z e^{-t^2}\, dt, \tag{2.1.1}$$

evaluated along an arbitrary path joining the origin to the point $t = z$. The form of this path does not matter, since the integrand is an entire function of the complex variable t, and in fact we can assume that the integration is along the line segment joining the points $t = 0$ and $t = z$. According to a familiar theorem of complex variable theory,[1] $\Phi(z)$ is an entire function and hence can be expanded in a convergent power series for any value of z. To find this expansion, we need only replace e^{-t^2} by its power series in (2.1.1), and then integrate term by term (this is always permissible for power series[2]), obtaining

$$\Phi(z) = \frac{2}{\sqrt{\pi}} \int_0^z \sum_{k=0}^{\infty} \frac{(-1)^k t^{2k}}{k!}\, dt = \frac{2}{\sqrt{\pi}} \sum_{k=0}^{\infty} \frac{(-1)^k z^{2k+1}}{k!(2k+1)}, \qquad |z| < \infty. \tag{2.1.2}$$

---

[1] If $f(t)$ is analytic in a simply connected domain D, then the integral

$$\varphi(z) = \int_a^z f(t)\, dt,$$

evaluated along any rectifiable path contained in D, defines an analytic function in D. See A. I. Markushevich, *op. cit.*, Theorem 13.5, p. 282. The theorem remains true if $f(a) = \infty$ or $a = \infty$, provided that the improper integral exists.

[2] *Ibid.*, Theorems 16.3 and 15.4, pp. 348 and 325.

It follows from (2.1.2) that $\Phi(z)$ is an odd function of z. For real values of its argument, $\Phi(z)$ is a real monotonically increasing function, whose graph is shown in Figure 1. At zero we have $\Phi(0) = 0$, and as z increases, $\Phi(z)$ rapidly approaches the limiting value $\Phi(\infty) = 1$, since

$$\int_0^\infty e^{-t^2}\,dt = \frac{\sqrt{\pi}}{2}. \tag{2.1.3}$$

The difference between $\Phi(z)$ and this limit can be written in the form

$$1 - \Phi(z) = \frac{2}{\sqrt{\pi}}\int_0^\infty e^{-t^2}\,dt - \frac{2}{\sqrt{\pi}}\int_0^z e^{-t^2}\,dt = \frac{2}{\sqrt{\pi}}\int_z^\infty e^{-t^2}\,dt. \tag{2.1.4}$$

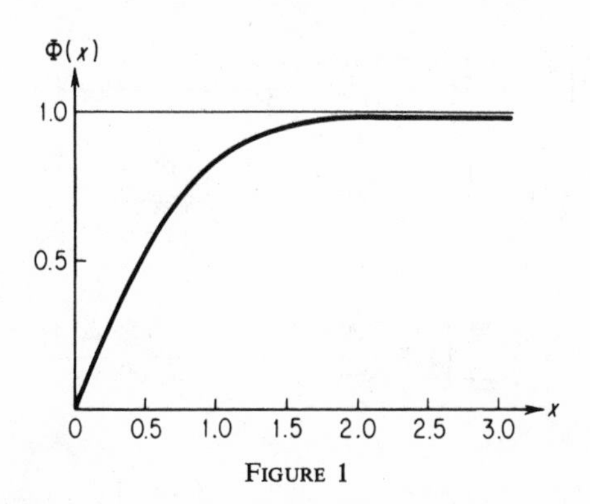

FIGURE 1

The probability integral is encountered in many branches of applied mathematics, e.g., probability theory, the theory of errors, the theory of heat conduction, and various branches of mathematical physics (see Secs. 2.5–2.7). In the literature, one often finds two functions related to the probability integral, i.e., the *error function*

$$\operatorname{Erf} z = \int_0^z e^{-t^2}\,dt = \frac{\sqrt{\pi}}{2}\,\Phi(z), \tag{2.1.5}$$

and its *complement*

$$\operatorname{Erfc} z = \int_z^\infty e^{-t^2}\,dt = \frac{\sqrt{\pi}}{2}\,[1 - \Phi(z)]. \tag{2.1.6}$$

Many more complicated integrals can be expressed in terms of the probability integral. For example, by differentiation of the parameter z it can be shown that

$$\frac{2}{\pi}\int_0^\infty \frac{e^{-zt^2}}{1 + t^2}\,dt = e^z[1 - \Phi(\sqrt{z})]. \tag{2.1.7}$$

2.2. Asymptotic Representation of the Probability Integral for Large $|z|$

To find an asymptotic representation of the function $\Phi(z)$ for large $|z|$, we apply repeated integration by parts to the integral in (2.1.4), obtaining

$$\int_z^\infty e^{-t^2}\,dt = -\frac{1}{2}\int_z^\infty \frac{1}{t}\,d(e^{-t^2}) = \frac{e^{-z^2}}{2z} - \frac{1}{2}\int_z^\infty \frac{e^{-t^2}}{t^2}dt$$

$$= \frac{e^{-z^2}}{2z} - \frac{e^{-z^2}}{2^2z^3} + \frac{1\cdot 3}{z^2}\int_z^\infty \frac{e^{-t^2}}{t^4}\,dt$$

$$= e^{-z^2}\left[\frac{1}{2z} - \frac{1}{2^2z^3} + \frac{1\cdot 3}{2^3z^5} - \frac{1\cdot 3\cdot 5}{2^4z^7} + \cdots + (-1)^n\frac{1\cdot 3\cdots(2n-1)}{2^{n+1}z^{2n+1}}\right]$$

$$+ (-1)^{n+1}\frac{1\cdot 3\cdots(2n+1)}{2^{n+1}}\int_z^\infty \frac{e^{-t^2}}{t^{2n+2}}\,dt.$$

It follows that

$$1 - \Phi(z) = \frac{e^{-z^2}}{\sqrt{\pi}\,z}\left[1 + \sum_{k=1}^{n}(-1)^k\frac{1\cdot 3\cdots(2k-1)}{(2z^2)^k} + r_n(z)\right], \tag{2.2.1}$$

where

$$r_n(z) = (-1)^{n+1}\frac{1\cdot 3\cdots(2n+1)}{2^n}\,ze^{z^2}\int_z^\infty \frac{e^{-t^2}}{t^{2n+2}}\,dt. \tag{2.2.2}$$

Now let

$$|\arg z| \leqslant \frac{\pi}{2} - \delta,$$

where δ is an arbitrarily small positive number, and choose the path of integration in (2.2.2) to be the infinite line segment beginning at the point $t = z$ and parallel to the real axis. If $z = x + iy = re^{i\varphi}$, then this segment has the equation $t = u + iy$ ($x \leqslant u < \infty$), and on the segment we have

$$|e^{z^2-t^2}| = e^{x^2-u^2}, \qquad |t|^{-(2n+3)} \leqslant |z|^{-(2n+3)}, \qquad |t| \leqslant u\sec\varphi.$$

Therefore

$$|r_n(z)| \leqslant \frac{1\cdot 3\cdots(2n+1)}{2^n|z|^{2n+2}}\sec\varphi\int_x^\infty e^{x^2-u^2}u\,du = \frac{1\cdot 3\cdots(2n+1)}{(2|z|^2)^{n+1}}\sec\varphi,$$

which implies

$$|r_n(z)| \leqslant \frac{1\cdot 3\cdots(2n+1)}{(2|z|^2)^{n+1}}\sec\varphi \leqslant \frac{1\cdot 3\cdots(2n+1)}{(2|z|^2)^{n+1}\sin\delta}. \tag{2.2.3}$$

It follows from (2.2.3) that as $|z| \to \infty$ the product $z^{2n}r_n(z)$ converges uniformly to zero in the indicated sector, i.e.,

$$1 - \Phi(z) \approx \frac{e^{-z^2}}{\sqrt{\pi}z}\left[1 + \sum_{n=1}^{\infty}(-1)^n \frac{1\cdot 3\cdots(2n-1)}{(2z^2)^n}\right], \qquad |z| \to \infty, \quad |\arg z| \leqslant \frac{\pi}{2} - \delta. \tag{2.2.4}$$

Thus the series on the right is the asymptotic series (see Sec. 1.4) of the function $1 - \Phi(z)$, and a bound on the error committed in approximating $1 - \Phi(z)$ by the sum of a finite number of terms of the series is given by (2.2.3). For positive real z this error does not exceed the first neglected term in absolute value.

An asymptotic representation of the probability integral in the sector

$$\frac{\pi}{2} + \delta \leqslant \arg z \leqslant \frac{3\pi}{2} - \delta$$

can be obtained from (2.2.1) by using the relation $\Phi(z) = -\Phi(-z)$, but the construction of an asymptotic representation in the sector

$$\frac{\pi}{2} - \delta \leqslant \arg z \leqslant \frac{\pi}{2} + \delta$$

requires a separate argument [cf. (2.3.5)].

2.3. The Probability Integral of Imaginary Argument. The Function $F(z)$

In the applications, one often encounters the case where the argument of the probability integral is a complex number. We now examine the particularly simple case where $z = ix$ is a pure imaginary. Choosing a segment of the imaginary axis as the path of integration, and making the substitution $t = iu$, we find from (2.1.1) that

$$\frac{\Phi(ix)}{i} = \frac{2}{\sqrt{\pi}}\int_0^x e^{u^2}\,du. \tag{2.3.1}$$

The integral in the right increases without limit as $x \to \infty$, and therefore it is more convenient to consider the function

$$F(z) = e^{-z^2}\int_0^z e^{u^2}\,du, \tag{2.3.2}$$

which remains bounded for all real z. In the general case of complex z, $F(z)$ is an entire function, and the choice of the path of integration in (2.3.2) is completely arbitrary.

To expand $F(z)$ in power series, we note that $F(z)$ satisfies the linear differential equation

$$F'(z) + 2zF(z) = 1, \tag{2.3.3}$$

with initial condition $F(0) = 0$. Substituting the series

$$F(z) = \sum_{k=0}^{\infty} a_k z^k$$

into (2.3.3), and comparing coefficients of identical powers of z, we obtain the recurrence relation

$$a_0 = 0, \quad a_1 = 1, \quad (k + 1)a_{k+1} + 2a_{k-1} = 0.$$

After some simple calculations, this leads to the expansion

$$F(z) = \sum_{k=0}^{\infty} \frac{(-1)^k 2^k z^{2k+1}}{1 \cdot 3 \cdots (2k + 1)}, \qquad |z| < \infty. \tag{2.3.4}$$

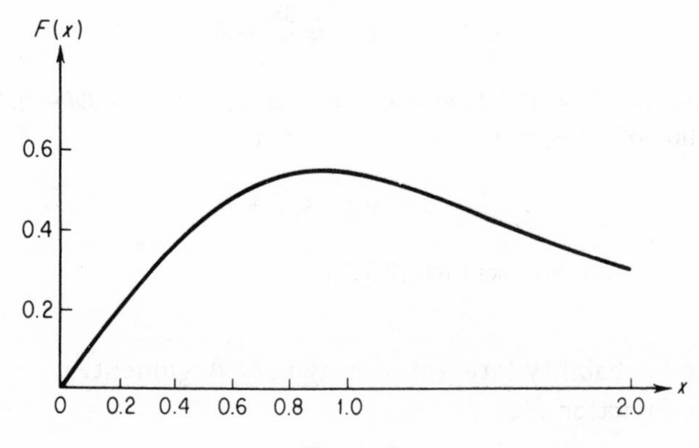

FIGURE 2

To study the behavior of $F(z)$ as $z \to \infty$ for real z, we apply L'Hospital's rule twice to the ratio

$$\frac{2z \displaystyle\int_0^z e^{u^2}\, du}{e^{z^2}},$$

and then use (2.3.2) to deduce that

$$\lim_{z \to \infty} 2zF(z) = 1,$$

i.e.,

$$F(z) \approx \frac{1}{2z}, \qquad z \to \infty. \tag{2.3.5}$$

In Figure 2 we show the graph of the function $F(z)$ for real $z \geqslant 0$. The maximum of the function occurs at $z = 0.924\ldots$ and equals $F_{\max} = 0.541\ldots$. The function $F(z)$ comes up in the theory of propagation of electromagnetic waves along the earth's surface, and in other problems of mathematical physics.

2.4. The Probability Integral of Argument $\sqrt{i}x$. The Fresnel Integrals

Another interesting case from the standpoint of the applications occurs when the argument of the probability integral is the complex number

$$z = \sqrt{i}x = \frac{x}{\sqrt{2}}(1 + i),$$

where x is real. In this case, we choose the path of integration in (2.1.1) to be a segment of the bisector of the angle between the real and imaginary axes. Then, using the formula $t = \sqrt{i}u$ to introduce the new real variable u, we find from (1.1.1) that

$$\frac{\Phi(\sqrt{i}x)}{\sqrt{i}} = \frac{2}{\sqrt{\pi}}\int_0^x e^{-iu^2}\,du = \frac{2}{\sqrt{\pi}}\int_0^x \cos u^2\,du - i\frac{2}{\sqrt{\pi}}\int_0^x \sin u^2\,du. \tag{2.4.1}$$

The integrals on the right can be expressed in terms of the functions

$$C(z) = \int_0^z \cos\frac{\pi t^2}{2}\,dt, \qquad S(z) = \int_0^z \sin\frac{\pi t^2}{2}\,dt, \tag{2.4.2}$$

where the integration is along any path joining the origin to the point $t = z$. The functions $C(z)$ and $S(z)$ are known as the *Fresnel integrals*. Since the integrands in (2.4.2) are entire functions of the complex variable t, the choice of the path of integration does not matter, and both $C(z)$ and $S(z)$ are entire functions of z.

For real $z = x$, the Fresnel integrals are real, with the graphs shown in Figure 3. Both $C(x)$ and $S(x)$ vanish for $x = 0$, and have an oscillatory character, as follows from the formulas

$$C'(x) = \cos\frac{\pi x^2}{2}, \qquad S'(x) = \sin\frac{\pi x^2}{2},$$

which show that $C(x)$ has extrema at $x = \pm\sqrt{2n+1}$, while $S(x)$ has extrema at $x = \pm\sqrt{2n}$ $(n = 0, 1, 2, \ldots)$. The largest maxima are $C(1) = 0.779893\ldots$ and $S(\sqrt{2}) = 0.713972\ldots$, respectively. As $x \to \infty$, each of the functions approaches the limit

$$C(\infty) = S(\infty) = \tfrac{1}{2},$$

as implied by the familiar formula[3]

$$\int_0^\infty \cos t^2\,dt = \int_0^\infty \sin t^2\,dt = \frac{\sqrt{\pi}}{2\sqrt{2}}. \tag{2.4.3}$$

---

[3] D. V. Widder, *op. cit.*, p. 382.

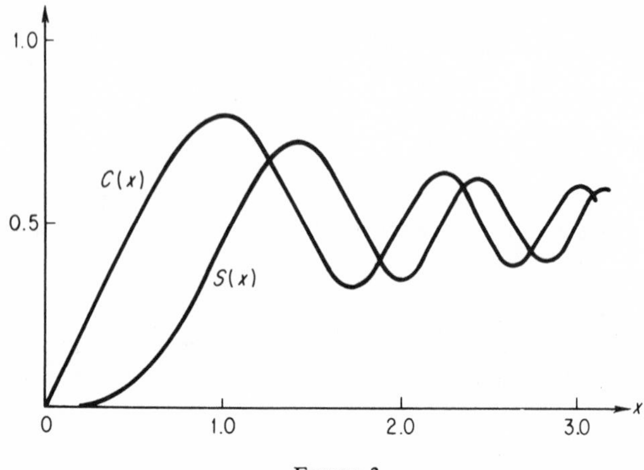

FIGURE 3

Replacing the trigonometric functions in the integrands in (2.4.2) by their power series expansions, and integrating term by term, we obtain the following series expansions for the Fresnel integrals, which converge for arbitrary z:

$$C(z) = \int_0^z \sum_{k=0}^{\infty} \frac{(-1)^k}{(2k)!} \left(\frac{\pi t^2}{2}\right)^{2k} dt = \sum_{k=0}^{\infty} \frac{(-1)^k}{(2k)!} \left(\frac{\pi}{2}\right)^{2k} \frac{z^{4k+1}}{4k+1},$$

$$S(z) = \int_0^z \sum_{k=0}^{\infty} \frac{(-1)^k}{(2k+1)!} \left(\frac{\pi t^2}{2}\right)^{2k+1} dt = \sum_{k=0}^{\infty} \frac{(-1)^k}{(2k+1)!} \left(\frac{\pi}{2}\right)^{2k+1} \frac{z^{4k+3}}{4k+3}. \tag{2.4.4}$$

The relation between the Fresnel integrals and the probability integral is given by the formula

$$\begin{aligned} C(z) \pm iS(z) &= \int_0^z e^{\pm \pi i t^2/2}\, dt = \sqrt{\frac{2}{\pi}}\, e^{\pm \pi i/4} \int_0^{\sqrt{\pi/2}\, z e^{\mp \pi i/4}} e^{-u^2} du \\ &= \frac{1}{\sqrt{2}}\, e^{\pm \pi i/4} \Phi\left(\sqrt{\frac{\pi}{2}}\, z e^{\mp \pi i/4}\right), \end{aligned} \tag{2.4.5}$$

which implies

$$C(z) = \frac{1}{2\sqrt{2}} \left[e^{\pi i/4} \Phi\left(\sqrt{\frac{\pi}{2}}\, z e^{-\pi i/4}\right) + e^{-\pi i/4} \Phi\left(\sqrt{\frac{\pi}{2}}\, z e^{\pi i/4}\right)\right],$$

$$S(z) = \frac{1}{2i\sqrt{2}} \left[e^{\pi i/4} \Phi\left(\sqrt{\frac{\pi}{2}}\, z e^{-\pi i/4}\right) - e^{-\pi i/4} \Phi\left(\sqrt{\frac{\pi}{2}}\, z e^{\pi i/4}\right)\right]. \tag{2.4.6}$$

Using (2.4.6), we can derive the properties of $C(z)$ and $S(z)$ from the corresponding properties of the probability integral. In particular, the results of

Sec. 2.2 lead to the following asymptotic representations of the Fresnel integrals, valid for large $|z|$ in the sector $|\arg z| \leqslant \frac{1}{4}\pi - \delta$,

$$
\begin{aligned}
C(z) &= \frac{1}{2} - \frac{1}{\pi z}\left[B(z)\cos\frac{\pi z^2}{2} - A(z)\sin\frac{\pi z^2}{2}\right],\\
S(z) &= \frac{1}{2} - \frac{1}{\pi z}\left[A(z)\cos\frac{\pi z^2}{2} + B(z)\sin\frac{\pi z^2}{2}\right],
\end{aligned}
\tag{2.4.7}
$$

where

$$A(z) = \sum_{k=0}^{N} \frac{(-1)^k \alpha_{2k}}{(\pi z^2)^{2k}} + O(|z|^{-4N-4}),$$

$$B(z) = \sum_{k=0}^{N} \frac{(-1)^k \alpha_{2k+1}}{(\pi z^2)^{2k+1}} + O(|z|^{-4N-6}),$$

$$\alpha_k = 1\cdot 3\cdots(2k-1), \quad \alpha_0 = 1.$$

The Fresnel integrals come up in various branches of physics and engineering, e.g., diffraction theory, theory of vibrations (see Sec. 2.7), etc. Many integrals of a more complicated type can be expressed in terms of the functions $C(z)$ and $S(z)$.

2.5. Application to Probability Theory

By a *normal* (or *Gaussian*) random variable with mean m and standard deviation σ is meant a random variable ξ such that the probability of ξ lying in the interval $[x, x + dx]$ is given by the expression[4,5]

$$\frac{1}{\sqrt{2\pi}\sigma} e^{-(x-m)^2/2\sigma^2}\,dx. \tag{2.5.1}$$

Then the probability

$$\mathbf{P}\{a \leqslant \xi - m \leqslant b\} \tag{2.5.2}$$

that $\xi - m$ lies in the interval $[a, b]$ is just the integral

$$
\begin{aligned}
\frac{1}{\sqrt{2\pi}\sigma}\int_{a+m}^{b+m} e^{-(x-m)^2/2\sigma^2}\,dx &= \frac{1}{\sqrt{\pi}}\int_{a/\sqrt{2}\sigma}^{b/\sqrt{2}\sigma} e^{-t^2}\,dt\\
&= \frac{1}{2}\left[\Phi\left(\frac{b}{\sqrt{2}\sigma}\right) - \Phi\left(\frac{a}{\sqrt{2}\sigma}\right)\right],
\end{aligned}
\tag{2.5.3}
$$

---

[4] As usual, $[a, b]$ denotes the *closed* interval $a \leqslant x \leqslant b$, and (a, b) the *open* interval $a < x < b$.

[5] See W. Feller, *An Introduction to Probability Theory and Its Applications, Vol. 1*, second edition, John Wiley and Sons, Inc., New York (1957). If $x_1, \ldots, x_n$ are the results of measurements of ξ, where n is large, then

$$m \approx \frac{1}{n}\sum_{k=1}^{n} x_k, \qquad \sigma^2 \approx \frac{1}{n}\sum_{k=1}^{n} (x - m)^2.$$

where $\Phi(x)$ is the probability integral. As one would expect, (2.5.2) equals 1 if $a = -\infty$, $b = \infty$.

Setting $a = -\delta$, $b = \delta$, we obtain the probability that $|\xi - m|$ does not exceed δ:

$$\mathbf{P}\{|\xi - m| \leqslant \delta\} = \Phi\left(\frac{\delta}{\sqrt{2}\sigma}\right). \tag{2.5.4}$$

Then the probability that $|\xi - m|$ exceeds δ is just

$$\mathbf{P}\{|\xi - m| > \delta\} = 1 - \Phi\left(\frac{\delta}{\sqrt{2}\sigma}\right). \tag{2.5.5}$$

The value $\delta = \delta_p$ for which (2.5.4) and (2.5.5) are equal is called the *probable error*, and clearly satisfies the equation

$$\Phi\left(\frac{\delta_p}{\sqrt{2}\sigma}\right) = \frac{1}{2}.$$

Using a table of the function $\Phi(x)$ to solve this equation,[6] we find that

$$\delta_p = 0.67449\sigma.$$

Example. *With standard deviation* 1 mm, *a machine produces parts of average length* 10 cm. *Find the probability that a part is of length* 10 cm *to within a tolerance of* 1 mm.

The required probability is

$$\mathbf{P}\{|\xi - 10| \leqslant 0.1\} = \Phi\left(\frac{1}{\sqrt{2}}\right) \approx 0.683,$$

i.e., some 68 percent of the parts satisfy the specified tolerance. In this case, the probable error is approximately 0.7 mm.

2.6. Application to the Theory of Heat Conduction. Cooling of the Surface of a Heated Object

Consider the following problem in the theory of heat conduction: An object occupying the half-space $x \geqslant 0$ is initially heated to temperature T_0. It then cools off by radiating heat through its surface $x = 0$ into the surrounding medium which is at zero temperature. We want to find the temperature $T(x, t)$ of the object as a function of position x and time t.

Let the object have thermal conductivity k, heat capacity c, density ρ and

---

[6] See E. Jahnke and F. Emde, *Tables of Higher Functions*, sixth edition, revised by F. Lösch, McGraw-Hill Book Co., New York (1960), p. 31.

emissivity λ, and let $\tau = kt/c\rho$. Then our problem reduces to the solution of the equation of heat conduction

$$\frac{\partial T}{\partial \tau} = \frac{\partial^2 T}{\partial x^2}, \tag{2.6.1}$$

subject to the initial condition

$$T|_{\tau=0} = T_0 \tag{2.6.2}$$

and the boundary conditions[7]

$$\left(\frac{\partial T}{\partial x} - hT\right)\Big|_{x=0} = 0, \qquad T|_{x\to\infty} = T_0, \tag{2.6.3}$$

where $h = \lambda/k > 0$.

To solve the problem, we introduce the *Laplace transform* $\bar{T} = \bar{T}(x, p)$ of $T = T(x, \tau)$, defined by the formula

$$\bar{T} = \int_0^\infty e^{-p\tau}\, T\, d\tau, \qquad \operatorname{Re} p > 0. \tag{2.6.4}$$

A system of equations determining $\bar{T}$ can be obtained from (2.6.1–3) if we multiply the first and third equations by $e^{-p\tau}$ and integrate from 0 to ∞, taking the second equation into account. The result is

$$\begin{aligned} &\frac{d^2\bar{T}}{dx^2} = p\bar{T} - T_0, \\ &\frac{d\bar{T}}{dx} - h\bar{T}|_{x=0} = 0, \qquad \bar{T}|_{x\to\infty} = \frac{T_0}{p}. \end{aligned} \tag{2.6.5}$$

The system (2.6.5) has the solution

$$\bar{T} = \frac{T_0}{p}\left(1 - \frac{h}{h + \sqrt{p}}\, e^{-\sqrt{p}x}\right), \qquad \operatorname{Re} p > 0, \quad \operatorname{Re} \sqrt{p} > 0. \tag{2.6.6}$$

We can now solve for T by inverting (2.6.4). This can be done either by using a table of Laplace transforms,[8] or by applying the *Fourier-Mellin inversion theorem*,[9] which states that

$$T = \frac{1}{2\pi i}\int_{a-i\infty}^{a+i\infty} e^{p\tau}\, \bar{T}\, dp, \tag{2.6.7}$$

where a is a constant greater than the real part of all the singular points of $\bar{T}$.

---

[7] For the derivation of equations (2.6.1, 3), see G. P. Tolstov, *Fourier Series* (translated by R. A. Silverman), Prentice-Hall, Inc., Englewood Cliffs, N.J. (1962), Chap. 9, Secs. 20 and 24.

[8] See A. Erdélyi, W. Magnus, F. Oberhettinger and F. G. Tricomi, *Tables of Integral Transforms, Volume 1* (of two volumes), Chaps. 4–5, McGraw-Hill Book Co., New York (1954). This two-volume set (based, in part, on notes left by Harry Bateman) will henceforth be referred to as the Bateman Manuscript Project, *Tables of Integral Transforms.*

[9] H. S. Carslaw and J. C. Jaeger, *Operational Methods in Applied Mathematics*, second edition, Oxford University Press, London (1953), Chap. 4, Secs. 28–31.

The quantity of greatest interest is the *surface temperature* of the object. Setting $x = 0$ in (2.6.6), we find that

$$\bar{T}|_{x=0} = \frac{T_0}{\sqrt{p}(\sqrt{p} + h)} = T_0\left(\frac{1}{p - h^2} - \frac{h}{p - h^2}\frac{1}{\sqrt{p}}\right). \tag{2.6.8}$$

The simplest way to solve (2.6.8) for the original function $T|_{x=0}$ is to use the *convolution theorem*,[10] which states that if $\bar{f}_1$ and $\bar{f}_2$ are the Laplace transforms of f_1 and f_2, then $\bar{f} = \bar{f}_1\bar{f}_2$ is the Laplace transform of the function

$$f(\tau) = \int_0^\tau f_1(t)f_2(\tau - t)\,dt. \tag{2.6.9}$$

Since it is easily verified that

$$\bar{f}_1 = \frac{h}{\sqrt{p}}, \qquad \bar{f}_2 = \frac{1}{p - h^2}$$

are the Laplace transforms of

$$f_1 = \frac{h}{\sqrt{\pi\tau}}, \qquad f_2 = e^{h^2\tau},$$

(2.6.9) implies

$$T|_{x=0} = T_0\left(e^{h^2\tau} - \frac{h}{\sqrt{\pi}}\int_0^\infty e^{h^2(\tau - t)}\frac{dt}{\sqrt{t}}\right) = T_0e^{h^2\tau}\left(1 - \frac{2}{\sqrt{\pi}}\int_0^{h\sqrt{\tau}} e^{-s^2}\,ds\right),$$

i.e.,

$$T|_{x=0} = T_0e^{h^2\tau}[1 - \Phi(h\sqrt{\tau})], \tag{2.6.10}$$

where $\Phi(x)$ is the probability integral. It follows from the asymptotic formula (2.2.1) that for large τ the surface temperature falls off like $1/\sqrt{\tau}$:

$$T|_{x=0} \approx \frac{T_0}{h\sqrt{\pi\tau}}, \qquad \tau \to \infty. \tag{2.6.11}$$

The temperature inside the object ($x \neq 0$) can also be expressed in closed form in terms of the probability integral.

2.7. Application to the Theory of Vibrations. Transverse Vibrations of an Infinite Rod under the Action of a Suddenly Applied Concentrated Force

Consider an infinite rod of linear density ρ and Young's modulus E, lying along the positive x-axis. Let I be the moment of inertia of a cross section of the rod about a horizontal axis through the center of mass of the section, and let $\tau = \sqrt{EI/\rho}\,t$. Suppose the end $x = 0$ satisfies a sliding condition, while

[10] H. S. Carslaw and J. C. Jaeger, *op. cit.*, Chap. 4, Sec. 33.

the end $x = \infty$ is clamped, and suppose a constant force Q is suddenly applied at the end $x = 0$. Then the displacement $u = u(x, t)$ at an arbitrary point $x \geqslant 0$ of the rod is described by the system of equations[11]

$$\frac{\partial^2 u}{\partial \tau^2} + \frac{\partial^4 u}{\partial x^4} = 0,$$

$$u\big|_{\tau=0} = \frac{\partial u}{\partial \tau}\bigg|_{\tau=0} = 0, \tag{2.7.1}$$

$$\frac{\partial u}{\partial x}\bigg|_{x=0} = 0, \qquad \frac{\partial^3 u}{\partial x^3}\bigg|_{x=0} = \frac{Q}{EI}, \qquad u\big|_{x\to\infty} = 0, \quad \frac{\partial u}{\partial x}\bigg|_{x\to\infty} = 0.$$

To solve this system, we use the Laplace transform, as in the preceding section. Writing

$$\bar{u} = \int_0^\infty e^{-p\tau}\, u\, d\tau, \qquad \operatorname{Re} p > 0, \tag{2.7.2}$$

we obtain the following equations for $\bar{u}$:

$$\frac{d^4\bar{u}}{dx^4} + p^2\bar{u} = 0,$$

$$\frac{d\bar{u}}{dx}\bigg|_{x=0} = 0, \qquad \frac{d^3\bar{u}}{dx^3}\bigg|_{x=0} = \frac{Q}{EIp}, \tag{2.7.3}$$

$$\bar{u}\big|_{x\to\infty} = 0, \qquad \frac{d\bar{u}}{dx}\bigg|_{x\to\infty} = 0.$$

Simple calculations then show that

$$\bar{u} = \frac{Q}{2EIp^2 i}\left(\frac{e^{-\sqrt{-pi}\,x}}{\sqrt{-pi}} - \frac{e^{-\sqrt{pi}\,x}}{\sqrt{pi}}\right), \qquad \operatorname{Re} p > 0, \quad \operatorname{Re}\sqrt{\pm pi} > 0. \tag{2.7.4}$$

To find u, we again use the convolution theorem. Since[12]

$$\bar{f}_1 = \frac{Q}{EIp^2}, \qquad \bar{f}_2 = \frac{1}{2i}\left(\frac{e^{-\sqrt{-pi}\,x}}{\sqrt{-pi}} - \frac{e^{-\sqrt{pi}\,x}}{\sqrt{pi}}\right)$$

are the Laplace transforms of

$$f_1 = \frac{Q}{EI}\tau, \qquad f_2 = \frac{1}{\sqrt{2\pi\tau}}\left(\sin\frac{x^2}{4\tau} + \cos\frac{x^2}{4\tau}\right),$$

(2.6.9) implies

$$u = \frac{Q}{EI\sqrt{2\pi}}\int_0^\tau \left(\sin\frac{x^2}{4t} + \cos\frac{x^2}{4t}\right)\frac{\tau - t}{\sqrt{t}}\,dt = \frac{Qx\tau}{EI}\, f\!\left(\frac{x}{2\sqrt{\tau}}\right), \tag{2.7.5}$$

---

[11] See R. E. D. Bishop and D. C. Johnson, *The Mechanics of Vibration*, Cambridge University Press, London (1960), p. 285.

[12] Bateman Manuscript Project, *Tables of Integral Transforms, Vol. 1*, formula (27), p. 146 or formula (6), p. 246.

where

$$f(x) = \frac{1}{\sqrt{2\pi}} \int_x^\infty (\sin y^2 + \cos y^2) \frac{1 - (x^2/y^2)}{y^2} \, dy. \tag{2.7.6}$$

The function $f(x)$ can be expressed in terms of the Fresnel integrals $C(z)$ and $S(z)$, introduced in Sec. 2.4. In fact, integrating (2.7.6) by parts twice, we find that

$$f(x) = \left(1 + \frac{2}{3}x^2\right)\left[\frac{1}{2} - C\left(\sqrt{\frac{2}{\pi}}\,x\right)\right] - \left(1 - \frac{2}{3}x^2\right)\left[\frac{1}{2} - S\left(\sqrt{\frac{2}{\pi}}\,x\right)\right]$$
$$+ \frac{2}{3\sqrt{2\pi}}\left[(1 + x^2)\frac{\sin x^2}{x} + (1 - x^2)\frac{\cos x^2}{x}\right]. \tag{2.7.7}$$

PROBLEMS

1. Show that the functions

$$\varphi(z) = \frac{\sqrt{\pi}}{2} e^{z^2} \Phi(z)$$

satisfies the differential equation $\varphi' - 2z\varphi = 1$, and use this fact to derive the expansion

$$\Phi(z) = \frac{2z}{\sqrt{\pi}} e^{-z^2} \sum_{k=0}^{\infty} \frac{(2z^2)^k}{1 \cdot 3 \cdots (2k+1)}, \qquad |z| < \infty.$$

2. Using formula (2.4.5) and the result of Problem 1, derive the following expansions of the Fresnel integrals

$$C(x) = x\left[\alpha(x) \cos\frac{\pi x^2}{2} + \beta(x) \sin\frac{\pi x^2}{2}\right],$$

$$S(x) = x\left[\alpha(x) \sin\frac{\pi x^2}{2} - \beta(x) \cos\frac{\pi x^2}{2}\right],$$

where

$$\alpha(x) = \sum_{k=0}^{\infty} \frac{(-1)^k(\pi x^2)^{2k}}{1 \cdot 3 \cdots (4k+1)}, \qquad \beta(x) = \sum_{k=0}^{\infty} \frac{(-1)^k(\pi x^2)^{2k+1}}{1 \cdot 3 \cdots (4k+3)}.$$

3. Use integration by parts to show that

$$\int \Phi(x)\, dx = x\Phi(x) + \frac{1}{\sqrt{\pi}} e^{-x^2} + C.$$

4. Let $\overline{\Phi}$ be the Laplace transform of the probability integral, i.e.,

$$\overline{\Phi}(p) = \int_0^\infty e^{-px}\, \Phi(x)\, dx.$$

Prove that

$$\overline{\Phi}(p) = \frac{1}{4} e^{p^2/4}\left[1 - \Phi\left(\frac{p}{2}\right)\right].$$

5. Derive the integral representations

$$F(z) = \int_0^\infty e^{-t^2} \sin 2zt\, dt, \qquad \Phi(z) = \frac{2}{\pi} \int_0^\infty e^{-t^2} \frac{\sin 2zt}{t}\, dt.$$

Hint. Replace $\sin 2zt$ by its power series expansion and integrate term by term.

6. Derive the following integral representations for the square of the probability integral:

$$\Phi^2(z) = 1 - \frac{4}{\pi} \int_0^1 \frac{e^{-z^2(1+t^2)}}{1+t^2}\, dt,$$

$$[1 - \Phi(z)]^2 = \frac{4}{\pi} \int_1^\infty \frac{e^{-z^2(1+t^2)}}{1+t^2}\, dt, \qquad |\arg z| \leqslant \frac{\pi}{4}.$$

Hint. Represent $\Phi^2(z)$ as a double integral over the region $0 \leqslant s \leqslant z$, $0 \leqslant t \leqslant z$, and transform to polar coordinates.

7. Derive the formulas

$$1 - \Phi(z) = \frac{2}{\sqrt{\pi}} e^{-z^2} \int_0^\infty e^{-t^2 - 2zt}\, dt,$$

$$[1 - \Phi(z)]^2 = \frac{4}{\sqrt{\pi}} e^{-2z^2} \int_0^\infty e^{-t^2 - 2\sqrt{2}zt} \Phi(t)\, dt.$$

Hint. The second formula is obtained from the first after introducing new variables $\alpha = s + t$, $\beta = st$ in the double integral over the region $0 \leqslant s < \infty$, $0 \leqslant t \leqslant s$.

8. Prove that

$$C^2(z) \pm S^2(z) = \frac{2}{\pi} \int_0^1 \frac{\sin \frac{\pi z^2}{2}(1 \mp t^2)}{1 \mp t^2}\, dt.$$

9. Prove that

$$C(x) = \int_0^{\pi x^2/2} J_{-1/2}(t)\, dt, \qquad S(x) = \int_0^{\pi x^2/2} J_{1/2}(t)\, dt,$$

where $J_\nu(x)$ is the Bessel function of order ν (see Sec. 5.8).

3

THE EXPONENTIAL INTEGRAL AND RELATED FUNCTIONS

3.1. The Exponential Integral and Its Basic Properties

The *exponential integral* is defined by

$$\mathrm{Ei}(z) = \int_{-\infty}^{z} \frac{e^t}{t}\, dt, \qquad |\arg(-z)| < \pi, \tag{3.1.1}$$

where the integration is along any path L in the t-plane with a cut along the positive real axis (see Figure 4). Since the integrand is an analytic function in

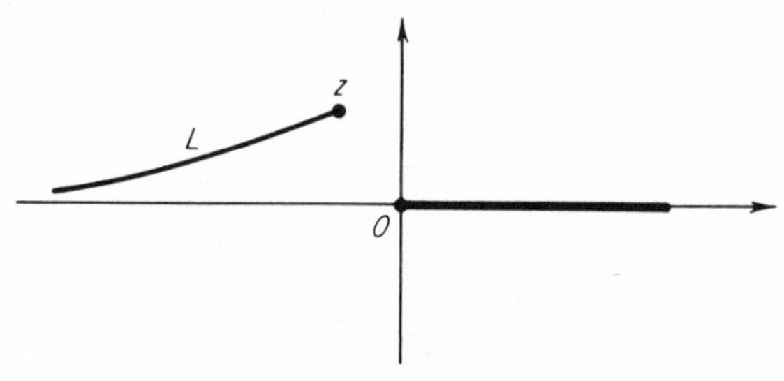

FIGURE 4

the resulting simply connected domain, the integral is path-independent and $\mathrm{Ei}(z)$ is an analytic function of z (cf. footnote 1, p. 16). A possible choice of the path of integration is the infinite line segment

$$-\infty < \mathrm{Re}\, t \leqslant \mathrm{Re}\, z, \qquad \mathrm{Im}\, t = \mathrm{Im}\, z, \tag{3.1.2}$$

passing through the point z and parallel to the real axis.

If we replace z by $-z$ and t by $-t$, formula (3.1.1) becomes

$$-\operatorname{Ei}(-z)=\int_z^\infty \frac{e^{-t}}{t}\,dt, \qquad |\arg z| < \pi, \tag{3.1.3}$$

where the function $-\operatorname{Ei}(-z)$ is analytic in the plane with a cut along the *negative* real axis. The graph of this function for $z = x > 0$ is shown in Figure 5. It will be noted that $-\operatorname{Ei}(-x)$ decreases monotonically from the value $-\operatorname{Ei}(0) = +\infty$ to the value $-\operatorname{Ei}(-\infty) = 0$, and in fact, its derivative is

$$\frac{d}{dx}[-\operatorname{Ei}(-x)] = -\frac{e^{-x}}{x} < 0 \qquad \text{if } x > 0.$$

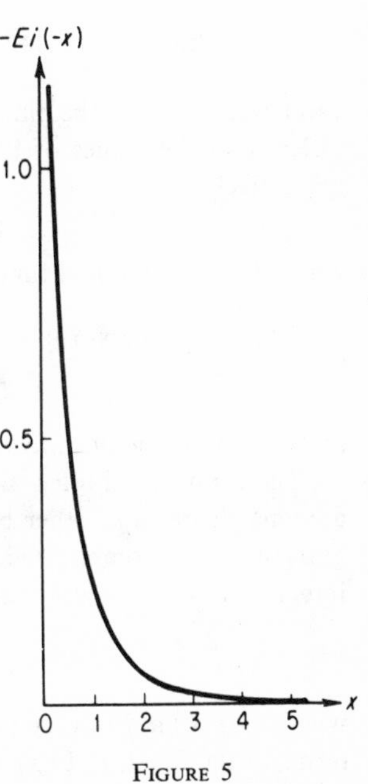

FIGURE 5

To derive a series expansion of the exponential integral, we represent (3.1.1) in the form

$$\operatorname{Ei}(z) = \int_{-\infty}^{-1} \frac{e^t}{t}\,dt + \int_{-1}^{0} \frac{e^t - 1}{t}\,dt + \int_0^z \frac{e^t - 1}{t}\,dt + \int_{-1}^{z} \frac{dt}{t},$$

and observe that the sum of the first two integrals is an absolute constant, which we denote by C. Setting $t = -u^{-1}$ in the first integral and $t = -u$ in the second, we find that

$$C = \int_0^1 \frac{1 - e^{-u} - e^{-1/u}}{u}\,du. \tag{3.1.4}$$

Comparison of (3.1.4) and (1.3.20) shows that C coincides with Euler's constant:

$$C = \gamma = 0.5772157\ldots$$

Thus we have[1]

$$\operatorname{Ei}(z) = \gamma + \log(-z) + \int_0^z \frac{e^t - 1}{t}\,dt, \qquad |\arg(-z)| < \pi. \tag{3.1.5}$$

The integral on the right, whose integrand is an entire function, is itself an entire function of the complex variable z, and can therefore be expanded in a power series which converges in the whole plane. To obtain this series, we

---

[1] In this book $\log z$ always means the single-valued branch of the logarithm defined by

$$\log z = \log|z| + i \arg z, \qquad |\arg z| < \pi.$$

Similarly, z^ν (ν arbitrary) means $e^{\nu \log z}$, and so on.

need only expand the integrand in powers of t and integrate term by term. The result is

$$\int_0^z \frac{e^t - 1}{t}\,dt = \int_0^z \sum_{k=1}^{\infty} \frac{t^{k-1}}{k!}\,dt = \sum_{k=1}^{\infty} \frac{z^k}{k!k}, \qquad |z| < \infty,$$

and therefore the desired expansion of the exponential integral is

$$\mathrm{Ei}(z) = \gamma + \log(-z) + \sum_{k=1}^{\infty} \frac{z^k}{k!k}, \qquad |\arg(-z)| < \pi, \tag{3.1.6}$$

valid everywhere in the plane cut along the positive real axis. It follows from (3.1.6) that the values of $\mathrm{Ei}(z)$ on the upper and lower edges of the cut are respectively

$$\mathrm{Ei}(x \pm i0) = \mathrm{Ei}_1(x) \mp \pi i, \qquad x > 0,$$

where $\mathrm{Ei}_1(x)$ is the real function defined by

$$\mathrm{Ei}_1(x) = \tfrac{1}{2}[\mathrm{Ei}(x + i0) + \mathrm{Ei}(x - i0)] = \gamma + \log x + \sum_{k=1}^{\infty} \frac{x^k}{k!k}, \qquad x > 0, \tag{3.1.7}$$

and known as the *modified exponential integral.*[2]

The exponential integral is often encountered in the applications, e.g., in antenna theory and other branches of physics and engineering. Many integrals of a more complicated type can be expressed in terms of the exponential integral. For example, the integral

$$\int e^z f(z)\,dz,$$

where $f(z)$ is an arbitrary rational function, can be written in finite form in terms of the function $\mathrm{Ei}(z)$ and elementary functions (see Problem 9, p. 42).

3.2. Asymptotic Representation of the Exponential Integral for Large $|z|$

To find an asymptotic representation of the function $\mathrm{Ei}(x)$ for large $|z|$, we apply repeated integration by parts to formula (3.1.1), obtaining

$$\begin{aligned}\int_{-\infty}^z \frac{e^t}{t}\,dt &= \int_{-\infty}^z \frac{1}{t}\,d(e^t) = \frac{e^z}{z} + \int_{-\infty}^z \frac{e^t}{t^2}\,dt \\ &= \frac{e^z}{z} + \frac{e^z}{z^2} + 1\cdot 2\int_{-\infty}^z \frac{e^t}{t^3}\,dt \\ &= e^z\left[\frac{1}{z} + \frac{1}{z^2} + \frac{1\cdot 2}{z^3} + \cdots + \frac{1\cdot 2\cdots n}{z^{n+1}}\right] + 1\cdot 2\cdots(n+1)\int_{-\infty}^z \frac{e^t}{t^{n+2}}\,dt.\end{aligned}$$

---

[2] Since (3.1.1) does not define $\mathrm{Ei}(z)$ for $z = x > 0$, one can formally extend the definition of the exponential integral by defining $\mathrm{Ei}(x) \equiv \mathrm{Ei}_1(x)$ for $x > 0$.

It follows that

$$\mathrm{Ei}(z) = \frac{e^z}{z}\left[\sum_{k=0}^{n} \frac{k!}{z^k} + r_n(z)\right], \tag{3.2.1}$$

where

$$r_n(z) = (n+1)!ze^{-z}\int_{-\infty}^{z} \frac{e^t}{t^{n+2}}\,dt, \qquad |\arg(-z)| < \pi. \tag{3.2.2}$$

To estimate the remainder $r_n(z)$, we choose the line segment (3.1.2) as the path of integration. Suppose $|\arg(-z)| \leqslant \pi - \delta$, where δ is an arbitrarily small positive number, and let $z = x + iy$. Then along the segment $t = \sigma + iy$ $(-\infty < \sigma \leqslant x)$ we have

$$|e^{t-z}| = e^{\sigma - x}, \qquad |t| \geqslant |z| \sin\delta,$$

and hence

$$|r_n(z)| \leqslant \frac{(n+1)!}{|z|^{n+1}(\sin\delta)^{n+2}}\int_{-\infty}^{x} e^{\sigma-x}\,d\sigma = \frac{(n+1)!}{(\sin\delta)^{n+2}}|z|^{-n-1} = O(|z|^{-n-1}). \tag{3.2.3}$$

Therefore we have the asymptotic representation

$$\mathrm{Ei}(z) = \frac{e^z}{z}\left[\sum_{k=0}^{n} \frac{k!}{z^k} + O(|z|^{-n-1})\right], \qquad |\arg(-z)| \leqslant \pi - \delta. \tag{3.2.4}$$

It follows from (3.2.4) that the divergent series

$$\frac{e^z}{z}\sum_{k=0}^{\infty} \frac{k!}{z^k}$$

is the asymptotic series for $\mathrm{Ei}(z)$ in the sector $|\arg(-z)| \leqslant \pi - \delta$.

It should be noted that if $\mathrm{Re}\, z \leqslant 0$, i.e., in the sector $|\arg(-z)| \leqslant \pi/2$, we have the sharper estimate

$$|r_n(z)| \leqslant \frac{(n+1)!}{|z|^{n+1}}. \tag{3.2.5}$$

In this case, the error committed in approximating $\mathrm{Ei}(z)$ by the sum of a finite number of terms of the asymptotic series does not exceed the first neglected term in absolute value.

3.3. The Exponential Integral of Imaginary Argument. The Sine and Cosine Integrals

If $z = ix$ is a pure imaginary, the function $\mathrm{Ei}(z)$ can be expressed in terms of two real functions $\mathrm{Si}(x)$ and $\mathrm{Ci}(x)$, known as the *sine integral* and the *cosine integral*, respectively. These functions, which are interesting in their own right, are defined for arbitrary complex z by the integrals

$$\mathrm{Si}(z) = \int_0^z \frac{\sin t}{t}\,dt, \qquad \mathrm{Ci}(z) = \int_\infty^z \frac{\cos t}{t}\,dt, \qquad |\arg z| < \pi. \tag{3.3.1}$$

The choice of the path of integration in the first integral is entirely arbitrary, but in the second integral it is required that the path of integration L lie in the plane cut along the negative real axis, as shown schematically in Figure 6.

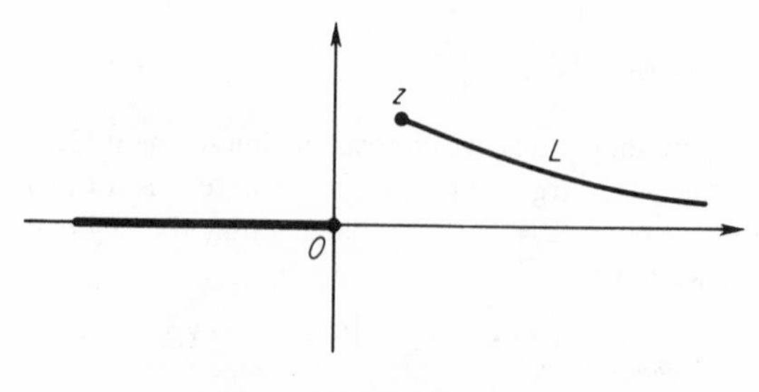

FIGURE 6

For the usual reason (cf. footnote 1, p. 16), $\mathrm{Si}(z)$ is an entire function, while $\mathrm{Ci}(z)$ is analytic in the plane cut along the negative real axis.

For real $z = x > 0$, both functions are real, with the graphs shown in

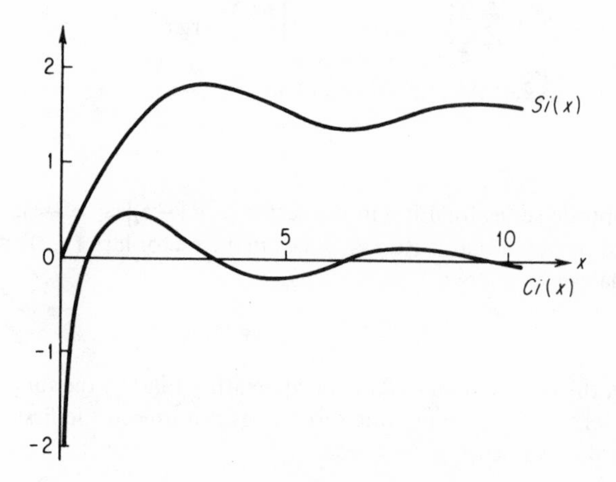

FIGURE 7

Figure 7. Moreover, $\mathrm{Si}(x)$ and $\mathrm{Ci}(x)$ have an oscillatory character, as follows from the formulas

$$\frac{d}{dx}\,\mathrm{Si}(x) = \frac{\sin x}{x}, \qquad \frac{d}{dx}\,\mathrm{Ci}(x) = \frac{\cos x}{x},$$

which show that $\mathrm{Si}(x)$ has extrema at the points $x = n\pi$ $(n = 0, 1, 2, \ldots)$, while $\mathrm{Ci}(x)$ has extrema at the points $x = (n + \frac{1}{2})\pi$. For $x < 0$,

$$\mathrm{Si}(x) = -\mathrm{Si}(|x|),$$

whereas $\mathrm{Ci}(x)$ is not defined. For large and small values of the argument, we have the limiting values

$$\begin{aligned} \mathrm{Si}(\infty) &= \frac{\pi}{2}, \qquad \mathrm{Ci}(\infty) = 0, \\ \mathrm{Si}(0) &= 0, \qquad \mathrm{Ci}(+0) = -\infty. \end{aligned} \tag{3.3.2}$$

To establish the relation between the functions $\mathrm{Ei}(ix)$, $\mathrm{Si}(x)$ and $\mathrm{Ci}(x)$, we substitute $z = ix$ $(x > 0)$ into (3.1.1). First we note that the integration along the original path L can be replaced by integration along the imaginary axis. In fact, consider the integral of the function e^t/t along the closed contour consisting of an arc C_R of the circle of radius R with center at the origin, the

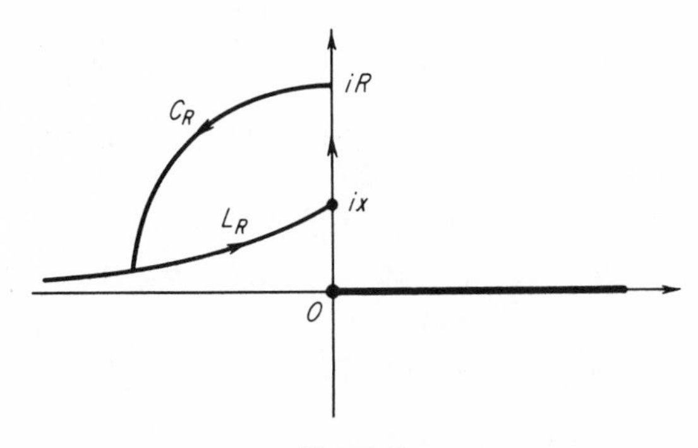

FIGURE 8

arc L_R of the curve L lying inside this circle, and the segment of the imaginary axis joining the points ix and iR (see Figure 8). According to Cauchy's integral theorem,

$$\int_{L_R} \frac{e^t}{t}\,dt + \int_{ix}^{iR} \frac{e^t}{t}\,dt + \int_{C_R} \frac{e^t}{t}\,dt = 0.$$

But as $R \to \infty$, the integral along L_R approaches $\mathrm{Ei}(ix)$, while the integral along C_R vanishes.[3] Therefore

$$\mathrm{Ei}(ix) = -\int_{ix}^{i\infty} \frac{e^t}{t}\,dt = \int_{\infty}^{x} \frac{e^{iu}}{u}\,du = \int_{\infty}^{x} \frac{\cos u}{u}\,du + i\int_{\infty}^{x} \frac{\sin u}{u}\,du,$$

[3] On the arc C_R we have $t = Re^{i\theta}$, $\pi/2 \leqslant \theta < \pi$, and hence

$$\left|\int_{C_R} \frac{e^t}{t}\,dt\right| \leqslant \int_{\pi/2}^{\pi} e^{R\cos\theta}\,d\theta = \int_0^{\pi/2} e^{-R\sin\chi}\,d\chi \leqslant \int_0^{\pi/2} e^{-2R\chi/\pi}\,d\chi = \frac{\pi}{2}\frac{1-e^{-R}}{R},$$

where we use the inequality $\sin\chi \geqslant (2\chi/\pi)$, valid for $0 \leqslant \chi \leqslant \pi/2$ [see A. I. Markushevich, *op. cit.*, formula (13.20), p. 272]. It follows that

$$\int_{C_R} \frac{e^t}{t}\,dt \to 0$$

as $R \to \infty$.

i.e.,

$$\mathrm{Ei}(ix) = \mathrm{Ci}(x) - i\left[\frac{\pi}{2} - \mathrm{Si}(x)\right], \qquad x > 0, \tag{3.3.3}$$

and similarly,

$$\mathrm{Ei}(-ix) = \mathrm{Ci}(x) + i\left[\frac{\pi}{2} - \mathrm{Si}(x)\right], \qquad x > 0. \tag{3.3.4}$$

We have proved formulas (3.3.3–4) for $x > 0$, but it is easily seen by using the principle of analytic continuation that they hold in a larger region, and in fact,

$$\begin{aligned} \mathrm{Ei}(-ze^{-\pi i/2}) &= \mathrm{Ci}(z) - i\left[\frac{\pi}{2} - \mathrm{Si}(z)\right], \qquad -\frac{\pi}{2} < \arg z < \pi, \\ \mathrm{Ei}(-ze^{\pi i/2}) &= \mathrm{Ci}(z) + i\left[\frac{\pi}{2} - \mathrm{Si}(z)\right], \qquad -\pi < \arg z < \frac{\pi}{2}. \end{aligned} \tag{3.3.5}$$

To prove (3.3.5), we merely note that both sides are analytic functions of z in the indicated sectors, and that these functions coincide for $z = x > 0$. From (3.3.5) we deduce the useful formulas

$$\begin{aligned} \mathrm{Ci}(z) &= \frac{1}{2}[\mathrm{Ei}(-ze^{\pi i/2}) + \mathrm{Ei}(-ze^{-\pi i/2})], \qquad |\arg z| < \frac{\pi}{2}, \\ \mathrm{Si}(z) &= \frac{\pi}{2} - \frac{1}{2i}[\mathrm{Ei}(-ze^{\pi i/2}) - \mathrm{Ei}(-ze^{-\pi i/2})], \qquad |\arg z| < \frac{\pi}{2}, \end{aligned} \tag{3.3.6}$$

which express $\mathrm{Ci}(z)$ and $\mathrm{Si}(z)$ in terms of the exponential integral.

The functions $\mathrm{Si}(z)$ and $\mathrm{Ci}(z)$ have simple series expansions. The expansion of $\mathrm{Si}(z)$ is found by substituting the power series for $\sin t$ into (3.3.1) and then integrating term by term. The result is

$$\mathrm{Si}(z) = \int_0^z \sum_{k=0}^{\infty} \frac{(-1)^k t^{2k}}{(2k+1)!}\,dt = \sum_{k=0}^{\infty} \frac{(-1)^k z^{2k+1}}{(2k+1)!(2k+1)}, \qquad |z| < \infty. \tag{3.3.7}$$

The derivation of the expansion of $\mathrm{Ci}(z)$ is somewhat more complicated. The simplest approach is to use the relation between the functions $\mathrm{Ci}(z)$ and $\mathrm{Ei}(-ze^{\pm\pi i/2})$, together with the expansion (3.1.6). In this way, we find that[4]

$$\mathrm{Ci}(z) = \gamma + \log z + \sum_{k=1}^{\infty} \frac{(-1)^k z^{2k}}{(2k)!2k}, \qquad |\arg z| < \pi. \tag{3.3.8}$$

In particular, (3.3.8) leads to the following values of the function $\mathrm{Ci}(z)$ on the upper and lower edges of the cut $[-\infty, 0]$:[5]

$$\mathrm{Ci}(-x \pm i0) = \mathrm{Ci}(x) \pm \pi i, \qquad x > 0. \tag{3.3.9}$$

[4] The original restriction $|\arg z| < \pi/2$ is easily eliminated by using the principle of analytic continuation.

[5] For simplicity of notation, we will always regard infinite branch cuts as passing through the point at infinity, as in the familiar representation of the extended complex plane by the Riemann sphere (see A. I. Markushevich, *op. cit.*, Chap. 5).

Finally, by using (3.2.4) and (3.3.6), we can derive asymptotic representations of the functions $\mathrm{Ci}(z)$ and $\frac{1}{2}\pi - \mathrm{Si}(z)$ for large $|z|$ in the sector $|\arg z| < \pi/2$. It is easily verified that

$$\begin{aligned} \mathrm{Ci}(z) &= \frac{\sin z}{z} P(z) - \frac{\cos z}{z} Q(z), \\ \frac{\pi}{2} - \mathrm{Si}(z) &= \frac{\cos z}{z} P(z) + \frac{\sin z}{z} Q(z), \end{aligned} \tag{3.3.10}$$

where

$$P(z) = \sum_{k=0}^{n} \frac{(-1)^k (2k)!}{z^{2k}} + O(|z|^{-2n-2}),$$

$$Q(z) = \sum_{k=0}^{n} \frac{(-1)^k (2k+1)!}{z^{2k+1}} + O(|z|^{-2n-3}).$$

3.4. The Logarithmic Integral

Another special function which is closely related to the exponential integral is the *logarithmic integral*. This function, which plays an important role in analysis, is defined by

$$\mathrm{li}(z) = \int_0^z \frac{dt}{\log t}, \qquad |\arg z| < \pi, \quad |\arg(1-z)| < \pi, \tag{3.4.1}$$

where the integral is along any path L belonging to the plane with two cuts along the segments $[-\infty, 0]$ and $[1, \infty]$ of the real axis (see Figure 9). By the

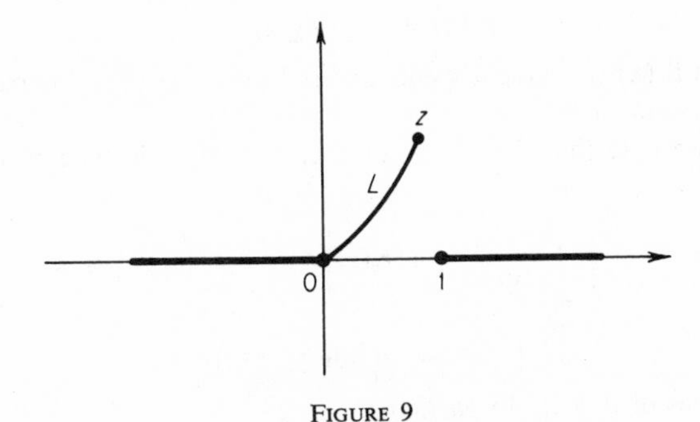

FIGURE 9

usual argument (cf. footnote 1, p. 16), $\mathrm{li}(z)$ is an analytic function in the cut plane. By introducing the new variable of integration $u = \log t$, we can easily express $\mathrm{li}(z)$ in terms of the exponential integral. In fact, the original cut t-plane is mapped onto the strip $|\mathrm{Im}\, u| < \pi$ in the u-plane, with a cut along the positive real axis, and (3.4.1) is transformed into the integral

$$\mathrm{li}(z) = \int_{-\infty}^{\log z} \frac{e^u}{u}\, du, \tag{3.4.2}$$

evaluated along any path belonging to this strip. Since the strip is a part of the domain of definition of the exponential integral (see Sec. 3.1), it follows from (3.4.2) that

$$\mathrm{li}(z) = \mathrm{Ei}\,(\log z), \tag{3.4.3}$$

where, as always, $\log z$ denotes the principal value of the logarithm (cf. footnote 1, p. 31).

Using (3.4.3), we can easily deduce the properties of the logarithmic integral from those of the exponential integral. For example, formula (3.1.6) implies the expansion

$$\mathrm{li}(z) = \gamma + \log(-\log z) + \sum_{k=1}^{\infty} \frac{(\log z)^k}{k!k}, \tag{3.4.4}$$

where z belongs to the plane with cuts along the segments $[-\infty, 0]$ and $[1, \infty]$. In particular, it follows from (3.4.4) that the values of $\mathrm{li}(z)$ on the upper and lower edges of the cut $[1, \infty]$ are

$$\mathrm{li}(x \pm i0) = \mathrm{li}_1(x) \mp \pi i, \qquad x > 1, \tag{3.4.5}$$

where $\mathrm{li}_1(x)$ denotes the real function

$$\mathrm{li}_1(x) = \tfrac{1}{2}[\mathrm{li}(x + i0) + \mathrm{li}(x - i0)] = \gamma + \log\log x + \sum_{k=1}^{\infty} \frac{(\log x)^k}{k!k}, \quad x > 1, \tag{3.4.6}$$

known as the *modified logarithmic integral*.[6] It follows from (3.1.7) and (3.4.6) that the modified exponential integral and the modified logarithmic integral are connected by the formula

$$\mathrm{li}_1(x) = \mathrm{Ei}_1\,(\log x). \tag{3.4.7}$$

The function $\mathrm{li}_1(x)$ is frequently encountered in analysis, and is particularly important in number theory.[7]

Finally, we note that the results of Sec. 3.2 imply the asymptotic representation

$$\mathrm{li}(z) = \frac{z}{\log z}\left[\sum_{k=0}^{n} \frac{k!}{(\log z)^k} + r_n(z)\right], \qquad \delta \leqslant |\arg z| \leqslant \pi - \delta, \tag{3.4.8}$$

where

$$|r_n(z)| = O(|\log z|^{-n-1})$$

for large values of $|\log z|$. In particular,

$$|r_n(z)| \leqslant \frac{(n+1)!}{|\log z|^{n+1}}$$

for $|z| < 1$, and in this case the sector is just $|\arg z| \leqslant \pi - \delta$.

---

[6] Since (3.4.1) does not define $\mathrm{li}(z)$ for $z = x > 1$, one can formally extend the definition of the logarithmic integral by defining $\mathrm{li}(x) \equiv \mathrm{li}_1(x)$ for $x > 1$.

[7] See A. E. Ingham, *The Distribution of Prime Numbers*, Cambridge Tracts in Mathematics and Mathematical Physics, No. 30, Cambridge University Press, London (1932).

3.5. Application to Electromagnetic Theory. Radiation of a Linear Half-Wave Oscillator[8]

As a simple example of the application of the special functions studied in this chapter, we consider the electromagnetic energy radiated by a linear oscillator of length $2l = \lambda/2$, driven by an alternating current I of frequency $\omega = 2\pi c/\lambda$ (c is the velocity of light and λ the wavelength), whose distribution along the conductor is give by

$$I = I_0 \cos \frac{\pi z}{2l} \cos \omega t, \qquad -l \leqslant z \leqslant l \tag{3.5.1}$$

(see Figure 10). Let $\mathbf{E}(t)$ and $\mathbf{H}(t)$ denote the time-dependent electric and magnetic field vectors, with complex amplitudes $\mathbf{E}$ and $\mathbf{H}$, so that

$$\mathbf{E}(t) = \operatorname{Re}\{\mathbf{E}e^{i\omega t}\}, \qquad \mathbf{H}(t) = \operatorname{Re}\{\mathbf{H}e^{i\omega t}\}. \tag{3.5.2}$$

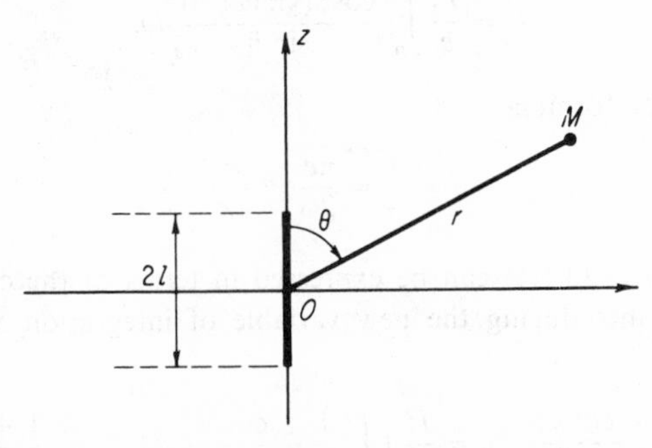

FIGURE 10

Then the power radiated by the oscillator, averaged over a period $T = \lambda/c$, is given by the formula[9]

$$P = \operatorname{Re}\left\{\frac{c}{8\pi}\int_S (\mathbf{E} \times \mathbf{H}^*)\cdot\mathbf{n}\, dS\right\}, \tag{3.5.3}$$

where S is an arbitrary surface surrounding the oscillator, $\mathbf{n}$ is the exterior normal to S, and $\mathbf{H}^*$ is the vector whose components are the complex conjugates of those of $\mathbf{H}$.[10]

In the present case, the vectors $\mathbf{E}$ and $\mathbf{H}$ have components $(E_r, E_\theta, 0)$ and $(0, 0, H)$ in a spherical coordinate system (r, θ, φ) [see Figure 10, where M is

---

[8] The necessary background information in electromagnetic theory, written in the system of units used here, can be found in G. Joos, *Theoretical Physics*, third edition, with the collaboration of I. Freeman, Blackie and Son, Ltd., London (1958).

[9] *Ibid.*, pp. 332, 341.

[10] As usual in vector algebra, the dot denotes the scalar product and the cross denotes the vector product.

the observation point], and for S we can choose a sphere $r = \rho$ of arbitrarily large radius ρ. Then (3.5.3) becomes

$$P = \operatorname{Re}\left\{\frac{c\rho^2}{4}\int_0^\pi E_\theta H^* \sin\theta\, d\theta\right\}, \tag{3.5.4}$$

where H^* is the complex conjugate of H. In (3.5.4) we can replace the exact values of E_θ and H by their asymptotic expressions for large r. Using the well-known formulas for the components of the electromagnetic field of an elementary dipole,[11] and integrating with respect to z, we easily find that

$$H \approx E_\theta \approx \frac{I_0 ik}{c\rho} e^{-ik\rho} \sin\theta \int_{-l}^{l} \cos\frac{\pi z}{2l} e^{ikz\cos\theta}\, dz = \frac{2I_0 i}{c\rho} e^{-ik\rho} \frac{\cos\left(\frac{1}{2}\pi\cos\theta\right)}{\sin\theta}$$

for sufficiently large ρ, where $k = \omega/c$. It follows that

$$P = \frac{I_0^2}{c}\int_0^\pi \frac{\cos^2\left(\frac{1}{2}\pi\cos\theta\right)}{\sin\theta}\, d\theta, \tag{3.5.5}$$

where we use the formula

$$l = \frac{\lambda}{4} = \frac{\pi c}{2\omega} = \frac{\pi}{2k}.$$

The integral in (3.5.5) can be expressed in terms of the cosine integral $\mathrm{Ci}(x)$. In fact, introducing the new variable of integration $x = \cos\theta$, we have

$$\begin{aligned} P &= \frac{I_0^2}{c}\int_0^1 \frac{1+\cos\pi x}{1-x^2}\, dx = \frac{I_0^2}{2c}\left(\int_0^1 \frac{1+\cos\pi x}{1-x}\, dx + \int_0^1 \frac{1+\cos\pi x}{1+x}\right) dx \\ &= \frac{I_0^2}{2c}\left(\int_0^1 \frac{1-\cos\pi y}{y}\, dy + \int_1^2 \frac{1-\cos\pi y}{y}\, dy\right) = \frac{I_0^2}{2c}\int_0^2 \frac{1-\cos\pi y}{y}\, dy \\ &= \frac{I_0^2}{2c}\int_0^{2\pi} \frac{1-\cos z}{z}\, dz. \end{aligned} \tag{3.5.6}$$

Finally, using the result of Problem 3, p. 41, we find that

$$P = \frac{I_0^2}{2c}[\gamma + \log 2\pi - \mathrm{Ci}(2\pi)], \tag{3.5.7}$$

where $\mathrm{Ci}(x)$ is the integral cosine and γ is Euler's constant.

The same method can be used to calculate the average power radiated by antennas with more complicated configurations. It is remarkable that the results can still be expressed in terms of sine and cosine integrals.

---

[11] G. Joos, *op. cit.*, pp. 338, 340.

PROBLEMS

1. Verify the integral representation

$$-\mathrm{Ei}(-z) = e^{-z} \int_0^\infty \frac{e^{-zt}}{1+t}\, dt, \qquad |\arg z| \leqslant \frac{\pi}{2}.$$

2. Verify the following integral representation for the square of the exponential integral:

$$[\mathrm{Ei}(-z)]^2 = 2e^{-2z} \int_0^\infty e^{-2zt} \frac{\log(1+2t)}{1+t}\, dt, \qquad |\arg z| \leqslant \frac{\pi}{2}.$$

Hint. Represent the left-hand side as a double integral over the region $0 \leqslant s < \infty$, $0 \leqslant t \leqslant s$, and introduce the new variables $\alpha = s + t$, $\beta = st$.

3. Prove that

$$\mathrm{Ci}(z) = \gamma + \log z - \int_0^z \frac{1 - \cos t}{t}\, dt, \qquad |\arg z| < \pi.$$

4. Starting from (3.1.1) and the definition of the modified exponential integral $\mathrm{Ei}_1(x)$, show that

$$\mathrm{Ei}_1(x) = \lim_{\varepsilon \to 0} \left(\int_{-\infty}^{-\varepsilon} \frac{e^t}{t}\, dt + \int_\varepsilon^x \frac{e^t}{t}\, dt \right), \qquad x > 0,$$

i.e., show that $\mathrm{Ei}_1(x)$ is the Cauchy principal value of the integral

$$\int_{-\infty}^x \frac{e^t}{t}\, dt.$$

5. Verify that

$$\mathrm{Ei}_1(x) = \gamma + \log x + \int_0^x \frac{e^t - 1}{t}\, dt.$$

6. Using L'Hospital's rule, show in turn that

$$\lim_{x \to +\infty} e^{-x}\, \mathrm{Ei}_1(x) = 0, \qquad \lim_{x \to +\infty} x e^{-x}\, \mathrm{Ei}_1(x) = 1,$$

and then deduce the asymptotic formula

$$\mathrm{Ei}_1(x) \approx \frac{e^x}{x}, \qquad x \to +\infty.$$

7. Using (3.4.7) and the result of the preceding problem, deduce the asymptotic formula

$$\mathrm{li}_1(x) \approx \frac{x}{\log x}, \qquad x \to +\infty.$$

Comment. This formula plays an important role in number theory.

8. Prove the formula

$$\mathrm{li}_1(x) = \lim_{\varepsilon \to 0} \left(\int_0^{1-\varepsilon} \frac{dt}{\log t} + \int_{1+\varepsilon}^x \frac{dt}{\log t} \right), \qquad x > 1.$$

Hint. Use (3.4.7) and the result of Problem 4.

9. Consider the integral

$$\int f(z)e^z\,dz, \tag{i}$$

where $f(z)$ is an arbitrary rational function, and the path of integration does not pass through any singular points of the integrand. By separating out the polynomial part of $f(z)$ and then expanding the remainder in partial fractions, the evaluation of (i) can be reduced to the evaluation of integrals of the form

$$\int z^n e^z\,dz, \tag{ii}$$

$$\int \frac{e^z}{(z-a)^n}\,dz, \tag{iii}$$

where n is a positive integer. By repeated integration by parts, (ii) can be expressed in terms of elementary functions, and the problem of evaluating (iii) can be reduced to the problem of evaluating the integral

$$\int \frac{e^z}{z-a}\,dz. \tag{iv}$$

Then the substitution $u = z - a$ reduces (iv) to an exponential integral (generally with a complex argument).

Using the method just described, prove that

$$\int_{-\infty}^{x} \frac{e^t}{t^2(t-1)}\,dt = \frac{e^x}{x} - 2\mathrm{Ei}(x) + e\mathrm{Ei}(x-1), \qquad x < 0.$$

10. As usual, let $\bar{f}$ denote the Laplace transform of f (see p. 25). Prove that

$$\overline{\mathrm{Si}}(x) = \frac{1}{p}\arctan\frac{1}{p}, \qquad -\overline{\mathrm{Ei}}(-x) = \frac{1}{p}\log(1+p),$$

where the arc tangent and the logarithm have their principal values.

4

ORTHOGONAL POLYNOMIALS

4.1. Introductory Remarks

A system of real functions $f_n(x)$ $(n = 0, 1, 2, \ldots)$ is said to be *orthogonal with weight* $\rho(x)$ on the interval $[a, b]$ if

$$\int_a^b \rho(x) f_m(x) f_n(x)\, dx = 0 \tag{4.1.1}$$

for every $m \neq n$, where $\rho(x)$ is a fixed nonnegative function which does not depend on the indices m and n. For example, the system of functions $\cos nx$ $(n = 0, 1, 2, \ldots)$ is orthogonal with weight 1 on the interval $[0, \pi]$, since

$$\int_0^\pi \cos mx \cos nx\, dx = 0 \qquad \text{if} \quad m \neq n.$$

Orthogonal systems play an important role in analysis, mainly because functions belonging to very general classes can be expanded in series of orthogonal functions, e.g., Fourier series, Fourier-Bessel series, etc.

An important class of orthogonal systems consists of orthogonal *polynomials* $p_n(x)$ $(n = 0, 1, 2, \ldots)$, where n is the degree of the polynomial $p_n(x)$. This class contains many special functions commonly encountered in the applications, e.g., Legendre, Hermite, Laguerre, Chebyshev and Jacobi polynomials. In addition to the orthogonality property (4.1.1), these functions have many other general properties. For example, they are the integrals of differential equations of a simple form, and can be defined as the coefficients in expansions in powers of t of suitably chosen functions $w(x, t)$, called *generating functions*. Orthogonal polynomials are of great importance in mathematical physics, approximation theory, the theory of mechanical quadratures, etc., and are the subject of an enormous literature, in which the contributions of Russian mathematicians like Adamov, Akhiezer, Bernstein, Chebyshev, Sonine, Steklov and Uspensky play a prominent role.

This chapter is devoted to the theory of Legendre, Hermite and Laguerre polynomials, which have extremely diverse applications to physics and engineering. For the convenience of readers primarily concerned with applications, each of these three kinds of polynomials is treated independently. Those interested in studying the subject from a more general point of view are referred to the books by Jackson, Sansone, Szegö and Tricomi cited in the Bibliography on p. 300.[1] In Problems 21–22, p.96–97, we also touch upon the theory of Jacobi and Chebyshev polynomials.

4.2. Definition and Generating Function of the Legendre Polynomials

The Legendre polynomials are defined by *Rodrigues' formula*

$$P_n(x) = \frac{1}{2^n n!}\frac{d^n}{dx^n}(x^2-1)^n, \qquad n = 0, 1, 2, \ldots \tag{4.2.1}$$

for arbitrary real or complex values of the variable x. Thus the first few Legendre polynomials are

$$P_0(x) = 1, \qquad P_1(x) = x, \qquad P_2(x) = \tfrac{1}{2}(3x^2-1),$$
$$P_3(x) = \tfrac{1}{2}(5x^3-3x), \ldots$$

The general expression for the nth Legendre polynomial is obtained from (4.2.1) by using the familiar binomial expansion

$$(x^2-1)^n = \sum_{k=0}^{n} \frac{(-1)^k n!}{k!(n-k)!} x^{2n-2k},$$

which implies

$$P_n(x) = \sum_{k=0}^{[n/2]} \frac{(-1)^k(2n-2k)!}{2^n k!(n-k)!(n-2k)!} x^{n-2k}, \tag{4.2.2}$$

where the symbol $[\nu]$ denotes the largest integer $\leqslant \nu$. It will be shown in Sec. 4.5 that the Legendre polynomials are orthogonal with weight 1 on the

---

[1] See also A. Erdélyi, W. Magnus, F. Oberhettinger and F. G. Tricomi, *Higher Transcendental Functions, Volume 2* (of three volumes), Chap. 10, McGraw-Hill Book Co., New York (1953). This three-volume set (based, in part, on notes left by Harry Bateman) will henceforth be referred to as the Bateman manuscript Project, *Higher Transcendental Functions.*

interval $[-1, 1]$.[2] As already noted, these orthogonal polynomials play an important role in the applications, particularly, in mathematical physics (see Secs. 8.3–4, 8.7–8, 8.13–14).

The properties of the Legendre polynomials can be derived very simply if we first prove that the function

$$w(x, t) = (1 - 2xt + t^2)^{-1/2}$$

(where the value of the square root is taken to be 1 for $t = 0$) is the *generating function* of the Legendre polynomials, i.e., that the expansion

$$w(x, t) = (1 - 2xt + t^2)^{-1/2} = \sum_{n=0}^{\infty} P_n(x)t^n \tag{4.2.3}$$

holds for sufficiently small $|t|$. Let r_1 and r_2 be the roots of the quadratic equation $1 - 2xt + t^2 = 0$, and let

$$r = \min\{|r_1|, |r_2|\}. \tag{4.2.4}$$

Then $w(x, t)$, regarded as a function of t, is analytic in the disk $|t| < r$.[3] It follows from a familiar theorem of complex variable theory[4] that

$$w(x, t) = (1 - 2xt + t^2)^{-1/2} = \sum_{n=0}^{\infty} c_n(x)t^n, \qquad |t| < r,$$

where the coefficients $c_n(x)$ can be written as contour integrals

$$c_n(x) = \frac{1}{2\pi i}\int_C (1 - 2xt + t^2)^{-1/2}t^{-n-1}\,dt, \tag{4.2.5}$$

evaluated along any closed contour C surrounding the point $t = 0$ and lying inside the disk $|t| < r$. If we make the substitution

$$1 - ut = (1 - 2xt + t^2)^{1/2},$$

then (4.2.5) transforms into the following integral of a rational function evaluated along a closed contour C' surrounding the point $u = x$:[5]

$$c_n(x) = \frac{1}{2\pi i}\int_{C'} \frac{(u^2 - 1)^n}{2^n(u - x)^{n+1}}\,du. \tag{4.2.6}$$

---

[2] This property can be proved directly, by starting from the definition (4.2.1), but our approach will be different. In fact, it can be shown that if $p_n(x)$ $(n = 0, 1, 2, \ldots)$ is an arbitrary system of polynomials orthogonal with weight 1 on the interval $[-1, 1]$, then $p_n(x) = \gamma_n P_n(x)$, where γ_n is independent of x. See G. E. Shilov, *An Introduction to the Theory of Linear Spaces* (translated by R. A. Silverman), Prentice-Hall, Inc., Englewood Cliffs, N.J. (1961), Sec. 58.

[3] In the case of greatest practical importance, x is a real number belonging to the interval $[-1, 1]$, and then $r = 1$.

[4] A. I. Markushevich, *op. cit.*, Theorem 16.7, p. 361.

[5] The point $u = x$ corresponds to the point $t = 0$, and the closed contour C' corresponds to the closed contour C, since the square root returns to its original value after making a circuit around C.

This integral can be evaluated by residue theory. In fact, using the familiar rule,[6] we find that

$$c_n(x) = \frac{1}{2^n n!}\left[\frac{d^n(u^2-1)^n}{du^n}\right]_{u=x} \equiv P_n(x),$$

thereby verifying (4.2.3).

To illustrate the utility of the generating function for deriving properties of the Legendre polynomials, we successively set $x = 1, -1, 0$ in (4.2.3), each time expanding the left-hand side in powers of t. As a result, we obtain the important formulas

$$\begin{gathered} P_n(1) = 1, \qquad P_n(-1) = (-1)^n, \\ P_{2n}(0) = (-1)^n \frac{1\cdot 3\cdots(2n-1)}{2\cdot 4\cdots 2n}, \qquad P_{2n+1}(0) = 0. \end{gathered} \tag{4.2.7}$$

4.3. Recurrence Relations and Differential Equation for the Legendre Polynomials

We further illustrate the use of the expansion (4.2.3) by deriving some recurrence relations satisfied by the Legendre polynomials. First we substitute the series (4.2.3) into the identity

$$(1 - 2xt + t^2)\frac{\partial w}{\partial t} + (t - x)w = 0.$$

Since power series can be differentiated term by term, this gives

$$(1 - 2xt + t^2)\sum_{n=0}^{\infty} nP_n(x)t^{n-1} + (t - x)\sum_{n=0}^{\infty} P_n(x)t^n = 0.$$

Setting the coefficient of t^n equal to zero, we find that

$$(n+1)P_{n+1}(x) - 2nxP_n(x) + (n-1)P_{n-1}(x) + P_{n-1}(x) - xP_n(x) = 0,$$

or

$$(n+1)P_{n+1}(x) - (2n+1)xP_n(x) + nP_{n-1}(x) = 0, \quad n = 1, 2, \ldots, \tag{4.3.1}$$

which is a recurrence relation connecting three Legendre polynomials with consecutive indices. One can use this relation to calculate the Legendre polynomials step by step, starting from $P_0(x) = 1$, $P_1(x) = x$.

Similarly, the identity

$$(1 - 2xt + t^2)\frac{\partial w}{\partial x} - tw = 0$$

---

[6] See F. B. Hildebrand, *Advanced Calculus for Applications*, Prentice-Hall, Inc., Englewood Cliffs, N.J. (1962), p. 548.

leads to[7]

$$(1 - 2xt + t^2) \sum_{n=0}^{\infty} P_n'(x)t^n - \sum_{n=0}^{\infty} P_n(x)t^{n+1} = 0,$$

which implies

$$P_{n+1}'(x) - 2xP_n'(x) + P_{n-1}'(x) - P_n(x) = 0, \qquad n = 1, 2, \ldots \tag{4.3.2}$$

Differentiating (4.3.1), we first eliminate $P_{n-1}'(x)$ and then $P_{n+1}'(x)$ from the resulting equation and (4.3.2). This gives two further recurrence relations[8]

$$P_{n+1}'(x) - xP_n'(x) = (n + 1)P_n(x), \qquad n = 0, 1, 2, \ldots, \tag{4.3.3}$$

$$xP_n'(x) - P_{n-1}'(x) = nP_n(x), \qquad n = 1, 2, \ldots \tag{4.3.4}$$

Adding (4.3.3) and (4.3.4), we obtain the more symmetric formula

$$P_{n+1}'(x) - P_{n-1}'(x) = (2n + 1)P_n(x), \qquad n = 1, 2, \ldots \tag{4.3.5}$$

Finally, replacing n by $n - 1$ in (4.3.3), and eliminating $P_{n-1}'(x)$ from the resulting equation and (4.3.4), we find that

$$(1 - x^2)P_n'(x) = nP_{n-1}(x) - nxP_n(x), \qquad n = 1, 2, \ldots \tag{4.3.6}$$

This last formula allows us to express the derivative of a Legendre polynomial in terms of Legendre polynomials. If we differentiate (4.3.6) with respect to x and again use (4.3.4) to eliminate $P_{n-1}'(x)$, we arrive at the formula

$$[(1 - x^2)P_n'(x)]' + n(n + 1)P_n(x) = 0, \qquad n = 0, 1, 2, \ldots, \tag{4.3.7}$$

which shows that the Legendre polynomial $u = P_n(x)$ is a particular integral of the second-order linear differential equation

$$[(1 - x^2)u']' + n(n + 1)u = 0. \tag{4.3.8}$$

This equation is often encountered in mathematical physics, and plays an

---

[7] To justify differentiating (4.2.3) term by term with respect to x, it is sufficient to prove that (4.2.3) converges uniformly in the domain $|x| < a$, for arbitrary finite $a > 0$ and sufficiently small $|t|$. (Here we rely on Weierstrass' theorem, cited in footnote 5, p. 2.) Let $|t| < b$, where $b = \sqrt{a^2 + 1} - a$. Then, according to (4.2.3), the series

$$\sum_{n=0}^{\infty} \frac{P_n(ia)}{i^n} |t|^n$$

converges to $(1 - 2a|t| - |t|^2)^{-1/2}$. The uniform convergence of (4.2.3) for $|x| < a$, $|t| < b$ now follows from the inequality

$$|P_n(x)t^n| \leqslant \frac{P_n(ia)}{i^n} |t|^n,$$

implied by (4.2.2).

[8] In some cases, the validity of a recurrence relation for small n does not follow from the general argument, but then one can always verify the relation by direct substitution of $P_0(x) = 1$, $P_1(x) = x, \ldots$

important role in the theory of Legendre polynomials. By making changes of variables in (4.3.8), we can easily derive many other equations whose integrals can be expressed in terms of Legendre polynomials. Thus, for example, the equation

$$\frac{1}{\sin\theta}\frac{d}{d\theta}\left(\sin\theta\frac{du}{d\theta}\right) + n(n+1)u = 0 \tag{4.3.9}$$

is satisfied by the function $u = P_n(\cos\theta)$, the equation

$$\frac{d^2u}{d\theta^2} + \left[(n+\tfrac{1}{2})^2 + \frac{1}{4\sin^2\theta}\right]u = 0 \tag{4.3.10}$$

is satisfied by the function $u = \sqrt{\sin\theta}\, P_n(\cos\theta)$, and so on.

4.4. Integral Representations of the Legendre Polynomials

The Legendre polynomials have simple representations in terms of definite integrals with the variable x as parameter. To obtain the first of these representations, we assume that x is a real or complex number, and choose the path of integration C' in formula (4.2.6) to be a circle of radius $\sqrt{|x^2-1|}$ with center at the point $u = x$.[9] Then

$$u = x + \sqrt{x^2-1}\,e^{i\varphi}, \qquad -\pi \leqslant \varphi \leqslant \pi,$$

and (4.2.6) becomes

$$P_n(x) = \frac{1}{2\pi}\int_{-\pi}^{\pi}\left[\frac{x^2 + 2x\sqrt{x^2-1}\,e^{i\varphi} + (x^2-1)e^{2i\varphi} - 1}{2\sqrt{x^2-1}\,e^{i\varphi}}\right]^n d\varphi,$$

which reduces to

$$P_n(x) = \frac{1}{\pi}\int_0^{\pi}[x + \sqrt{x^2-1}\cos\varphi]^n\,d\varphi. \tag{4.4.1}$$

Formula (4.4.1) is called *Laplace's integral.* Here the choice of the value of the square root $\sqrt{x^2-1}$ does not matter, since after raising the expression in brackets to the nth power and integrating the result term by term, odd powers of the square root vanish.

From (4.4.1) we can derive an important inequality satisfied by Legendre polynomials. Let x be a real number such that $-1 \leqslant x \leqslant 1$. Then

$$|x + \sqrt{x^2-1}\cos\varphi| = \sqrt{x^2 + (1-x^2)\cos^2\varphi} \leqslant 1,$$

and hence

$$|P_n(x)| \leqslant 1, \qquad -1 \leqslant x \leqslant 1. \tag{4.4.2}$$

Another important integral representation of the Legendre polynomials

---

[9] According to Cauchy's integral theorem, replacing the contour C' by any other closed Jordan curve surrounding the point $u = x$ does not change the value of the integral.

can be deduced from (4.4.1) by assuming that x is a real number such that $-1 < x < 1$. In this case, setting

$$x = \cos\theta, \qquad 0 < \theta < \pi,$$

we can write (4.4.1) in the form

$$P_n(\cos\theta) = \frac{1}{\pi}\int_0^{\pi} (\cos\theta + i\sin\theta\cos\varphi)^n \, d\varphi.$$

If we introduce a new complex variable of integration $t = \cos\theta + i\sin\theta\cos\varphi$ this formula becomes

$$P_n(\cos\theta) = \frac{1}{\pi i}\int_{e^{-i\theta}}^{e^{i\theta}} \frac{t^n \, dt}{\sqrt{1 - 2t\cos\theta + t^2}}, \tag{4.4.3}$$

where the integral is evaluated along the line segment AB joining the points $t = e^{\pm i\theta}$ (see Figure 11), and the choice of the square root is determined by

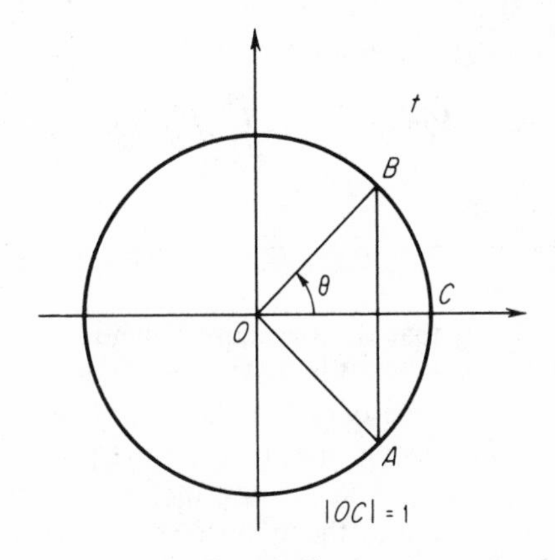

FIGURE 11

the condition that its value at the point $t = \cos\theta$ be $\sin\theta$. According to Cauchy's integral theorem, the integration along AB can be replaced by integration along the arc ACB of the unit circle, since the integrand is analytic in the region between the arc and the chord. Making this change, and writing $t = e^{i\psi}$, we find that

$$P_n(\cos\theta) = \frac{1}{\pi}\int_{-\theta}^{\theta} \frac{e^{i(n+\frac{1}{2})\psi}}{\sqrt{2\cos\psi - 2\cos\theta}} \, d\psi,$$

which becomes

$$P_n(\cos\theta) = \frac{2}{\pi}\int_0^{\theta} \frac{\cos(n+\frac{1}{2})\psi}{\sqrt{2\cos\psi - 2\cos\theta}} \, d\psi, \qquad 0 < \theta < \pi, \quad n = 0, 1, 2, \ldots, \tag{4.4.4}$$

after taking the real part. This integral representation is known as the *Mehler-Dirichlet formula.*

4.5. Orthogonality of the Legendre Polynomials

One of the most important properties of the Legendre polynomials is their orthogonality on the interval $[-1, 1]$, which follows from the differential equation (4.3.7). To prove this property, we subtract the differential equation for the nth polynomial multiplied by $P_m(x)$ from the differential equation for the mth polynomial multiplied by $P_n(x)$. This gives

$$[(1 - x^2)P_m'(x)]'P_n(x) - [(1 - x^2)P_n'(x)]'P_m(x) + [m(m + 1) - n(n + 1)]P_m(x)P_n(x) = 0,$$

or

$$\{(1 - x^2)[P_m'(x)P_n(x) - P_n'(x)P_m(x)]\}' + (m - n)(m + n + 1)P_m(x)P_n(x) = 0.$$

Integrating the last equation over the interval $[-1, 1]$ and noting that the integral of the first term vanishes, we find that

$$(m - n)(m + n - 1)\int_{-1}^{1} P_m(x)P_n(x)\,dx = 0,$$

i.e.,

$$\int_{-1}^{1} P_m(x)P_n(x)\,dx = 0 \quad \text{if} \quad m \neq n. \tag{4.5.1}$$

Formula (4.5.1) shows that the Legendre polynomials are orthogonal with weight $\rho(x) = 1$ on the interval $[-1, 1]$.

The orthogonality property (4.5.1) plays an important role in the theory of expansions of functions in series of Legendre polynomials (see Sec. 4.7). In this theory, we will also need to know the value of the integral (4.5.1) for $m = n$, which can be found by the following device (brought to our attention by V. L. Kan): We replace n by $n - 1$ in the recurrence relation (4.3.1) and multiply the result by $(2n + 1)P_n(x)$. Then from this equation we subtract (4.3.1) multiplied by $(2n - 1)P_{n-1}(x)$, obtaining

$$n(2n + 1)P_n^2(x) + (n - 1)(2n + 1)P_{n-2}(x)P_n(x) - (n + 1)(2n - 1)P_{n-1}(x)P_{n+1}(x) - n(2n - 1)P_{n-1}^2(x) = 0,$$
$$n = 2, 3, \ldots$$

Finally, integrating this relation over the interval $[-1, 1]$, and taking account of (4.5.1), we find that

$$\int_{-1}^{1} P_n^2(x)\,dx = \frac{2n - 1}{2n + 1}\int_{-1}^{1} P_{n-1}^2(x)\,dx, \qquad n = 2, 3, \ldots$$

Repeated application of this formula gives

$$\int_{-1}^{1} P_n^2(x)\,dx = \frac{3}{2n + 1}\int_{-1}^{1} P_1^2(x)\,dx = \frac{2}{2n + 1}.$$

Direct calculation shows that this result is also valid for $n = 0, 1$, and hence

$$\int_{-1}^{1} P_n^2(x)\,dx = \frac{2}{2n+1}, \qquad n = 0, 1, 2, \ldots \tag{4.5.2}$$

It follows from (4.5.1–2) that the functions

$$\varphi_n(x) = \sqrt{n + \tfrac{1}{2}}\,P_n(x), \qquad n = 0, 1, 2, \ldots$$

form an orthonormal system on the interval $[-1, 1]$.[10]

4.6. Asymptotic Representation of the Legendre Polynomials for Large n

The Legendre polynomials $P_n(x)$ $(-1 < x < 1)$ have a simple asymptotic representation which describes their behavior for large values of the degree n. To obtain this representation, we use a general method due to Steklov.[11] Our starting point is the differential equation (4.3.10) satisfied by the function

$$u(\theta) = \sqrt{\sin\theta}\,P_n(\cos\theta).$$

Writing this equation in the form

$$u'' + (n + \tfrac{1}{2})^2 u = -\frac{u}{4\sin^2\theta}, \tag{4.6.1}$$

taking account of the initial conditions

$$u\left(\frac{\pi}{2}\right) = P_n(0), \qquad u'\left(\frac{\pi}{2}\right) = -P_n'(0),$$

and regarding the right-hand side of (4.6.1) as a known function, we find that[12]

$$\begin{aligned} u(\theta) = P_n(0)\cos\left[\left(n+\frac{1}{2}\right)\left(\frac{\pi}{2}-\theta\right)\right] &+ \frac{P_n'(0)}{n+\frac{1}{2}}\sin\left[\left(n+\frac{1}{2}\right)\left(\frac{\pi}{2}-\theta\right)\right] \\ &+ \frac{1}{4(n+\frac{1}{2})}\int_{\theta}^{\pi/2} u(\varphi)\sin\left[(n+\tfrac{1}{2})(\theta-\varphi)\right]\frac{d\varphi}{\sin^2\varphi}. \end{aligned} \tag{4.6.2}$$

Equation (4.6.2) can be regarded as an integral equation for the function $u(\theta)$.

---

[10] A system of functions $\varphi_n(x)$ $(n = 0, 1, 2, \ldots)$ is said to be *orthonormal* on the interval $[a, b]$ if

$$\int_a^b \varphi_m(x)\varphi_n(x)\,dx = \begin{cases} 0, & m \neq n, \\ 1, & m = n. \end{cases}$$

[11] V. A. Steklov, *Sur les expressions asymptotiques de certaines fonctions, définies par les équations différentielles linéaires du second ordre, et leurs applications au problème du développement d'une fonction arbitraire en séries procédant suivant les-dites fonctions*, Communications de la Société Mathématique de Kharkow, (2), **10**, 97 (1907).

[12] See E. A. Coddington, *An Introduction to Ordinary Differential Equations*, Prentice-Hall, Inc., Englewood Cliffs, N.J. (1961), Theorem 11, p. 123.

Next, using formulas (4.2.6), (4.3.6), and the relations (1.2.1, 4, 6) involving the gamma function, we obtain

$$P_{2m}(0) = (-1)^m \frac{\Gamma(m+\frac{1}{2})}{\sqrt{\pi}\,\Gamma(m+1)}, \qquad P_{2m+1}(0) = 0,$$

$$P'_{2m}(0) = 0, \qquad P'_{2m+1}(0) = (-1)^m \frac{2\,\Gamma(m+\frac{3}{2})}{\sqrt{\pi}\,\Gamma(m+1)}.$$

It follows that equation (4.6.2) can be written in the form

$$u(\theta) = \alpha_n\left\{\sin\left[(n+\tfrac{1}{2})\theta + \frac{\pi}{4}\right] + r_n(\theta)\right\}, \tag{4.6.3}$$

where α_n denotes the first or the second of the expressions

$$\frac{\Gamma\left(\dfrac{n}{2}+\dfrac{1}{2}\right)}{\sqrt{\pi}\,\Gamma\left(\dfrac{n}{2}+1\right)}, \qquad \frac{2\Gamma\left(\dfrac{n}{2}+1\right)}{\sqrt{\pi}\left(n+\dfrac{1}{2}\right)\Gamma\left(\dfrac{n}{2}+\dfrac{1}{2}\right)},$$

depending on whether n is even or odd, and

$$r_n(\theta) = \frac{1}{4\alpha_n(n+\frac{1}{2})}\int_\theta^{\pi/2} u(\varphi)\sin\left[(n+\tfrac{1}{2})(\theta-\varphi)\right]\frac{d\varphi}{\sin^2\varphi}. \tag{4.6.4}$$

Now suppose that the variable θ is confined to the interval $\delta \leqslant \theta \leqslant \pi - \delta$, where δ is a fixed positive number, and let M_n denote the maximum modulus of $u(\theta)$ in this interval. Then it follows from (4.6.3) and (4.6.4) that for every θ in $[\delta, \pi - \delta]$,

$$|u(\theta)| \leqslant \alpha_n + \frac{\pi M_n}{4(2n+1)}\csc^2\delta,$$

and hence

$$M_n \leqslant \alpha_n + \frac{\pi M_n}{4(2n+1)}\csc^2\delta.$$

Solving this last inequality for M_n, we obtain

$$M_n \leqslant \alpha_n\left[1 - \frac{\pi}{4(2n+1)}\csc^2\delta\right]^{-1}, \qquad 2n+1 > \frac{\pi}{4}\csc^2\delta,$$

which implies the estimate

$$|r_n(\theta)| \leqslant \frac{\pi\csc^2\delta}{4(2n+1)}\left[1 - \frac{\pi}{4(2n+1)}\csc^2\delta\right]^{-1}, \qquad 2n+1 > \frac{\pi}{4}\csc^2\delta.$$

Thus $r_n(\theta) = O(n^{-1})$ uniformly in the interval $[\delta, \pi - \delta]$. Therefore (4.6.3) leads to the asymptotic formula

$$u(\theta) \approx \alpha_n \sin\left[(n+\tfrac{1}{2})\theta + \frac{\pi}{4}\right], \qquad n \to \infty \tag{4.6.5}$$

for all $\delta \leqslant \theta \leqslant \pi - \delta$.

Making some simple calculations based on Stirling's formula (1.4.25),[13] we find that

$$\alpha_n \approx \sqrt{\frac{2}{\pi n}}, \qquad n \to \infty,$$

and therefore (4.6.5) can be written in the simpler form

$$u(\theta) \approx \sqrt{\frac{2}{\pi n}} \sin\left[(n + \tfrac{1}{2})\theta + \frac{\pi}{4}\right], \qquad n \to \infty. \tag{4.6.6}$$

Recalling the definition of $u(\theta)$, we finally have the following asymptotic representation for the Legendre polynomials:

$$P_n(\cos\theta) \approx \sqrt{\frac{2}{\pi n \sin\theta}} \sin\left[(n + \tfrac{1}{2})\theta + \frac{\pi}{4}\right], \qquad n \to \infty, \quad \delta \leqslant \theta \leqslant \pi - \delta. \tag{4.6.7}$$

For more exact asymptotic representations, we refer the reader to Hobson's treatise.[14]

4.7. Expansion of Functions in Series of Legendre Polynomials

In the applications it is often necessary to expand a given real function $f(x)$, defined in the interval $(-1, 1)$, in a series of Legendre polynomials:

$$f(x) = \sum_{n=0}^{\infty} c_n P_n(x), \qquad -1 < x < 1. \tag{4.7.1}$$

The coefficients c_n can be determined formally by using the orthogonality property of the Legendre polynomials (see Sec. 4.5). In fact, multiplying the series (4.7.1) by $P_m(x)$, integrating term by term over the interval $[-1, 1]$ and using (4.5.1–2), we find that

$$\int_{-1}^{1} f(x)P_m(x)\,dx = \int_{-1}^{1} \sum_{n=0}^{\infty} c_n P_n(x)P_m(x)\,dx$$
$$= \sum_{n=0}^{\infty} c_n \int_{-1}^{1} P_m(x)P_n(x)\,dx = \frac{2}{2m+1} c_m,$$

which implies

$$c_n = (n + \tfrac{1}{2}) \int_{-1}^{1} f(x)P_n(x)\,dx, \qquad n = 0, 1, 2, \ldots \tag{4.7.2}$$

However, it is not known in advance whether $f(x)$ can be expanded in a series

---

[13] The fact that $\lim_{n\to\infty}\left(1 + \frac{x}{n}\right)^n = e^x$ is also used.

[14] E. W. Hobson, *The Theory of Spherical and Ellipsoidal Harmonics*, Cambridge University Press, London (1931).

of the form (4.7.1), or whether the term-by-term integration used to determine the coefficients c_n is legitimate. Therefore, it cannot be asserted without further study that the series (4.7.1) with the coefficients (4.7.2) actually converges and has the sum $f(x)$. In order to establish simple sufficient conditions for such convergence (see Theorem 1 below), we first prove the following

LEMMA. *If the real function* $\varphi(x)$ *is piecewise continuous*[15] *in* $(-1, 1)$ *and if the integral*

$$\int_{-1}^{1} \varphi^2(x)\, dx \tag{4.7.3}$$

is finite,[16] *then*

$$\lim_{n\to\infty} \sqrt{n+\tfrac{1}{2}} \int_{-1}^{1} \varphi(x) P_n(x)\, dx = 0. \tag{4.7.4}$$

Proof. First we write (4.7.4) as a sum of three integrals

$$\sqrt{n+\tfrac{1}{2}} \int_{-1}^{1} \cdots = \sqrt{n+\tfrac{1}{2}} \left[\int_{-1}^{-1+\delta} \cdots + \int_{-1+\delta}^{1-\delta} \cdots + \int_{1-\delta}^{1} \cdots\right]$$
$$= \mathscr{I}_1 + \mathscr{I}_2 + \mathscr{I}_3. \tag{4.7.5}$$

Then, using Schwarz's inequality[17] and formula (4.5.2), we find that

$$|\mathscr{I}_3| \leqslant \sqrt{n+\tfrac{1}{2}} \left[\int_{1-\delta}^{1} P_n^2(x)\, dx\right]^{1/2} \left[\int_{1-\delta}^{1} \varphi^2(x)\, dx\right]^{1/2}$$
$$\leqslant \sqrt{n+\tfrac{1}{2}} \left[\int_{-1}^{1} P_n^2(x)\, dx\right]^{1/2} \left[\int_{1-\delta}^{1} \varphi^2(x)\, dx\right]^{1/2} = \left[\int_{1-\delta}^{1} \varphi^2(x)\, dx\right]^{1/2},$$

and similarly,

$$|\mathscr{I}_1| \leqslant \left[\int_{-1}^{-1+\delta} \varphi^2(x)\, dx\right]^{1/2}.$$

It follows from these estimates and the existence of (4.7.3) that given any $\varepsilon > 0$, there is a $\delta = \delta(\varepsilon) > 0$, independent of n, such that

$$|\mathscr{I}_1| < \frac{\varepsilon}{3}, \qquad |\mathscr{I}_3| < \frac{\varepsilon}{3}. \tag{4.7.6}$$

---

[15] For the definition of piecewise continuous and piecewise smooth functions, see G. P. Tolstov, *op. cit.*, p. 18.

[16] If $\varphi(x)$ is defined only in $(-1, 1)$, then (4.7.3) means

$$\lim_{a,b\to 0+} \int_{-1+a}^{1-b} \varphi^2(x)\, dx.$$

If $\varphi(x)$ is piecewise continuous in the *closed* interval $[-1, 1]$, then the finiteness of (4.7.3) is obvious. In other words, we allow $\varphi(x)$ to become infinite at the end points -1 and 1, provided the integral (4.7.3) remains finite.

[17] According to Schwarz's inequality,

$$\left[\int_a^b f(x)g(x)\, dx\right]^2 \leqslant \int_a^b f^2(x)\, dx \int_a^b g^2(x)\, dx,$$

provided the integrals on the right exist. See G. P. Tolstov, *op. cit.*, p. 50.

Assuming that δ has been chosen in this way, we now use (4.6.3) to write

$$\begin{aligned}\mathscr{I}_2 &= \sqrt{n+\tfrac{1}{2}}\int_{\delta_1}^{\pi-\delta_1} \varphi(\cos\theta)\, P_n(\cos\theta)\sin\theta\, d\theta \\ &= \sqrt{n+\tfrac{1}{2}}\,\alpha_n\left[\frac{1}{\sqrt{2}}\int_{\delta_1}^{\pi-\delta_1}\varphi(\cos\theta)\sqrt{\sin\theta}\,\sin\left(n+\tfrac{1}{2}\right)\theta\, d\theta\right. \\ &\quad + \frac{1}{\sqrt{2}}\int_{\delta_1}^{\pi-\delta_1}\varphi(\cos\theta)\sqrt{\sin\theta}\,\cos\left(n+\tfrac{1}{2}\right)\theta\, d\theta \\ &\quad \left. + \int_{\delta_1}^{\pi-\delta_1}\varphi(\cos\theta)\sqrt{\sin\theta}\, r_n(\theta)\, d\theta\right],\end{aligned}$$

where $\delta_1 = \arccos(1-\delta)$. Since, by hypothesis, $\varphi(\cos\theta)\sqrt{\sin\theta}$ is piecewise continuous and hence absolutely integrable on $[\delta_1, \pi-\delta_1]$, the first two integrals on the right approach zero as $n\to\infty$.[18] Moreover, the last integral also approaches zero as $n\to\infty$, since $r_n(\theta) = O(n^{-1})$ uniformly in $[\delta_1, \pi-\delta_1]$, as shown in Sec. 4.6, where it was also proved that

$$\sqrt{n+\tfrac{1}{2}}\,\alpha_n \to \sqrt{\frac{2}{\pi}}$$

as $n\to\infty$. Therefore $\mathscr{I}_2 \to 0$ as $n\to\infty$, so that for a suitable choice of $N = N(\varepsilon)$, we have

$$|\mathscr{I}_2| < \frac{\varepsilon}{3} \tag{4.7.7}$$

for every $n > N$. Combining (4.7.7) and (4.7.6) we find that

$$|\mathscr{I}_1 + \mathscr{I}_2 + \mathscr{I}_3| < \varepsilon, \qquad n < N,$$

and the lemma is proved.

We are now ready to prove

THEOREM 1. *If the real function* $f(x)$ *is piecewise smooth in* $(-1, 1)$ *and if the integral*

$$\int_{-1}^{1} f^2(x)\, dx \tag{4.7.8}$$

is finite, then the series (4.7.1), *with coefficients* c_n *calculated from* (4.7.2), *converges to* $f(x)$ *at every continuity point of* $f(x)$.

Proof. First we note that the conditions imposed on $f(x)$ imply the existence of the integrals in the right-hand side of (4.7.2),[19] so that the coefficients c_n can actually be calculated. Let $S_m(x)$ denote the sum of

---

[18] G. P. Tolstov, *op. cit.*, p. 70.

[19] Apply Schwarz's inequality to the functions $f(x)$ and $P_n(x)$.

the first $m + 1$ terms of the series (4.7.1). Then it follows from (4.7.2) that

$$S_m(x) = \sum_{n=0}^{m} c_n P_n(x) = \sum_{n=0}^{m} (n + \tfrac{1}{2}) P_n(x) \int_{-1}^{1} f(y) P_n(y)\, dy$$
$$= \int_{-1}^{1} f(y) K_m(x, y)\, dy, \tag{4.7.9}$$

where

$$K_m(x, y) = \sum_{n=0}^{m} (n + \tfrac{1}{2}) P_n(x) P_n(y). \tag{4.7.10}$$

The "kernel" $K_m(x, y)$ can be calculated by the following device: We multiply the recurrence relation (4.3.1) by $P_n(y)$ and then from the resulting equation we subtract the same equation with x and y interchanged. This gives

$$(n + 1)[P_{n+1}(x)P_n(y) - P_{n+1}(y)P_n(x)]$$
$$- n[P_n(x)P_{n-1}(y) - P_n(y)P_{n-1}(x)]$$
$$= (2n + 1)(x - y)P_n(x)P_n(y).$$

Summing over n from 1 to m, and noting that $P_0(x) = 1$, $P_1(x) = x$, we obtain

$$(x - y) \sum_{n=1}^{m} (2n + 1) P_n(x) P_n(y)$$
$$= (m + 1)[P_{m+1}(x)P_m(y) - P_{m+1}(y)P_m(x)] - (x - y),$$

which implies

$$K_m(x, y) = \frac{m + 1}{2} \frac{P_{m+1}(x)P_m(y) - P_{m+1}(y)P_m(x)}{x - y}. \tag{4.7.11}$$

Integrating (4.7.10) with respect to y between the limits -1 and 1, and using (4.5.1–2),[20] we find that

$$\int_{-1}^{1} K_m(x, y)\, dy = 1. \tag{4.7.12}$$

Now suppose x is a point of $(-1, 1)$ at which $f(x)$ is continuous. Multiplying (4.7.12) by $f(x)$, subtracting the result from (4.7.9), and using (4.7.11), we obtain

$$S_m(x) - f(x) = \int_{-1}^{1} K_m(x, y)[f(y) - f(x)]\, dy$$
$$= \frac{m + 1}{2} P_m(x) \int_{-1}^{1} P_{m+1}(y) \varphi(x, y)\, dy \tag{4.7.13}$$
$$- \frac{m + 1}{2} P_{m+1}(x) \int_{-1}^{1} P_m(y) \varphi(x, y)\, dy,$$

[20] Since $P_0(y) = 1$, we have

$$\int_{-1}^{1} P_n(y)\, dy = \int_{-1}^{1} P_0(y) P_n(y)\, dy = \begin{cases} 0, & n \neq 0, \\ 2, & n = 0. \end{cases}$$

where

$$\varphi(x, y) = \frac{f(y) - f(x)}{y - x}.$$

Regarded as a function of y, $\varphi(x, y)$ is piecewise continuous in $(-1, 1)$, and moreover

$$\int_{-1}^{1} \varphi^2(x, y)\, dy \tag{4.7.14}$$

is finite. In fact, if $y \neq x$, the piecewise continuity of $\varphi(x, y)$ in $(-1, 1)$ follows from that of $f(y)$,while $\varphi(x, y)$ is piecewise continuous at $y = x$ since

$$\varphi(x, x - 0) = f'(x - 0), \qquad \varphi(x, x + 0) = f'(x + 0)$$

both exist if x is a continuity point of $f(x)$.[21] The fact that (4.7.14) is finite follows from (4.7.8) and the fact that $\varphi(x, y)$ is bounded in a neighborhood of $y = x$, where both $\varphi(x, x - 0)$ and $\varphi(x, x + 0)$ exist. Therefore, according to the lemma,

$$\lim_{m \to \infty} \sqrt{m + \tfrac{3}{2}} \int_{-1}^{1} P_{m+1}(y)\varphi(x, y)\, dy$$
$$= \lim_{m \to \infty} \sqrt{m + \tfrac{1}{2}} \int_{-1}^{1} P_m(y)\varphi(x, y)\, dy = 0.$$

Moreover, using (4.6.7), we see that each of the expressions

$$\frac{m + 1}{2\sqrt{m + \frac{3}{2}}} P_m(x), \qquad \frac{m + 1}{2\sqrt{m + \frac{1}{2}}} P_{m+1}(x)$$

remains bounded as $m \to \infty$. It follows that the right-hand side of (4.7.13) goes to zero as $m \to \infty$, i.e.,

$$\lim_{m \to \infty} S_m(x) = f(x),$$

and the proof of Theorem 1 is complete.

Remark 1. The case where x is a discontinuity point of $f(x)$ is also of interest. It can be shown that in this case, under the same conditions as in Theorem 1, the series (4.7.1) converges to the limit[22]

$$\lim_{m \to \infty} S_m(x) = \tfrac{1}{2}[f(x + 0) + f(x - 0)]. \tag{4.7.15}$$

Remark 2. Theorem 1 gives sufficient conditions for expanding $f(x)$ in a

---

[21] Cf. G. P. Tolstov, *op. cit.*, p. 73.

[22] This should be compared with the similar situation encountered in the theory of Fourier series (*ibid.*, p. 75 ff.).

series of the form (4.7.1). These conditions can be considerably weakened. A theorem which is valid for a larger class of functions can be found in Hobson's book.[23]

4.8. Examples of Expansions in Series of Legendre Polynomials

We now give some simple examples illustrating the technique of expanding functions in series of Legendre polynomials:

Example 1. Let $f(x)$ be a polynomial of degree m:

$$f(x) = \sum_{n=0}^{m} a_n x^n.$$

Then (4.7.1) takes the form

$$f(x) = \sum_{n=0}^{m} c_n P_n(x). \tag{4.8.1}$$

In this case, there is no need to calculate the integrals (4.7.2), since the coefficients c_n can easily be found by solving the system of linear equations obtained when the explicit expressions for the Legendre polynomials are substituted into (4.8.1) and coefficients of identical powers of x in both sides of the equation are equated. Thus, for example,

$$x^2 = c_0 P_0(x) + c_1 P_1(x) + c_2 P_2(x) = c_0 + c_1 x + \tfrac{1}{2} c_2 (3x^2 - 1),$$

so that

$$c_0 = \tfrac{1}{3}, \qquad c_1 = 0, \qquad c_2 = \tfrac{2}{3}.$$

Therefore

$$x^2 = \tfrac{1}{3} P_0(x) + \tfrac{2}{3} P_2(x),$$

an expansion which is valid for all x.

Example 2. Suppose $f(x)$ is the function

$$f(x) = \begin{cases} 0, & -1 \leqslant x < \alpha, \\ 1, & \alpha < x \leqslant 1. \end{cases}$$

According to Theorem 1, $f(x)$ can be expanded in a series of the form (4.7.1), with coefficients

$$c_n = (n + \tfrac{1}{2}) \int_{\alpha}^{1} P_n(x)\, dx.$$

Using (4.3.5) and noting that $P_n(1) = 1$, we find that

$$c_n = -\tfrac{1}{2}[P_{n+1}(\alpha) - P_{n-1}(\alpha)], \qquad c_0 = \tfrac{1}{2}(1 - \alpha),$$

[23] E. W. Hobson, *op. cit.*, p. 329.

which leads to the required expansion

$$f(x) = \tfrac{1}{2}(1 - \alpha) - \tfrac{1}{2} \sum_{n=1}^{m} [P_{n+1}(\alpha) - P_{n-1}(\alpha)]P_n(x), \quad -1 < x < 1. \quad (4.8.2)$$

Next, we verify that the relation (4.7.15) holds at the discontinuity point $x = \alpha$. Letting $S_m(x)$ denote the sum of the first $m + 1$ terms of the series (4.8.2), we have

$$\begin{aligned} S_m(\alpha) &= \tfrac{1}{2}(1 - \alpha) - \tfrac{1}{2} \sum_{n=1}^{m} [P_{n+1}(\alpha)P_n(\alpha) - P_n(\alpha)P_{n-1}(\alpha)] \\ &= \tfrac{1}{2} - \tfrac{1}{2}P_{m+1}(\alpha)P_m(\alpha). \end{aligned}$$

Since, according to (4.6.7), $P_n(\alpha) \to 0$ as $n \to \infty$,

$$\lim_{m \to \infty} S_m(\alpha) = \tfrac{1}{2} = \tfrac{1}{2}[f(\alpha + 0) + f(\alpha - 0)],$$

in keeping with the general theory.

Example 3. Finally, let

$$f(x) = \sqrt{\frac{1 - x}{2}}.$$

This function satisfies the conditions of Theorem 1, and hence can be expanded in a series of the form (4.7.1). The coefficients c_n can be calculated by the following method, which is often useful: We multiply the expansion (4.2.3) by $f(x)$ and integrate over the interval $[-1, 1]$. After some elementary calculations, we obtain

$$\frac{1}{2t}\left[1 + t - \frac{(1 - t)^2}{2\sqrt{t}} \log \frac{1 + \sqrt{t}}{1 - \sqrt{t}}\right] = \sum_{n=0}^{\infty} t^n \int_{-1}^{1} \sqrt{\frac{1 - x}{2}}\, P_n(x)\, dx, \quad |t| < 1, \tag{4.8.3}$$

where the term-by-term integration is justified by the uniform convergence of the series (4.2.3) in the interval $[-1, 1]$, which follows from the estimate (4.4.2). Expanding the left-hand side of (4.8.3) in powers of t, we find that

$$\frac{4}{3} - 4 \sum_{n=1}^{\infty} \frac{t^n}{(4n^2 - 1)(2n + 3)} = \sum_{n=0}^{\infty} t^n \int_{-1}^{1} \sqrt{\frac{1 - x}{2}}\, P_n(x)\, dx,$$

which implies

$$\int_{-1}^{1} \sqrt{\frac{1 - x}{2}}\, P_0(x)\, dx = \frac{4}{3},$$

$$\int_{-1}^{1} \sqrt{\frac{1 - x}{2}}\, P_n(x)\, dx = -\frac{4}{(4n^2 - 1)(2n + 3)}.$$

We now use (4.7.2) to write the required expansion in the form

$$\sqrt{\frac{1-x}{2}} = \frac{2}{3}P_0(x) - 2\sum_{n=1}^{\infty}\frac{P_n(x)}{(2n-1)(2n+3)}, \qquad -1 < x < 1. \tag{4.8.4}$$

4.9. Definition and Generating Function of the Hermite Polynomials

Another important class of orthogonal polynomials encountered in the applications, especially in mathematical physics,[24] consists of the *Hermite polynomials* $H_n(x)$,[25] which can be defined by the formula

$$H_n(x) = (-1)^n e^{x^2}\frac{d^n e^{-x^2}}{dx^n}, \qquad n = 0, 1, 2, \ldots \tag{4.9.1}$$

According to (4.9.1), the first few Hermite polynomials are

$$H_0(x) = 1, \qquad H_1(x) = 2x, \qquad H_2(x) = 4x^2 - 2,$$
$$H_3(x) = 8x^3 - 12x, \ldots,$$

and in general,

$$H_n(x) = \sum_{k=0}^{[n/2]}\frac{(-1)^k n!}{k!(n-2k)!}(2x)^{n-2k}, \tag{4.9.2}$$

where $[\nu]$ denotes the largest integer $\leqslant \nu$. It will be shown later (see Sec. 4.13) that the Hermite polynomials are orthogonal with weight $\rho(x) = e^{-x^2}$ on the interval $(-\infty, \infty)$.

The Hermite polynomials (or more exactly, the Hermite polynomials multiplied by the constant factor $1/n!$) are the coefficients in the expansion

$$w(x, t) = e^{2xt-t^2} = \sum_{n=0}^{\infty}\frac{H_n(x)}{n!}t^n, \qquad |t| < \infty, \tag{4.9.3}$$

and hence $w(x, t)$ is called the *generating function* of the Hermite polynomials. To prove (4.9.3), we need only note that $w(x, t)$, regarded as a function of the complex variable t, is an entire function, and therefore has the Taylor series

$$w(x, t) = e^{2xt-t^2} = \sum_{n=0}^{\infty}\frac{1}{n!}\left[\frac{\partial^n w}{\partial t^n}\right]_{t=0} t^n, \qquad |t| < \infty,$$

---

[24] In problems involving the integration of Laplace's equation and Helmholtz' equation in parabolic coordinates, in quantum mechanics, etc. (see Secs. 10.7–8).

[25] Actually introduced in 1859 by Chebyshev, some years before the publication of Hermite's work.

which immediately implies (4.9.3), since

$$\left(\frac{\partial^n w}{\partial t^n}\right)_{t=0} = e^{x^2}\left[\frac{\partial^n}{\partial t^n} e^{-(x-t)^2}\right]_{t=0} = (-1)^n e^{x^2}\left[\frac{d^n e^{-u^2}}{du^n}\right]_{u=x} \equiv H_n(x).$$

Formula (4.9.3) can be used to derive various properties of the Hermite polynomials. For example, setting $x = 0$ in (4.9.3), expanding e^{-t^2} in power series, and comparing coefficients of powers of t in both sides of the resulting equation, we find that

$$H_{2n}(0) = (-1)^n \frac{(2n)!}{n!}, \qquad H_{2n+1}(0) = 0. \tag{4.9.4}$$

There is another expansion closely related to (4.9.3), which we will prove in Sec. 4.11, i.e.,

$$W(x, y, t) = (1 - t^2)^{-1/2} e^{[2xyt - (x^2 + y^2)t^2]/(1-t^2)} = \sum_{n=0}^{\infty} \frac{H_n(x)H_n(y)}{2^n n!} t^n, \quad |t| < 1, \tag{4.9.5}$$

where the left-hand side can be regarded as the generating function of *products* of Hermite polynomials. Setting $y = x$ in (4.9.5), we obtain

$$W(x, x, t) = (1 - t^2)^{-1/2} e^{2x^2 t/(1+t)} = \sum_{n=0}^{\infty} \frac{H_n^2(x)}{2^n n!} t^n, \qquad |t| < 1. \tag{4.9.6}$$

Formulas (4.9.3, 4, 6) play an important role in the theory of Hermite polynomials.

4.10. Recurrence Relations and Differential Equation for the Hermite Polynomials

Substituting (4.9.3) into the identity

$$\frac{\partial w}{\partial t} - (2x - 2t)w = 0$$

(a power series can always be differentiated term by term), we find that

$$\sum_{n=0}^{\infty} \frac{H_{n+1}(x)}{n!} t^n - 2x \sum_{n=0}^{\infty} \frac{H_n(x)}{n!} t^n + 2 \sum_{n=0}^{\infty} \frac{H_n(x)}{n!} t^{n+1} = 0,$$

which gives

$$H_{n+1}(x) - 2xH_n(x) + 2nH_{n-1}(x) = 0, \qquad n = 1, 2, \ldots \tag{4.10.1}$$

when the coefficient of t^n is equated to zero. The recurrence relation (4.10.1), connecting three Hermite polynomials with consecutive indices, can be used to calculate the Hermite polynomials step by step, starting from $H_0(x) = 1$, $H_1(x) = 2x$.

We can derive another recurrence relation satisfied by the Hermite polynomials by substituting (4.9.3) into the identity[26]

$$\frac{\partial w}{\partial x} - 2tw = 0.$$

This gives

$$\sum_{n=0}^{\infty} \frac{H_n'(x)}{n!} t^n - 2 \sum_{n=0}^{\infty} \frac{H_n(x)}{n!} t^{n+1} = 0,$$

or

$$H_n'(x) = 2nH_{n-1}(x), \qquad n = 1, 2, \ldots \tag{4.10.2}$$

Formula (4.10.2) allows us to express the derivative of a Hermite polynomial in terms of another Hermite polynomial, and is very useful. Using the recurrence relations (4.10.1–2), we can easily derive a differential equation satisfied by the Hermite polynomials. In fact, eliminating $H_{n-1}(x)$ from these two relations, we obtain

$$H_{n+1}(x) - 2xH_n(x) + H_n'(x) = 0.$$

Then, differentiating this formula and using (4.10.2) again, we find that

$$H_n''(x) - 2xH_n'(x) + 2nH_n(x) = 0, \qquad n = 0, 1, 2, \ldots, \tag{4.10.3}$$

where the validity of (4.10.3) for $n = 0$ can be verified directly. It follows from (4.10.3) that the function $u = H_n(x)$ is a particular integral of the second-order linear differential equation

$$u'' - 2xu' + 2nu = 0. \tag{4.10.4}$$

By making changes of variables, we can easily derive other differential equations whose integrals can be expressed in terms of Hermite polynomials. For example, it is easy to see that

$$u = e^{-x^2/2} H_n(x)$$

is a particular solution of the equation

$$u'' + (2n + 1 - x^2)u = 0. \tag{4.10.5}$$

---

[26] The justification for differentiating (4.9.3) term by term with respect to x follows from the uniform convergence of (4.9.3) in the domain $|x| < a$ for arbitrary finite $a > 0$. According to (4.9.2),

$$|H_n(x)| \leqslant \frac{H_n(ia)}{i^n}, \qquad |x| < a,$$

so that (4.9.3) is majorized by the convergent series

$$\sum_{n=0}^{\infty} \frac{H_n(ia)}{i^n} \frac{|t|^n}{n!} = e^{2a|t| + |t|^2}$$

and hence converges uniformly for $|x| < a$ (cf. footnote 7, p. 47).

4.11. Integral Representations of the Hermite Polynomials

The Hermite polynomials have simple and useful representations in terms of definite integrals containing the variable x as parameter. To derive these representations, we start from the familiar integral

$$e^{-x^2} = \frac{2}{\sqrt{\pi}} \int_0^\infty e^{-t^2} \cos 2xt \, dt, \tag{4.11.1}$$

where x is an arbitrary real or complex number. Differentiating (4.11.1) $2n$ times with respect to x,[27] and comparing the result with (4.9.1), we find that

$$H_{2n}(x) = \frac{2^{2n+1}(-1)^n e^{x^2}}{\sqrt{\pi}} \int_0^\infty e^{-t^2} t^{2n} \cos 2xt \, dt, \qquad n = 0, 1, 2, \ldots \tag{4.11.2}$$

Similarly, for odd indices we have

$$H_{2n+1}(x) = \frac{2^{2n+2}(-1)^n e^{x^2}}{\sqrt{\pi}} \int_0^\infty e^{-t^2} t^{2n+1} \sin 2xt \, dt, \qquad n = 0, 1, 2, \ldots, \tag{4.11.3}$$

which can be combined with (4.11.2) into a single formula

$$H_n(x) = \frac{2^n(-i)^n e^{x^2}}{\sqrt{\pi}} \int_{-\infty}^\infty e^{-t^2+2itx} t^n \, dt, \qquad n = 0, 1, 2, \ldots \tag{4.11.4}$$

To illustrate the utility of these representations, we now derive formula (4.9.5). According to (4.11.4), for $|t| < 1$ we have

$$\begin{aligned}
\sum_{n=0}^\infty \frac{H_n(x)H_n(y)}{2^n n!} t^n &= \frac{e^{x^2+y^2}}{\pi} \sum_{n=0}^\infty \frac{(-1)^n}{n!}(2t)^n \\
&\qquad \times \int_{-\infty}^\infty \int_{-\infty}^\infty e^{-u^2-v^2+2iux+2ivy}(uv)^n \, du \, dv \\
&= \frac{e^{x^2+y^2}}{\pi} \int_{-\infty}^\infty \int_{-\infty}^\infty e^{-u^2-v^2+2iux+2ivy} \, du \, dv \sum_{n=0}^\infty \frac{(-1)^n(2uvt)^n}{n!} \\
&= \frac{e^{x^2+y^2}}{\pi} \int_{-\infty}^\infty \int_{-\infty}^\infty e^{-u^2-v^2+2iux+2ivy-2uvt} \, du \, dv.
\end{aligned} \tag{4.11.5}$$

After two applications of the familiar formula

$$\int_{-\infty}^\infty e^{-a^2s^2-2bs} \, ds = \frac{\sqrt{\pi}}{a} e^{b^2/a^2}, \qquad \operatorname{Re} a^2 > 0, \tag{4.11.6}$$

[27] To justify differentiating behind the integral sign, see E. C. Titchmarsh, *op. cit.* pp. 99–100, noting that the integral in (4.11.1) is uniformly convergent in the disk $|x| \leqslant a$ for arbitrary finite $a > 0$, since it is majorized by the absolutely convergent integral

$$\frac{2}{\sqrt{\pi}} \int_0^\infty e^{-t^2+2at} \, dt.$$

the right-hand side of (4.11.5) reduces to

$$W(x, y, t) = (1 - t^2)^{-1/2} e^{[2xyt - (x^2+y^2)t^2]/(1-t^2)}.$$

The legitimacy of the various formal calculations follows from the convergence of the expression

$$\frac{e^{|x|^2+|y|^2}}{\pi} \int_{-\infty}^{\infty} \int_{-\infty}^{\infty} e^{-u^2-v^2+2|u|\,|x|+2|v|\,|y|}\, du\, dv \sum_{n=0}^{\infty} \frac{(2|u|\,|v|\,|t|)^n}{n!}$$

for all $|t| < 1$.

4.12. Integral Equations Satisfied by the Hermite Polynomials

The Hermite polynomials satisfy simple integral equations with symmetric kernels. To derive these equations, we replace x by y in the expansion (4.9.3) of the generating function, multiply the result by $e^{ixy - \frac{1}{2}y^2}$ $(-\infty < x < \infty)$ and integrate over $(-\infty, \infty)$. This gives

$$\begin{aligned} \int_{-\infty}^{\infty} e^{2yt - t^2 + ixy - \frac{1}{2}y^2}\, dy &= \int_{-\infty}^{\infty} e^{ixy - \frac{1}{2}y^2}\, dy \sum_{n=0}^{\infty} \frac{H_n(y)}{n!}\, t^n \\ &= \sum_{n=0}^{\infty} \frac{t^n}{n!} \int_{-\infty}^{\infty} e^{ixy - \frac{1}{2}y^2} H_n(y)\, dy. \end{aligned} \tag{4.12.1}$$

Interchanging the order of integration and summation is permissible, since

$$\begin{aligned} \int_{-\infty}^{\infty} |e^{ixy - \frac{1}{2}y^2}|\, dy \sum_{n=0}^{\infty} \frac{|H_n(y)|}{n!} |t|^n &\leqslant \int_{-\infty}^{\infty} e^{-\frac{1}{2}y^2}\, dy \sum_{n=0}^{\infty} \frac{1}{n!} \left(\frac{|t|}{i}\right)^n H_n(i|y|) \\ &= \int_{-\infty}^{\infty} e^{-\frac{1}{2}y^2 + 2|y|\,|t| + |t|^2}\, dy < \infty, \end{aligned}$$

where we have used the inequality

$$|H_n(x)| \leqslant \frac{1}{i^n} H_n(i|x|),$$

implied by (4.9.2).

Evaluating the integral in the left-hand side of (4.12.1), we find that

$$\begin{aligned} \int_{-\infty}^{\infty} e^{2yt - t^2 - \frac{1}{2}y^2 + ixy}\, dy &= \sqrt{2\pi}\, e^{t^2 + 2ixt - \frac{1}{2}x^2} \\ &= \sqrt{2\pi}\, e^{-\frac{1}{2}x^2} \sum_{n=0}^{\infty} \frac{(it)^n}{n!} H_n(x). \end{aligned} \tag{4.12.2}$$

Comparing coefficients of identical powers of t in (4.12.1–2), we obtain the desired integral equation satisfied by the Hermite polynomials

$$e^{-x^2/2} H_n(x) = \frac{1}{i^n \sqrt{2\pi}} \int_{-\infty}^{\infty} e^{ixy} e^{-y^2/2} H_n(y)\, dy, \qquad n = 0, 1, 2, \ldots \tag{4.12.3}$$

If we consider separately the cases of even and odd n, bearing in mind that $H_{2m}(x)$ is an even function and $H_{2m+1}(x)$ an odd function (of the variable x), then (4.12.3) implies the following two integral equations with real kernels:

$$e^{-x^2/2}H_{2m}(x) = (-1)^m\sqrt{\frac{2}{\pi}}\int_0^\infty e^{-y^2/2}H_{2m}(y)\cos xy\,dy,$$

$$e^{-x^2/2}H_{2m+1}(x) = (-1)^m\sqrt{\frac{2}{\pi}}\int_0^\infty e^{-y^2/2}H_{2m+1}(y)\sin xy\,dy, \quad m = 0, 1, 2, \ldots \tag{4.12.4}$$

4.13. Orthogonality of the Hermite Polynomials

It is easy to show that the Hermite polynomials are orthogonal with weight e^{-x^2} on the interval $(-\infty, \infty)$, i.e.,

$$\int_{-\infty}^\infty e^{-x^2}H_m(x)H_n(x)\,dx = 0 \quad \text{if} \quad m \neq n. \tag{4.13.1}$$

In fact, setting $u_n = e^{-x^2/2}H_n(x)$ and using equation (4.10.5), we have

$$u_n'' + (2n + 1 - x^2)u_n = 0, \qquad u_m'' + (2m + 1 - x^2)u_m = 0.$$

Multiplying the first of these equations by u_m and the second by u_n, we see that

$$\frac{d}{dx}(u_n'u_m - u_m'u_n) + 2(n - m)u_mu_n = 0. \tag{4.13.2}$$

Then, integrating (4.13.2) over $(-\infty, \infty)$, we find that

$$(n - m)\int_{-\infty}^\infty u_mu_n\,dx = 0,$$

which implies (4.13.1).

The value of the integral (4.13.1) for $m = n$ can be found as follows: We replace the index n by $n - 1$ in the recurrence relation (4.10.1) and multiply the result by $H_n(x)$. Then from this equation we subtract (4.10.1) multiplied by $H_{n-1}(x)$. This gives

$$H_n^2(x) + 2(n - 1)H_n(x)H_{n-2}(x) - H_{n+1}(x)H_{n-1}(x) - 2nH_{n-1}^2(x) = 0,$$
$$n = 2, 3, \ldots \tag{4.13.3}$$

Multiplying (4.13.3) by e^{-x^2}, integrating over $(-\infty, \infty)$ and using the orthogonality property (4.13.1), we obtain

$$\int_{-\infty}^\infty e^{-x^2}H_n^2(x)\,dx = 2n\int_{-\infty}^\infty e^{-x^2}H_{n-1}^2(x)\,dx, \qquad n = 2, 3, \ldots$$

Repeated application of this formula gives[28]

$$\int_{-\infty}^{\infty} e^{-x^2} H_n^2(x)\,dx = 2^{n-1}n! \int_{-\infty}^{\infty} e^{-x^2} H_1^2(x)\,dx = 2^n n!\sqrt{\pi}, \qquad n = 2, 3, \ldots$$

Direct calculation shows that this result is also valid for $n = 0, 1$, and hence

$$\int_{-\infty}^{\infty} e^{-x^2} H_n^2(x)\,dx = 2^n n!\sqrt{\pi}, \qquad n = 0, 1, 2, \ldots \tag{4.13.4}$$

It follows from (4.13.1, 4) that the functions

$$\varphi_n(x) = (2^n n!\sqrt{\pi})^{-1/2} e^{-x^2/2} H_n(x), \qquad n = 0, 1, 2, \ldots$$

form an orthonormal system on the interval $(-\infty, \infty)$.

4.14. Asymptotic Representation of the Hermite Polynomials for Large *n*

The Hermite polynomials have a simple asymptotic representation which describes their behavior for large values of the degree n. This representation was first found by Adamov,[29] and plays an important role in the problem of expanding functions in series of Hermite polynomials (see Sec. 4.15). We again apply the general method used in Sec. 4.6 to solve the analogous problem for the Legendre polynomials. Our starting point is the differential equation (4.10.5) for the function $u = e^{-x^2/2}H_n(x)$. Writing this equation in the form

$$u'' + (2n + 1)u = x^2 u, \tag{4.14.1}$$

taking account of the initial conditions

$$u(0) = H_n(0), \qquad u'(0) = H_n'(0),$$

and regarding the right-hand side of (4.14.1) as a known function, we find that

$$\begin{aligned} u(x) &= H_n(0)\cos\sqrt{2n+1}\,x + H_n'(0)\frac{\sin\sqrt{2n+1}\,x}{\sqrt{2n+1}} \\ &\quad + \frac{1}{\sqrt{2n+1}}\int_0^x y^2 u(y)\sin[\sqrt{2n+1}\,(x-y)]\,dy. \end{aligned} \tag{4.14.2}$$

---

[28] Note that

$$\int_{-\infty}^{\infty} e^{-x^2} H_1^2(x)\,dx = 4\int_{-\infty}^{\infty} e^{-x^2} x^2\,dx = 2\sqrt{\pi}.$$

[29] A. A. Adamov, *On the asymptotic expansion of the polynomials* $e^{\alpha x^2/2}\, d^n(e^{-\alpha x^2/2})/dx^n$ *for large values of n* (in Russian), Annals of the Polytechnic Insitute of St. Petersburg, **5**, 127 (1906).

Next, using formulas (4.9.4), (4.10.2) and (1.2.1, 4), we obtain

$$H_{2m}(0) = (-1)^m \frac{\Gamma(2m+1)}{\Gamma(m+1)}, \qquad H_{2m+1}(0) = 0,$$

$$H'_{2m}(0) = 0, \qquad H'_{2m+1}(0) = 2(-1)^m \frac{\Gamma(2m+2)}{\Gamma(m+1)}.$$

It follows that equation (4.14.2) can be written in the form

$$u(x) = \alpha_n\left[\cos\left(\sqrt{2n+1}\,x - \frac{n\pi}{2}\right) + r_n(x)\right], \tag{4.14.3}$$

where α_n denotes the first or the second of the expressions

$$\frac{\Gamma(n+1)}{\Gamma\left(\frac{n}{2}+1\right)}, \qquad \frac{2\Gamma(n+1)}{\sqrt{2n+1}\,\Gamma\left(\frac{n}{2}+\frac{1}{2}\right)}, \tag{4.14.4}$$

depending on whether n is even or odd, and

$$r_n(x) = \frac{1}{\alpha_n\sqrt{2n+1}} \int_0^x y^2 u(y) \sin\left[\sqrt{2n+1}\,(x-y)\right] dy. \tag{4.14.5}$$

To estimate the remainder $r_n(x)$ for arbitrary real x, we use Schwarz's inequality (see footnote 17, p. 54). Taking account of (4.13.4), we have

$$\begin{aligned} |r_n(x)| &\leqslant \frac{1}{\alpha_n\sqrt{2n+1}} \left[\int_0^{|x|} y^4\,dy\right]^{1/2} \left[\int_0^{|x|} u^2(y)\,dy\right]^{1/2} \\ &= \frac{1}{a_n\sqrt{2n+1}} \left[\int_0^{|x|} y^4\,dy\right]^{1/2} \left[\int_0^{\infty} u^2(y)\,dy\right]^{1/2} \\ &= \frac{(2^n n!\sqrt{\pi})^{1/2}}{\alpha_n\sqrt{2n+1}} \frac{|x|^{5/2}}{\sqrt{2}\sqrt{5}} = \beta_n|x|^{5/2}. \end{aligned}$$

It follows from Stirling's formula (1.4.25) that

$$\alpha_n \approx 2^{(n+1)/2} n^{n/2} e^{-n/2}, \qquad 2^n n!\sqrt{\pi} \approx 2^{n+\frac{1}{2}} e^{-n} n^{n+\frac{1}{2}} \pi \tag{4.14.6}$$

as $n \to \infty$, and hence the product $\beta_n n^{1/4}$ is bounded for arbitrary $n \geqslant 0$. Therefore

$$|r_n(x)| \leqslant C|x|^{5/2} n^{-1/4}, \tag{4.14.7}$$

where C is some constant. This last inequality shows that for any finite x we have the asymptotic formula

$$u(x) \approx \alpha_n \cos\left(\sqrt{2n+1}\,x - \frac{n\pi}{2}\right), \qquad n \to \infty \tag{4.14.8}$$

or

$$H_n(x) \approx 2^{(n+1)/2} n^{n/2} e^{-n/2} e^{x^2/2} \cos\left(\sqrt{2n+1}\,x - \frac{n\pi}{2}\right), \qquad n \to \infty. \tag{4.14.9}$$

For more exact asymptotic representations of the Hermite polynomials $H_n(x)$ for large n, we refer the reader to the monographs by Szegö and Sansone, cited in the Bibliography (see p. 300).

4.15. Expansion of Functions in Series of Hermite Polynomials

We now show that a real function $f(x)$ defined in the infinite interval $(-\infty, \infty)$ can be expanded in a series of Hermite polynomials

$$f(x) = \sum_{n=0}^{\infty} c_n H_n(x), \qquad -\infty < x < \infty, \tag{4.15.1}$$

provided $f(x)$ satisfies certain general conditions. The coefficients c_n can be determined formally by using the orthogonality property of the Hermite polynomials (see Sec. 4.13). In fact, multiplying the series (4.15.1) by $e^{-x^2}H_m(x)$, integrating term by term over the interval $(-\infty, \infty)$, and using (4.13.1, 4), we find that

$$\int_{-\infty}^{\infty} e^{-x^2} f(x) H_m(x)\, dx = \sum_{n=0}^{\infty} c_n \int_{-\infty}^{\infty} e^{-x^2} H_m(x) H_n(x)\, dx = 2^m m! \sqrt{\pi}\, c_m,$$

which implies

$$c_n = \frac{1}{2^n n! \sqrt{\pi}} \int_{-\infty}^{\infty} e^{-x^2} f(x) H_n(x)\, dx, \qquad n = 0, 1, 2, \ldots \tag{4.15.2}$$

In the course of establishing simple sufficient conditions for the series (4.15.1) with these coefficients to actually converge and to have the sum $f(x)$, we will need the following

LEMMA. *If the real function $\varphi(x)$ defined in the infinite interval $(-\infty, \infty)$ is piecewise continuous in every finite subinterval $[-a, a]$ and if the integral*

$$\int_{-\infty}^{\infty} (1 + x^2) e^{-x^2} \varphi^2(x)\, dx \tag{4.15.3}$$

is finite, then

$$\lim_{n\to\infty} \frac{n^{1/4}}{(2^n n! \sqrt{\pi})^{1/2}} \int_{-\infty}^{\infty} e^{-x^2} H_n(x) \varphi(x)\, dx = 0. \tag{4.15.4}$$

Proof. First we write the integral (4.15.4) as a sum of three integrals

$$\frac{n^{1/4}}{(2^n n! \sqrt{\pi})^{1/2}} \int_{-\infty}^{\infty} \cdots = \frac{n^{1/4}}{(2^n n! \sqrt{\pi})^{1/2}} \left[\int_{-\infty}^{-a} \cdots + \int_{-a}^{a} \cdots + \int_{a}^{\infty} \cdots \right]$$

$$= \mathscr{I}_1 + \mathscr{I}_2 + \mathscr{I}_3. \tag{4.15.5}$$

Then, using Schwarz's inequality, we find that

$$|\mathscr{I}_1| \leqslant \frac{n^{1/4}}{(2^n n!\sqrt{\pi})^{1/2}} \int_{-\infty}^{-a} e^{-x^2}|H_n(x)|\,|\varphi(x)|\,dx$$

$$\leqslant \frac{n^{1/4}}{(2^n n!\sqrt{\pi})^{1/2}} \left[\int_{-\infty}^{-a} \frac{H_n^2(x)}{1+x^2} e^{-x^2}\,dx\right]^{1/2} \left[\int_{-\infty}^{-a} (1+x^2)\,e^{-x^2}\varphi^2(x)\,dx\right]^{1/2}.$$

$$\leqslant \left[\frac{\sqrt{n}}{2^n n!\sqrt{\pi}} \int_{-\infty}^{\infty} \frac{H_n^2(x)}{1+x^2}\,dx\right]^{1/2} \left[\int_{-\infty}^{-a} (1+x^2)e^{-x^2}\varphi^2(x)\,dx\right]^{1/2}, \tag{4.15.6}$$

and similarly,

$$|\mathscr{I}_3| \leqslant \left[\frac{\sqrt{n}}{2^n n!\sqrt{\pi}} \int_{-\infty}^{\infty} \frac{H_n^2(x)}{1+x^2} e^{-x^2}\,dx\right]^{1/2} \left[\int_{a}^{\infty} (1+x^2)e^{-x^2}\varphi^2(x)\,dx\right]^{1/2}. \tag{4.15.7}$$

Our next step is to show that the integral

$$\mathscr{I} = \frac{\sqrt{n}}{2^n n!\sqrt{\pi}} \int_{-\infty}^{\infty} \frac{H_n^2(x)}{1+x^2} e^{-x^2}\,dx \tag{4.15.8}$$

satisfies the condition

$$\mathscr{I} = O(1), \tag{4.15.9}$$

i.e., $\mathscr{I}$ is bounded for all n. To show this, we use the identity

$$\mathscr{I} = \sqrt{\frac{n}{\pi}} \int_{-\infty}^{\infty} \left(\frac{1-x^2}{1+x^2}\right)^n \frac{e^{-x^2}}{1+x^2}\,dx, \tag{4.15.10}$$

proved in Problem 8, p. 95. Writing (4.5.10) in the form

$$\mathscr{I} = 2\sqrt{\frac{n}{\pi}} \left[\int_0^1 \cdots + \int_1^{\infty} \cdots\right],$$

and making the change of variable $x \to x^{-1}$ in the second integral, we obtain

$$\mathscr{I} = 2\sqrt{\frac{n}{\pi}} \int_0^1 \left(\frac{1-x^2}{1+x^2}\right)^n \frac{e^{-x^2} + (-1)^n e^{-x^{-2}}}{1+x^2}\,dx. \tag{4.15.11}$$

Since

$$e^{-x^2} + (-1)^n e^{-x^{-2}} \leqslant 2, \qquad 0 \leqslant x \leqslant 1,$$

it follows from (4.15.11) that

$$\mathscr{I} \leqslant 4\sqrt{\frac{n}{\pi}} \int_0^1 \left(\frac{1-x^2}{1+x^2}\right)^n \frac{dx}{1+x^2}.$$

The integral on the right can be evaluated by making the substitution

$$\frac{1-x^2}{1+x^2} = \sqrt{1-t}.$$

Then

$$\frac{dx}{1+x^2} = \frac{dt}{4\sqrt{t(1-t)}},$$

and according to (1.5.2),

$$4\int_0^1 \left(\frac{1-x^2}{1+x^2}\right)^n \frac{dx}{1+x^2} = \int_0^1 t^{-1/2}(1-t)^{(n-1)/2}\,dt = B\left(\frac{1}{2}, \frac{n+1}{2}\right),$$

where $B(x, y)$ is the beta function. Using (1.5.6) and (1.2.5), we find that

$$B\left(\frac{1}{2}, \frac{n+1}{2}\right) = \frac{\sqrt{\pi}\,\Gamma\left(\frac{n+1}{2}\right)}{\Gamma\left(\frac{n}{2}+1\right)},$$

and hence

$$\mathscr{I} \leqslant \frac{\sqrt{n}\,\Gamma\left(\frac{n+1}{2}\right)}{\Gamma\left(\frac{n}{2}+1\right)}.$$

The estimate (4.15.9) is now an immediate consequence of Stirling's formula (1.4.25).

Since $\mathscr{I}$ is bounded, it follows from the existence of (4.15.3) that given any $\varepsilon > 0$, there is an $a = a(\varepsilon) > 0$, independent of n, such that

$$|\mathscr{I}_1| < \frac{\varepsilon}{3}, \qquad |\mathscr{I}_3| < \frac{\varepsilon}{3}. \tag{4.15.12}$$

Assuming that a has been chosen in this way, we now use (4.14.3) to write

$$\mathscr{I}_2 = \frac{\alpha_n n^{1/4}}{(2^n n!\sqrt{\pi})^{1/2}}\left[\int_{-a}^{a} e^{-x^2/2}\varphi(x)\cos\left(\sqrt{2n+1}\,x - \frac{n\pi}{2}\right)dx + \int_{-a}^{a} e^{-x^2/2}\varphi(x)r_n(x)\,dx\right].$$

Since $\varphi(x)e^{-x^2/2}$ is piecewise continuous and hence absolutely integrable in $[-a, a]$, the first of the integrals on the right approaches zero as $n \to \infty$. The second integral also approaches zero as $n \to \infty$, since, according to (4.14.7), the integrand is $O(n^{-1/4})$ uniformly in $[-a, a]$, while the factor in front of the brackets is bounded, as follows from (4.14.6). Therefore $\mathscr{I}_2 \to 0$ as $n \to \infty$, so that for a suitable choice of $N = N(\varepsilon)$, we have

$$|\mathscr{I}_2| < \frac{\varepsilon}{3} \tag{4.15.13}$$

for every $n > N$. Combining (4.15.13) and (4.15.12), we and that

$$|\mathscr{J}_1 + \mathscr{J}_2 + \mathscr{J}_3| < \varepsilon, \qquad n > N,$$

and the lemma is proved.

We are now ready to prove

THEOREM 2. *If the real function $f(x)$ defined in the infinite interval $(-\infty, \infty)$ is piecewise smooth in every finite interval $[-a, a]$, and if the integral*

$$\int_{-\infty}^{\infty} e^{-x^2} f^2(x)\, dx \tag{4.15.14}$$

is finite, then the series (4.15.1), *with coefficients c_n calculated from* (4.15.2), *converges to $f(x)$ at every continuity point of $f(x)$.*

Proof. First we note that the conditions imposed on $f(x)$ imply the existence of the integrals in the right-hand side of (4.15.2), so that the coefficients c_n can actually be calculated.[30] Let $S_m(x)$ denote the sum of the first $m + 1$ terms of the series (4.15.1). Then it follows from (4.1.52) that

$$\begin{aligned} S_m(x) = \sum_{n=0}^{m} c_n H_n(x) &= \sum_{n=0}^{m} H_n(x) \frac{1}{2^n n! \sqrt{\pi}} \int_{-\infty}^{\infty} e^{-y^2} f(y) H_n(y)\, dy \\ &= \int_{-\infty}^{\infty} e^{-y^2} f(y) K_m(x, y)\, dy, \end{aligned} \tag{4.15.15}$$

where

$$K_m(x, y) = \frac{1}{\sqrt{\pi}} \sum_{n=0}^{m} \frac{H_n(x) H_n(y)}{2^n n!}. \tag{4.15.16}$$

The "kernel" $K_m(x, y)$ can be calculated by the following device: We multiply the recurrence relation (4.10.1) by $H_n(y)$ and then from the resulting equation we subtract the same equation with x and y interchanged. This gives

$$\begin{aligned} [H_{n+1}(x)H_n(y) - H_{n+1}(y)H_n(x)] - 2n[H_n(x)H_{n-1}(y) - H_n(y)H_{n-1}(x)] \\ = 2(x - y)H_n(x)H_n(y), \qquad n = 1, 2, \ldots \end{aligned} \tag{4.15.17}$$

Dividing (4.15.17) by $2^n n!$, summing over n from 1 to m, and noting that $H_0(x) = 1$, $H_1(x) = 2x$, we obtain

$$2(x - y) \sum_{n=1}^{m} \frac{H_n(x)H_n(y)}{2^n n!} = \frac{H_{m+1}(x)H_m(y) - H_{m+1}(y)H_m(x)}{2^m m!} - 2(x - y),$$

which implies

$$K_m(x, y) = \frac{H_{m+1}(x)H_m(y) - H_{m+1}(y)H_m(x)}{(x - y)2^{m+1} m! \sqrt{\pi}}. \tag{4.15.18}$$

---

[30] Apply Schwarz's inequality to the functions $e^{-x^2/2} f(x)$ and $e^{-x^2/2} H_n(x)$.

We note that $K_m(x, y)$ satisfies the important identity

$$\int_{-\infty}^{\infty} e^{-y^2} K_m(x, y)\, dy = 1, \tag{4.15.19}$$

which is an immediate consequence of (4.15.16) and (4.13.1, 4).[31]

Now suppose x is a continuity point of $f(x)$, and consider the difference $S_m(x) - f(x)$, which, according to (4.15.15) and (4.15.18, 19), can be written in the form

$$\begin{aligned} S_m(x) - f(x) &= \int_{-\infty}^{\infty} e^{-y^2} K_m(x, y)[f(y) - f(x)]\, dy \\ &= \frac{H_m(x)}{2^{m+1} m! \sqrt{\pi}} \int_{-\infty}^{\infty} e^{-y^2} H_{m+1}(y) \varphi(x, y)\, dy \\ &\quad - \frac{H_{m+1}(x)}{2^{m+1} m! \sqrt{\pi}} \int_{-\infty}^{\infty} e^{-y^2} H_m(y) \varphi(x, y)\, dy, \end{aligned} \tag{4.15.20}$$

where

$$\varphi(x, y) = \frac{f(y) - f(x)}{y - x}.$$

Regarded as a function of y, $\varphi(x, y)$ is piecewise continuous in $(-\infty, \infty)$, for exactly the same reasons as given in the proof of Theorem 1, p. 55. Moreover, the integral

$$\int_{-\infty}^{\infty} (1 + y^2) e^{-y^2} \varphi(x, y)\, dy$$

is finite, since $\varphi(x, y)$ is bounded in any neighborhood of $y = x$ (see p. 57), and for sufficiently large $b > x$,

$$\begin{aligned} \int_{b}^{\infty} (1 + y^2) e^{-y^2} \varphi^2(x, y)\, dy &= \int_{b}^{\infty} (1 + y^2) e^{-y^2} \left[\frac{f(y) - f(x)}{y - x}\right]^2 dy \\ &= O(1) \int_{b}^{\infty} e^{-y^2} [f^2(y) + f^2(x)]\, dy, \end{aligned}$$

where the last integral is finite, because of (4.15.14). A similar estimate can be given for the interval $(-\infty, -b)$. Therefore, according to the lemma,

$$\begin{aligned} &\lim_{m \to \infty} \frac{(m + 1)^{1/4}}{[2^{m+1}(m + 1)! \sqrt{\pi}]^{1/2}} \int_{-\infty}^{\infty} e^{-y^2} H_{m+1}(y) \varphi(x, y)\, dy \\ &\quad = \lim_{m \to \infty} \frac{m^{1/4}}{(2^m m! \sqrt{\pi})^{1/2}} \int_{-\infty}^{\infty} e^{-y^2} H_m(y) \varphi(x, y)\, dy = 0. \end{aligned} \tag{4.15.21}$$

---

[31] Since $H_0(y) = 1$, we have

$$\int_{-\infty}^{\infty} e^{-y^2} H_n(y)\, dy = \int_{-\infty}^{\infty} e^{-y^2} H_0(y) H_n(y)\, dy = \begin{cases} 0, & n \neq 0, \\ \sqrt{\pi}, & n = 0. \end{cases}$$

On the other hand, according to (4.14.19) and Stirling's formula, each of the expressions

$$\frac{[2^{2m+1}(m+1)!\sqrt{\pi}]^{1/2}}{(m+1)^{1/4}} \frac{H_m(x)}{2^{m+1}m!\sqrt{\pi}}, \qquad \frac{(2^m m!\sqrt{\pi})^{1/2}}{m^{1/4}} \frac{H_{m+1}(x)}{2^{m+1}m!\sqrt{\pi}}$$

remains bounded as $m \to \infty$. It follows that the right-hand side of (4.15.20) goes to zero as $m \to \infty$, i.e.,

$$\lim_{m \to \infty} S_m(x) = f(x),$$

and the proof of Theorem 2 is complete.

Remark 1. The case where x is a discontinuity of $f(x)$ is also of interest. It can be shown that in this case, under the same conditions as in Theorem 2, the series (4.15.1) converges to the limit

$$\tfrac{1}{2}[f(x+0) + f(x-0)].$$

Remark 2. Other sufficient conditions for expanding a function $f(x)$ in a series of Hermite polynomials can be found in the books mentioned at the end of Sec. 4.1.[32]

4.16. Examples of Expansions in Series of Hermite Polynomials

In applying Theorem 2 to a given function $f(x)$, we have to evaluate the integral in (4.15.2). In most cases this is done by replacing $H_n(x)$ by its explicit expression (4.9.1) or by one of the integral representations given in Sec. 4.11. The following examples serve to illustrate the technique of expanding functions in series of Hermite polynomials:

Example 1. The function

$$f(x) = x^{2p}, \qquad p = 0, 1, 2, \ldots$$

satisfies the conditions of Theorem 2. In this case,

$$x^{2p} = \sum_{n=0}^{p} c_{2n} H_{2n}(x),$$

where

$$c_{2n} = \frac{1}{2^{2n}(2n)!\sqrt{\pi}} \int_{-\infty}^{\infty} e^{-x^2} x^{2p} H_{2n}(x)\, dx.$$

[32] See also J. Korous, *On expansion of functions of one real variable in a series of Hermite polynomials* (in Czech), Rozpravy České Akademie, (2), **37**, no. 11 (1928).

Substituting from (4.9.1) and integrating by parts n times, we find that

$$c_{2n} = \frac{1}{2^{2n}(2n)!\sqrt{\pi}} \int_{-\infty}^{\infty} x^{2p} \frac{d^{2n}}{dx^{2n}} (e^{-x^2})\, dx$$

$$= \frac{1}{2^{2n}(2n)!\sqrt{\pi}} \frac{(2p)!}{(2p-2n)!} \int_{-\infty}^{\infty} e^{-x^2} x^{2p-2n}\, dx$$

$$= \frac{1}{2^{2n}(2n)!\sqrt{\pi}} \frac{(2p)!}{(2p-2n)!} \Gamma(p - n + \tfrac{1}{2}).$$

According to the duplication formula (1.2.3) for the gamma function,

$$2^{2p-2n}\Gamma(p - n + \tfrac{1}{2})(p-n)! = \sqrt{\pi}(2p-2n)!,$$

and therefore the expression for c_{2n} simplifies to

$$c_{2n} = \frac{(2p)!}{2^{2p}(2n)!(p-n)!}.$$

Thus the desired expansion is

$$x^{2p} = \frac{(2p)!}{2^{2p}} \sum_{n=0}^{p} \frac{H_{2n}(x)}{(2n)!(p-n)!}, \qquad -\infty < x < \infty, \qquad p = 0, 1, 2, \ldots \tag{4.16.1}$$

In the same way, we find that

$$x^{2p+1} = \frac{(2p+1)!}{2^{2p+1}} \sum_{n=0}^{p} \frac{H_{2n+1}(x)}{(2n+1)!(p-n)!}, \quad -\infty < x < \infty, \quad p = 0, 1, 2, \ldots \tag{4.16.2}$$

Example 2. Let $f(x) = e^{ax}$, where a is an arbitrary real or complex number. Then the same method as used in Example 1 shows that

$$e^{ax} = \sum_{n=0}^{\infty} c_n H_n(x),$$

where

$$c_n = \frac{1}{2^n n!\sqrt{\pi}} \int_{-\infty}^{\infty} e^{-x^2+ax} H_n(x)\, dx = \frac{(-1)^n}{2^n n!\sqrt{\pi}} \int_{-\infty}^{\infty} e^{ax} \frac{d^n}{dx^n} (e^{-x^2})\, dx$$

$$= \frac{a^n}{2^n n!\sqrt{\pi}} \int_{-\infty}^{\infty} e^{ax-x^2}\, dx = \frac{a^n}{2^n n!} e^{a^2/4},$$

so that

$$e^{ax} = e^{a^2/4} \sum_{n=0}^{\infty} \frac{a^n}{2^n n!} H_n(x), \qquad -\infty < x < \infty. \tag{4.16.3}$$

We get the same result by setting $t = a/2$ in the expansion (4.9.3) of the generating function.

Example 3. Consider the function

$$f(x) = e^{-a^2x^2}, \qquad \operatorname{Re} a^2 > -1.$$

In this case,

$$e^{-a^2x^2} = \sum_{n=0}^{\infty} c_{2n} H_{2n}(x),$$

where

$$c_{2n} = \frac{1}{2^{2n}(2n)!\sqrt{\pi}} \int_{-\infty}^{\infty} e^{-(a^2+1)x^2} H_{2n}(x)\, dx.$$

To evaluate the integral, we replace $H_{2n}(x)$ by its integral representation (4.11.2). Making an appropriate change of variable and again using the duplication formula (1.2.3), we obtain (cf. footnote 12, p. 6)

$$\begin{aligned} c_{2n} &= \frac{2(-1)^n}{\pi(2n)!} \int_0^{\infty} e^{-t^2} t^{2n}\, dt \int_{-\infty}^{\infty} e^{-a^2x^2} \cos 2xt\, dx \\ &= \frac{2(-1)^n}{\sqrt{\pi}(2n)!a} \int_0^{\infty} e^{-t^2(1+a^{-2})} t^{2n}\, dt = \frac{(-1)^n}{\sqrt{\pi}(2n)!} \frac{a^{2n}}{(1+a^2)^{n+\frac{1}{2}}} \int_0^{\infty} e^{-s} s^{n-\frac{1}{2}}\, ds \\ &= \frac{(-1)^n}{\sqrt{\pi}(2n)!} \frac{a^{2n}}{(1+a^2)^{n+\frac{1}{2}}} \Gamma(n+\tfrac{1}{2}) = \frac{(-1)^n a^{2n}}{2^{2n} n!(1+a^2)^{n+\frac{1}{2}}}. \end{aligned}$$

With this value of c_{2n}, we have

$$e^{-a^2x^2} = \sum_{n=0}^{\infty} \frac{(-1)^n a^{2n}}{2^{2n} n!(1+a^2)^{n+\frac{1}{2}}} H_{2n}(x), \qquad -\infty < x < \infty, \quad \operatorname{Re} a^2 > -1. \tag{4.16.4}$$

Example 4. If

$$f(x) = \operatorname{sgn} x = \begin{cases} 1, & x > 0, \\ -1, & x < 0, \end{cases}$$

then

$$\operatorname{sgn} x = \sum_{n=0}^{\infty} c_{2n+1} H_{2n+1}(x),$$

where

$$\begin{aligned} c_{2n+1} &= \frac{1}{2^{2n+1}(2n+1)!\sqrt{\pi}} \int_{-\infty}^{\infty} e^{-x^2} H_{2n+1}(x) \operatorname{sgn} x\, dx \\ &= \frac{1}{2^{2n}(2n+1)!\sqrt{\pi}} \int_0^{\infty} e^{-x^2} H_{2n+1}(x)\, dx. \end{aligned}$$

Using the identity

$$e^{-x^2} H_n(x) = -\frac{d}{dx}[e^{-x^2} H_{n-1}(x)], \tag{4.16.5}$$

which follows from (4.10.1) and (4.10.2), we find that

$$c_{2n+1} = \frac{H_{2n}(0)}{2^{2n}(2n+1)!\sqrt{\pi}} = \frac{(-1)^n}{2^{2n}(2n+1)n!\sqrt{\pi}},$$

and hence

$$\operatorname{sgn} x = \frac{1}{\sqrt{\pi}} \sum_{n=0}^{\infty} \frac{(-1)^n}{2^{2n}(2n+1)n!} H_{2n+1}(x), \qquad -\infty < x < \infty. \tag{4.16.6}$$

Example 5. By integrating (or differentiating) these formulas with respect to the variable x or the parameter a, we can derive further expansions of the same type. For example, integrating (4.16.4) with respect to x over the interval $[0, x]$ and using (4.10.2), we obtain

$$\Phi(ax) = \frac{1}{\sqrt{\pi}} \sum_{n=0}^{\infty} \frac{(-1)^n a^{2n+1}}{2^{2n} n!(1+a^2)^{n+\frac{1}{2}}} \frac{H_{2n+1}(x)}{2n+1}, \qquad -\infty < x < \infty, \tag{4.16.7}$$

where $\Phi(x)$ is the probability integral. Another interesting expansion is obtained if we multiply the series (4.16.4) by $(1 + a^2)^{-1}$ and integrate with respect to a from 0 to ∞. This gives

$$e^{x^2}[1 - \Phi(x)] = \frac{2}{\pi} \sum_{n=0}^{\infty} \frac{(-1)^n}{2^{2n} n!} \frac{H_{2n}(x)}{2n+1}, \qquad 0 \leqslant x < \infty, \tag{4.16.8}$$

where we have used the identity (2.1.7).

Other examples of expansion of functions in Hermite polynomials are given in the problems at the end of the chapter (see p. 93).

4.17. Definition and Generating Function of the Laguerre Polynomials

Still another important class of orthogonal polynomials encountered in the applications, especially in mathematical physics,[33] consists of the *Laguerre polynomials* $L_n^\alpha(x)$,[34] defined by the formula

$$L_n^\alpha(x) = e^x \frac{x^{-\alpha}}{n!} \frac{d^n}{dx^n}(e^{-x} x^{n+\alpha}), \qquad n = 0, 1, 2, \ldots \tag{4.17.1}$$

---

[33] In problems involving the integration of Helmholtz's equation in parabolic coordinates, in the theory of the hydrogen atom, in the theory of propagation of electromagnetic waves along transmission lines, etc.

[34] The polynomials $L_n^\alpha(x)$ differ by only a constant factor from the polynomials $T_n^\alpha(x)$ investigated by N. Y. Sonine, *Recherches sur les fonctions cylindriques et le développement des fonctions continues en séries*, Math. Ann. **16**, 1 (1880). Laguerre studied only the special case $\alpha = 0$. In the literature, the polynomials $L_n^\alpha(x)$ are sometimes called the *generalized* Laguerre polynomials.

for arbitrary real $\alpha > -1$. According to (4.17.1), the first few Laguerre polynomials are

$$L_0^\alpha(x) = 1, \qquad L_1^\alpha(x) = 1 + \alpha - x,$$

$$L_2^\alpha(x) = \tfrac{1}{2}[(1 + \alpha)(2 + \alpha) - 2(2 + \alpha)x + x^2], \ldots,$$

and in general, using Leibniz's formula, we have

$$L_n^\alpha(x) = \sum_{k=0}^{n} \frac{\Gamma(n + \alpha + 1)}{\Gamma(k + \alpha + 1)} \frac{(-x)^k}{k!(n - k)!}, \tag{4.17.2}$$

where for all $k < n$ the ratio of gamma functions can be replaced by the product

$$(n + \alpha)(n + \alpha - 1)\cdots(n + \alpha - (n - k - 1)).$$

It will be shown below (see Sec. 4.21) that the Laguerre polynomials $L_n^\alpha(x)$ are orthogonal with weight $\rho(x) = x^\alpha e^{-x}$ on the interval $0 \leqslant x < \infty$. The polynomials $L_n^0(x) = L_n(x)$ form the simplest class of Laguerre polynomials. Another important class consists of the polynomials $L_n^{\pm 1/2}(x)$ which are simply related to the Hermite polynomials (see Sec. 4.19).

As the starting point for the theory of Laguerre polynomials, we begin with the following expansion

$$w(x, t) = (1 - t)^{-\alpha-1}e^{-xt/(1-t)} = \sum_{n=0}^{\infty} L_n^\alpha(x)t^n, \qquad |t| < 1 \tag{4.17.3}$$

of the *generating function* $w(x, t)$. To prove (4.17.3), we note that the left-hand side, regarded as a function of the complex variable t, is analytic in the disk $|t| < 1$, and hence must have an expansion of the form

$$w(x, t) = (1 - t)^{-\alpha-1}e^{-xt/(1-t)} = \sum_{n=0}^{\infty} c_n^\alpha(x)t^n, \qquad |t| < 1.$$

According to a familiar theorem from complex variable theory, the coefficients $c_n^\alpha(x)$ can be written as contour integrals

$$c_n^\alpha(x) = \frac{1}{2\pi i}\int_C (1 - t)^{-\alpha-1}e^{-xt/(1-t)}t^{-n-1}\,dt, \tag{4.17.4}$$

evaluated along any closed contour C surrounding the point $t = 0$ and lying inside the disk $|t| < 1$. Choosing a contour of sufficiently small size and introducing the new variable of integration $u = x/(1 - t)$, we find that

$$c_n^\alpha(x) = \frac{e^x x^{-\alpha}}{2\pi i}\int_{C'} \frac{e^{-u}u^{n+\alpha}}{(u - x)^{n+1}}\,du, \tag{4.17.5}$$

where C' is a small closed contour surrounding the point $u = x$. Evaluating this integral by residue theory, we obtain

$$c_n^\alpha(x) = \frac{e^x x^{-\alpha}}{n!}\left[\frac{d^n}{du^n}e^{-u}u^{n+\alpha}\right]_{u=x} \equiv L_n^\alpha(x),$$

thereby verifying (4.17.3).

There is another expansion closely related to (4.17.3), i.e.,

$$\begin{aligned} W(x, y, t) &= (1 - t)^{-1} e^{-(x+y)t/(1-t)} (xyt)^{-\alpha/2} I_\alpha\left[\frac{2(xyt)^{1/2}}{1-t}\right] \\ &= \sum_{n=0}^{\infty} \frac{n! L_n^\alpha(x) L_n^\alpha(y)}{\Gamma(n+\alpha+1)} t^n, \qquad |t| < 1, \quad \alpha > -1, \end{aligned} \tag{4.17.6}$$

where $I_\alpha(z)$ is the modified Bessel function of the first kind (defined in Sec. 5.7).[35] Here the function $W(x, y, t)$ can be regarded as a generating function of *products* of Laguerre polynomials. The following special case of (4.17.6), obtained by setting $y = x$, is important in the applications:

$$\begin{aligned} W(x, x, t) &= (1 - t)^{-1} e^{-2xt/(1-t)} x^{-\alpha} t^{-\alpha/2} I_\alpha\left(\frac{2xt^{1/2}}{1-t}\right) \\ &= \sum_{n=0}^{\infty} \frac{n! [L_n^\alpha(x)]^2}{\Gamma(n+\alpha+1)} t^n, \qquad |t| < 1, \quad \alpha > -1. \end{aligned} \tag{4.17.7}$$

4.18. Recurrence Relations and Differential Equation for the Laguerre Polynomials

Substituting (4.17.3) into the easily verified identity

$$(1 - t^2) \frac{\partial w}{\partial t} + [x - (1 - t)(1 + \alpha)]w = 0,$$

we find that

$$(1 - t^2) \sum_{n=0}^{\infty} n L_n^\alpha(x) t^{n-1} + [x - (1 - t)(1 + \alpha)] \sum_{n=0}^{\infty} L_n^\alpha(x) t^n = 0,$$

which gives

$$(n + 1)L_{n+1}^\alpha(x) + (x - \alpha - 2n - 1)L_n^\alpha(x) + (n + \alpha)L_{n-1}^\alpha(x) = 0, \qquad n = 1, 2, \ldots \tag{4.18.1}$$

when the coefficient of t^n is set equal to zero. Similarly, substituting (4.17.3) into the identity[36]

$$(1 - t) \frac{\partial w}{\partial x} + tw = 0,$$

[35] See E. Hille, *On Laguerre's series, I*, Proc. Nat. Acad. Sci., **12**, 261 (1926); *Part II*, ibid., **12**, 265 (1926); *Part III*, ibid., **12**, 348 (1926).

[36] The justification for differentiating (4.17.3) term by term with respect to x follows from the uniform convergence of (4.17.3) in the domain $|x| < a$ for arbitrary finite $a > 0$. According to (4.17.2),

$$|L_n^\alpha(x)| \leqslant L_n^\alpha(-a), \qquad |x| < a, \quad \alpha > -1,$$

so that (4.17.3) is majorized by the convergent series

$$\sum_{n=0}^{\infty} L_n^\alpha(-a)|t|^n = (1 - |t|)^{-\alpha-1} e^{a|t|/(1-|t|)},$$

and hence converges uniformly for $|x| < a$.

we obtain

$$(1 - t) \sum_{n=0}^{\infty} t^n \frac{dL_n^\alpha(x)}{dx} + \sum_{n=0}^{\infty} L_n^\alpha(x) t^{n+1} = 0,$$

which implies

$$\frac{dL_n^\alpha(x)}{dx} - \frac{dL_{n-1}^\alpha(x)}{dx} + L_{n-1}^\alpha(x) = 0, \qquad n = 1, 2, \ldots \tag{4.18.2}$$

Elimination of $L_{n-1}^\alpha(x)$ from (4.18.1–2) leads to the equation[37]

$$\begin{aligned}(x - n - 1) \frac{dL_n^\alpha(x)}{dx} &+ (n + 1) \frac{dL_{n+1}^\alpha(x)}{dx} \\ + (2n + 2 + \alpha - x)L_\alpha^n(x) &- (n + 1)L_{n+1}^\alpha(x) = 0, \quad n = 0, 1, 2, \ldots\end{aligned} \tag{4.18.3}$$

Finally, replacing n by $n - 1$ in (4.18.3) and using (4.18.2) to eliminate $(d/dx)L_{n-1}^\alpha(x)$, we obtain

$$x \frac{dL_n^\alpha(x)}{dx} = nL_n^\alpha(x) - (n + \alpha)L_{n-1}^\alpha(x), \qquad n = 1, 2, \ldots \tag{4.18.4}$$

Formula (4.18.4) allows us to expand the derivative of a Laguerre polynomial in terms of another Laguerre polynomial.

Recurrence relations of another type, involving Laguerre polynomials with different *superscripts* can be obtained by regarding the generating function as a function of the parameter α, and then writing equations connecting $w(x, t, \alpha)$ and $w(x, t, \alpha + 1)$. Thus, substituting (4.17.3) into the identity

$$(1 - t)w(x, t, \alpha + 1) = w(x, t, \alpha),$$

and comparing coefficients of identical powers of t in both sides of the resulting equation, we obtain

$$L_n^{\alpha+1}(x) - L_{n-1}^{\alpha+1}(x) = L_n^\alpha(x), \qquad n = 1, 2, \ldots \tag{4.18.5}$$

Similarly, substituting (4.17.3) into the identity

$$\frac{\partial w(x, t, \alpha)}{\partial x} = -tw(x, t, \alpha+1),$$

we obtain another formula of this type:

$$\frac{dL_n^\alpha(x)}{dx} = -L_{n-1}^{\alpha+1}(x), \qquad n = 1, 2, \ldots \tag{4.18.6}$$

Using the recurrence relations (4.18.2, 4), we can derive a differential equation satisfied by the Laguerre polynomials. In fact, differentiating

[37] In some cases, the validity of a recurrence relation for small n does not follow from the general argument, but then one can always verify the relation by direct substitution of $L_0^\alpha(x) = 1, L_1^\alpha(x) = 1 + \alpha - x, \ldots$

(4.18.4) with respect to x and then using (4.18.2, 4) to eliminate $(d/dx)L_{n-1}^{\alpha}(x)$ and $L_{n-1}^{\alpha}(x)$, we find that

$$x\frac{d^2L_n^{\alpha}(x)}{dx^2} + (\alpha + 1 - x)\frac{dL_n^{\alpha}(x)}{dx} + nL_n^{\alpha}(x) = 0, \qquad n = 0, 1, 2, \ldots \quad (4.18.7)$$

It follows from (4.18.7) that $u = L_n^{\alpha}(x)$ is a particular solution of the second-order linear differential equation

$$xu'' + (\alpha + 1 - x)u' + nu = 0. \quad (4.18.8)$$

Equation (4.18.8) is encountered in mathematical physics and plays an important role in the theory of Laguerre polynomials. By making changes of variables, we can easily derive other differential equations whose integrals can be expressed in terms of Laguerre polynomials. For example, it is easy to see that the differential equations

$$xu'' + (\alpha + 1 - 2\nu)u' + \left[n + \frac{\alpha + 1}{2} - \frac{x}{4} + \frac{\nu(\nu - \alpha)}{x}\right]u = 0 \quad (4.18.9)$$

and

$$u'' + \left[4n + 2\alpha + 2 - x^2 + \frac{\frac{1}{4} - \alpha^2}{x^2}\right]u = 0 \quad (4.18.10)$$

have the particular solutions

$$u = e^{-x/2}x^{\nu}L_n^{\alpha}(x)$$

and

$$u = e^{-x^2/2}x^{\alpha + \frac{1}{2}}L_n^{\alpha}(x^2),$$

respectively.

4.19. An Integral Representation of the Laguerre Polynomials. Relation between the Laguerre and Hermite Polynomials

The Laguerre polynomials have a simple representation in terms of definite integrals containing the variable x as parameter. To obtain this representation, we assume that x is a positive real number. Then

$$e^{-x}x^{n+\alpha} = \int_0^{\infty} (\sqrt{xt})^{n+\alpha} J_{n+\alpha}(2\sqrt{xt})e^{-t}\, dt, \quad (4.19.1)$$

where $J_{\nu}(x)$ is the Bessel function of order ν.[38] Differentiating (4.19.1) with

[38] Here we anticipate some results on Bessel functions, proved in Chap. 5. Formula (4.19.1) is a special case of formula (5.15.2), obtained by setting

$$a = 1, \quad b = 2\sqrt{x}, \quad x = \sqrt{t}, \quad \nu = n + \alpha.$$

respect to x and taking account of the identity

$$\frac{d}{du} u^{\nu/2} J_\nu(2\sqrt{u}) = u^{(\nu-1)/2} J_{\nu-1}(2\sqrt{u}),$$

obtained by setting $z = 2\sqrt{u}$ in the first of the formulas (5.3.6), we find that

$$\frac{d^m}{dx^m}(e^{-x}x^{m+\alpha}) = \int_0^\infty (\sqrt{xt})^{n-m+\alpha} J_{n-m+\alpha}(2\sqrt{xt}) e^{-t} t^m \, dt, \quad m = 0, 1, 2 \ldots, \tag{4.19.2}$$

where it is easy to justify the differentiation behind the integral sign. Setting $m = n$ in (4.19.2) and taking account of (4.17.1), we obtain the desired integral representation of the Laguerre polynomials:

$$L_n^\alpha(x) = \frac{e^x x^{-\alpha/2}}{n!} \int_0^\infty t^{n+\frac{1}{2}\alpha} J_\alpha(2\sqrt{xt}) e^{-t} \, dt, \qquad \alpha > 1, \quad n = 0, 1, 2, \ldots \tag{4.19.3}$$

Although this formula has been derived under the assumption that x is a positive real number, it can easily be extended to arbitrary complex values of x by using the principle of analytic continuation.

We now set $\alpha = \pm\frac{1}{2}$ in (4.19.3) and use the familiar formulas (5.8.1–2) from the theory of Bessel functions. Then we have

$$\begin{aligned} L_n^{-1/2}(x) &= \frac{e^x}{n!\sqrt{\pi}} \int_0^\infty e^{-t} t^{n-\frac{1}{2}} \cos(2\sqrt{xt}) \, dt \\ &= \frac{e^x}{n!} \frac{2}{\sqrt{\pi}} \int_0^\infty e^{-u^2} u^{2n} \cos(2\sqrt{x}u) \, du, \\ L_n^{1/2}(x) &= \frac{e^x}{n!\sqrt{\pi x}} \int_0^\infty e^{-t} t^n \sin(2\sqrt{xt}) \, dt \\ &= \frac{e^x}{n!\sqrt{x}} \frac{2}{\sqrt{\pi}} \int_0^\infty e^{-u^2} u^{2n+1} \sin(2\sqrt{x}u) \, du, \end{aligned} \tag{4.19.4}$$

which, taken together with (4.11.2–3), imply

$$\begin{aligned} L_n^{-1/2}(x) &= \frac{(-1)^n}{2^{2n} n!} H_{2n}(\sqrt{x}), \\ L_n^{1/2}(x) &= \frac{(-1)^n}{2^{2n+1} n!} \frac{H_{2n+1}(\sqrt{x})}{\sqrt{x}}. \end{aligned} \tag{4.19.5}$$

These formulas establish a connection between two classes of orthogonal polynomials, and allow us to regard the theory of Hermite polynomials as a special branch of the theory of Laguerre polynomials.[39]

---

[39] One can also prove the formulas (4.19.5) directly from the expansions (4.17.2) and (4.9.2).

4.20. An Integral Equation Satisfied by the Laguerre Polynomials

The Laguerre polynomials satisfy a simple integral equation with a symmetric kernel. To obtain this equation, we replace x by y in the expansion

$$(1-t)^{-\alpha-1}e^{-xt/(1-t)} = \sum_{n=0}^{\infty} L_n^\alpha(x)t^n, \qquad |t| < 1, \quad \alpha > -1, \tag{4.20.1}$$

multiply the result by

$$e^{-y/2}y^{\alpha/2}J_\alpha(\sqrt{xy}),$$

where $J_\alpha(z)$ is the Bessel function of order α, and then integrate from 0 to ∞. This gives

$$\begin{aligned}(1-t)^{-\alpha-1}\int_0^\infty & e^{-y(1+t)/2(1-t)}y^{\alpha/2}J_\alpha(\sqrt{xy})\,dy \\ &= \sum_{n=0}^{\infty} t^n \int_0^\infty e^{-y/2}y^{\alpha/2}J_\alpha(\sqrt{xy})L_n^\alpha(y)\,dy,\end{aligned} \tag{4.20.2}$$

provided that the process of term-by-term integration is permissible. To prove the legitimacy of this process, suppose $|t| < \frac{1}{3}$. Then, using the inequalities[40]

$$|L_n^\alpha(x)| \leqslant L_n^\alpha(-x), \quad |J_\alpha(x)| \leqslant I_\alpha(x), \qquad x \geqslant 0, \quad \alpha > -1,$$

where $I_\alpha(x)$ is the modified Bessel function of the first kind (see Sec. 5.7), we have

$$\begin{aligned}\int_0^\infty & |J_\alpha(\sqrt{xy})|e^{-y/2}y^{\alpha/2}\sum_{n=0}^{\infty}|t|^n|L_n^\alpha(y)|dy \\ &\leqslant \int_0^\infty I_\alpha(\sqrt{xy})e^{-y/2}y^{\alpha/2}\sum_{n=0}^{\infty}|t|^n L_n^\alpha(-y)\,dy \\ &= (1-|t|)^{-\alpha-1}\int_0^\infty I_\alpha(\sqrt{xy})y^{\alpha/2}e^{-y(1-3|t|)/2(1-|t|)}\,dy,\end{aligned}$$

where, in evaluating the sum, (4.20.1) has been used again. For $|t| < \frac{1}{3}$, $\alpha > -1$ the last integral on the right converges, as can be verified by considering the asymptotic behavior of the function $I_\alpha(x)$ for large and small x (see Chap. 5). Therefore the right-hand side of (4.20.2) is absolutely convergent, which guarantees the validity of reversing the order of summation and integration.[41]

---

[40] The first inequality follows from (4.17.2), the second from the power series expansions of the appropriate Bessel functions (see Chap. 5).

[41] E. C. Titchmarsh, *op. cit.*, p. 45.

We now set $\sqrt{y} = u$ in the left-hand side of (4.20.2) and use formula (5.15.2). This gives

$$(1 - t)^{-\alpha-1} \int_0^\infty e^{-y(1+t)/2(1-t)} y^{\alpha/2} J_\alpha(\sqrt{xy})\, dy$$
$$= 2(1 + t)^{-\alpha-1} x^{\alpha/2} e^{-x(1-t)/2(1+t)} = 2x^{\alpha/2} e^{-x/2} \sum_{n=0}^\infty L_n^\alpha(x)(-t)^n,$$

for $|t| < 1$.[42] Thus, for all $|t| < \frac{1}{3}$, we have the identity

$$2e^{-x/2} x^{\alpha/2} \sum_{n=0}^\infty (-1)^n L_n^\alpha(x) t^n = \sum_{n=0}^\infty t^n \int_0^\infty e^{-y/2} y^{\alpha/2} J_\alpha(\sqrt{xy}) L_n^\alpha(y)\, dy,$$

and then, comparing coefficients of identical powers of t, we obtain the desired integral equation

$$e^{-x/2} x^{\alpha/2} L_n^\alpha(x) = \frac{(-1)^n}{2} \int_0^\infty J_\alpha(\sqrt{xy}) e^{-y/2} y^{\alpha/2} L_n^\alpha(y)\, dy, \qquad \alpha > -1, \quad n = 0, 1, 2, \ldots \tag{4.20.3}$$

For $\alpha = \pm\frac{1}{2}$ this equation reduces to the corresponding integral equations (4.11.4–5) for the Hermite polynomials.

4.21. Orthogonality of the Laguerre Polynomials

We now prove one of the most important properties of the Laguerre polynomials, i.e., their orthogonality with weight $e^{-x} x^\alpha$ on the interval $0 \leqslant x < \infty$. Setting

$$u_n(x) = e^{-x/2} x^{\alpha/2} L_n^\alpha(x)$$

and recalling (4.18.9), we see that $u_n(x)$ and $u_m(x)$ satisfy the differential equations

$$(xu_n')' + \left(n + \frac{\alpha + 1}{2} - \frac{x}{4} - \frac{\alpha^2}{4x}\right) u_n = 0,$$

$$(xu_m')' + \left(m + \frac{\alpha + 1}{2} - \frac{x}{4} - \frac{\alpha^2}{4x}\right) u_m = 0.$$

Subtracting the second of these equations multiplied by u_n from the first multiplied by u_m, and integrating from 0 to ∞, we obtain

$$x(u_n' u_m - u_m' u_n)\Big|_0^\infty + (n - m) \int_0^\infty u_m u_n\, dx = 0.$$

---

[42] For such t,

$$\operatorname{Re} \frac{1 + t}{1 - t} > 0,$$

and hence the convergence condition is satisfied.

For $\alpha > -1$ the first term vanishes at both limits,[43] and hence

$$\int_0^\infty u_m(x)u_n(x)\,dx = 0 \qquad \text{if} \quad m \neq n$$

or

$$\int_0^\infty e^{-x}x^\alpha L_m^\alpha(x)L_n^\alpha(x)\,dx = 0 \qquad \text{if} \quad m \neq n, \quad \alpha > -1. \qquad (4.21.1)$$

The value of the integral (4.21.1) for $m = n$ can be found as follows: We replace the index n by $n - 1$ in the recurrence relation (4.18.1) and multiply the result by $L_n^\alpha(x)$. Then from this equation we subtract (4.18.1) multiplied by $L_{n-1}^\alpha(x)$, obtaining

$$\begin{aligned} n[L_n^\alpha(x)]^2 &- (n+\alpha)[L_{n-1}^\alpha(x)]^2 - (n+1)L_{n+1}^\alpha(x)L_{n-1}^\alpha(x) \\ &+ 2L_n^\alpha(x)L_{n-1}^\alpha(x) + (n+\alpha-1)L_n^\alpha(x)L_{n-2}^\alpha(x) = 0, \qquad n = 2, 3, \ldots \end{aligned}$$

Multiplying this equation by $e^{-x}x^\alpha$, integrating from 0 to ∞, and using the orthogonality property (4.21.1), we find that

$$n\int_0^\infty e^{-x}x^\alpha[L_n^\alpha(x)]^2\,dx = (n+\alpha)\int_0^\infty e^{-x}x^\alpha[L_{n-1}^\alpha(x)]^2\,dx, \qquad n = 2, 3, \ldots$$

Repeated application of this formula gives[44]

$$\begin{aligned} \int_0^\infty e^{-x}x^\alpha[L_n^\alpha(x)]^2\,dx &= \frac{(n+\alpha)(n+\alpha-1)\cdots(\alpha+2)}{n(n-1)\cdots 3\cdot 2}\int_0^\infty e^{-x}x^\alpha[L_1^\alpha(x)]^2\,dx \\ &= \frac{\Gamma(n+\alpha+1)}{n!}, \qquad n = 2, 3, \ldots \end{aligned}$$

It follows by direct substitution that this formula is also valid for $n = 0, 1$, and hence

$$\int_0^\infty e^{-x}x^\alpha[L_n^\alpha(x)]^2\,dx = \frac{\Gamma(n+\alpha+1)}{n!}, \qquad \alpha > -1, \quad n = 0, 1, 2, \ldots \qquad (4.21.2)$$

Obviously, the functions

$$\varphi_n(x) = \left[\frac{n!}{\Gamma(n+\alpha+1)}\right]^{1/2} e^{-x/2}x^{\alpha/2}L_n^\alpha(x), \qquad n = 0, 1, 2, \ldots$$

form an orthonormal system on the interval $0 \leqslant x < \infty$.

Formulas (4.21.1–2) play an important role in the problem of expanding functions in series of Laguerre polynomials (see Sec. 4.23).

---

[43] Substituting for u_m and u_n, we easily verify that this term is $O(x^{1+\alpha})$ as $x \to 0$.

[44] Direct calculation shows that

$$\int_0^\infty e^{-x}x^\alpha[L_1^\alpha(x)]^2\,dx = \int_0^\infty e^{-x}x^\alpha(\alpha+1-x)^2\,dx = (\alpha+1)\Gamma(\alpha+1).$$

4.22. Asymptotic Representation of the Laguerre Polynomials for Large n

Like the other orthogonal polynomials, the Laguerre polynomials have a simple asymptotic representation which describes their behavior for large values of the degree n. To obtain this representation, we write

$$u = e^{-x/2}L_n^\alpha(x), \tag{4.22.1}$$

and note that u is the solution of the differential equation

$$xu'' + (\alpha + 1)u' + \left(n + \frac{\alpha + 1}{2}\right)u = \frac{xu}{4} \tag{4.22.2}$$

which is analytic in a neighborhood of the point $x = 0$ and satisfies the initial condition

$$u(0) = L_n^\alpha(0) = \frac{\Gamma(n + \alpha + 1)}{n!\Gamma(\alpha + 1)}. \tag{4.22.3}$$

The rest of the argument is somewhat dependent on whether α is positive or negative, but since this difference is not of a fundamental nature, we will only consider the case $\alpha \geqslant 0$.

Regarding the right-hand side of (4.22.2) as a known function, we find that

$$u(x) = A_1u_1(x) + A_2u_2(x) + \frac{\pi}{4N}\int_0^x (Ny)^{\alpha+1}u(y)[u_1(y)u_2(x) - u_1(x)u_2(y)]\,dy, \tag{4.22.4}$$

where

$$u_1(x) = (\sqrt{Nx})^{-\alpha}J_\alpha(2\sqrt{Nx}), \qquad u_2(x) = (\sqrt{Nx})^{-\alpha}\,Y_\alpha(2\sqrt{Nx}),$$

$$N = n + \frac{\alpha + 1}{2},$$

and $J_\alpha(x)$, $Y_\alpha(x)$ are the Bessel functions of the first and second kinds, respectively (see Chap. 5).[45] Taking account of the asymptotic behavior of the Bessel functions, described by formulas (5.16.1, 2), we find that as $x \to 0$,

$$u_1(x) \to \frac{1}{\Gamma(\alpha + 1)}, \qquad u_2(x) \to \infty,$$

---

[45] Here u_1 and u_2 are a pair of linearly independent solutions of the homogenous equation

$$u'' + \frac{\alpha + 1}{x}u' + \frac{N}{x}u = 0,$$

with Wronskian

$$W\{u_1, u_2\} = \frac{N}{\pi}(Nx)^{-\alpha-1}.$$

See equations (5.4.11–12) and (5.9.2)

while the integral is $O(x^2)$.[46] Therefore the values of the constants of integration are

$$A_1 = \frac{\Gamma(n + \alpha + 1)}{n!}, \qquad A_2 = 0, \tag{4.22.5}$$

and (4.22.4) can be written in the form

$$u(x) = A_1[u_1(x) + r_n(x)], \tag{4.22.6}$$

where

$$r_n(x) = \frac{\pi}{4A_1N}\int_0^x (Ny)^{\alpha+1}u(y)[u_1(y)u_2(x) - u_1(x)u_2(y)]\,dy. \tag{4.22.7}$$

It will now be shown that for fixed $x \geqslant 0$ the size of the remainder in (4.22.6) is small compared to the first term. In proving this, we distinguish two cases: (a) $0 \leqslant x \leqslant N^{-1}$ and (b) $x > N^{-1}$. First we find an upper bound (denoted by M_n) for the absolute value of $|u(x)|$ in the interval $0 \leqslant x \leqslant N^{-1}$. According to Sec. 5.16, for $0 \leqslant x \leqslant N^{-1}$ we have

$$u_1(x) = O(1), \qquad u_2(x) = \begin{cases} O(N^{-\alpha}x^{-\alpha}), & \alpha > 0, \\ O\left(\log \dfrac{1}{Nx}\right), & \alpha = 0. \end{cases} \tag{4.22.8}$$

Therefore, if $\alpha > 0$, it follows from (4.22.4–5) that

$$\begin{aligned} |u(x)| &\leqslant A_1O(1) + M_nN^{-1}\int_0^x (Ny)^{\alpha+1}[O(N^{-\alpha}x^{-\alpha}) + O(N^{-\alpha}y^{-\alpha})]\,dy \\ &= A_1O(1) + M_nx^2O(1) = A_1O(1) + M_nO(N^{-2}), \end{aligned}$$

which implies that

$$M_n = A_1O(1) \tag{4.22.9}$$

for large n, a result which remains valid for $\alpha = 0$. Using (4.22.9), we find that

$$|r_n(x)| \leqslant x^2O(1) = O(N^{-2}) \tag{4.22.10}$$

for $0 \leqslant x \leqslant N^{-1}$, $\alpha > 0$, whereas

$$|r_n(x)| \leqslant x^2 \log(N^{-1}x^{-1})O(1) = O(N^{-2}) \tag{4.22.11}$$

for $0 \leqslant x \leqslant N^{-1}$, $\alpha = 0$.

To estimate $r_n(x)$ for $x > N^{-1}$, we write (4.22.7) as a sum of integrals:

$$r_n(x) = \frac{\pi}{4A_1N}\left[\int_0^{1/N} \cdots + \int_{1/N}^x \cdots\right] = \mathcal{J}_1 + \mathcal{J}_2. \tag{4.22.12}$$

According to Sec. 5.16, in the interval $N^{-1} < x < \infty$ we have

$$u_1(x) = O(N^{-\frac{1}{2}\alpha-\frac{1}{4}}x^{-\frac{1}{2}\alpha-\frac{1}{4}}), \qquad u_2(x) = O(N^{-\frac{1}{2}\alpha-\frac{1}{4}}x^{-\frac{1}{2}\alpha-\frac{1}{4}}). \tag{4.22.13}$$

---

[46] Except in the case $\alpha = 0$, where the integral is $O\left(x^2 \log \dfrac{1}{x}\right)$.

Therefore, if $\alpha > 0$, we find as before that

$$|\mathscr{I}_1| \leqslant N^{-1}\int_0^{1/N}(Ny)^{\alpha+1}(Nx)^{-\frac{1}{2}\alpha-\frac{1}{4}}[O(1)+O(N^{-\alpha}y^{-\alpha})]\,dy$$
$$= N^{-2}O(N^{-\frac{1}{2}\alpha-\frac{1}{4}}x^{-\frac{1}{2}\alpha-\frac{1}{4}}), \tag{4.22.14}$$

a result which remains valid for $\alpha = 0$, and moreover

$$|\mathscr{I}_2| \leqslant (A_1N)^{-1}O(N^{-\frac{1}{2}\alpha-\frac{1}{4}}x^{-\frac{1}{2}\alpha-\frac{1}{4}})\int_{1/N}^{x}(Ny)^{\alpha+1}|u(y)|(Ny)^{-\frac{1}{2}\alpha-\frac{1}{4}}\,dy$$
$$\leqslant A_1^{-1}N^{\frac{1}{2}\alpha-\frac{1}{4}}O(N^{-\frac{1}{2}\alpha-\frac{1}{4}}x^{-\frac{1}{2}\alpha-\frac{1}{4}})\int_0^x y^{\frac{1}{2}\alpha+\frac{3}{4}}|u(y)|dy.$$

Using Schwarz's inequality and formula (4.21.2), we have

$$\int_0^x y^{\frac{1}{2}\alpha+\frac{3}{4}}|u(y)|dy \leqslant \left[\int_0^x y^{\frac{3}{2}}\,dy\right]^{1/2}\left[\int_0^\infty y^\alpha u^2(y)\,dy\right]^{1/2}$$
$$= A_1^{1/2}x^{\frac{5}{4}}(\alpha+\tfrac{5}{2})^{-1/2}.$$

and hence

$$|\mathscr{I}_2| \leqslant A_1^{-1/2}N^{\frac{1}{2}\alpha-\frac{1}{4}}x^{\frac{5}{4}}O(N^{-\frac{1}{2}\alpha-\frac{1}{4}}x^{-\frac{1}{2}\alpha-\frac{1}{4}}),$$

which becomes

$$|\mathscr{I}_2| \leqslant N^{-1/4}x^{\frac{5}{4}}O(N^{-\frac{1}{2}\alpha-\frac{1}{4}}x^{-\frac{1}{2}\alpha-\frac{1}{4}}) \tag{4.22.15}$$

since $A_1 = O(N^\alpha)$, according to (4.22.5). It follows from (4.22.12, 14, 15) that

$$|r_n(x)| \leqslant O(N^{-\frac{1}{2}\alpha-\frac{1}{4}}x^{-\frac{1}{2}\alpha-\frac{1}{4}})[N^{-1/4}x^{\frac{5}{4}}+N^{-2}O(1)]. \tag{4.22.16}$$

A comparison of (4.22.8) with (4.22.10–11), and of (4.22.13) with (4.22.16), shows that the size of the remainder term in (4.22.6) is small compared to $u_1(x)$ for all $0 \leqslant x \leqslant a$ and arbitrary finite $a > 0$, provided that n is large. Therefore, finally, we have the asymptotic formula

$$u(x) \approx A_1u_1(x), \qquad n\to\infty \tag{4.22.17}$$

or

$$L_n^\alpha(x) \approx \frac{\Gamma(n+\alpha+1)}{n!}e^{x/2}(Nx)^{-\alpha/2}J_\alpha(2\sqrt{Nx}), \qquad n\to\infty, \quad N = n+\frac{\alpha+1}{2}. \tag{4.22.18}$$

In the interval $0 < \delta \leqslant x \leqslant a$ we can replace the Bessel function by its asymptotic representation (5.16.1). This reduces (4.22.18) to the simpler form

$$L_n^\alpha(x) \approx \pi^{-1/2}e^{x/2}n^{\frac{1}{2}\alpha-\frac{1}{4}}x^{-\frac{1}{2}\alpha-\frac{1}{4}}\cos\left(2\sqrt{nx}-\frac{\alpha\pi}{2}-\frac{\pi}{4}\right), \qquad n\to\infty. \tag{4.22.19}$$

4.23. Expansion of Functions in Series of Laguerre Polynomials

One of the most important properties of the Laguerre polynomials is the fact that a real function $f(x)$ defined in the infinite interval $(0, \infty)$ can be expanded in a series of the form

$$f(x) = \sum_{n=0}^{\infty} c_n L_n^{\alpha}(x), \qquad 0 < x < \infty, \tag{4.23.1}$$

provided $f(x)$ satisfies certain general conditions. The coefficients c_n can be determined formally by using the orthogonality property of the Laguerre polynomials (see Sec. 4.21). In fact, multiplying (4.23.1) by $e^{-x}x^{\alpha}L_n^{\alpha}(x)$ and integrating term by term over the interval $(0, \infty)$, we find that

$$c_n = \frac{n!}{\Gamma(n+\alpha+1)} \int_0^{\infty} e^{-x}x^{\alpha}f(x)L_n^{\alpha}(x)\,dx. \tag{4.23.2}$$

This expansion is valid if $f(x)$ is piecewise smooth in every finite interval $[x_1, x_2]$ and suitably well-behaved near the points $x = 0$ and $x = \infty$. In particular, we have

THEOREM 3. *If the real function $f(x)$, defined in the infinite interval $(0, \infty)$, is piecewise smooth in every finite subinterval $[x_1, x_2]$, where $0 < x_1 < x_2 < \infty$, and if the integral*

$$\int_0^{\infty} e^{-x}x^{\alpha}f^2(x)\,dx$$

is finite, then the series (4.23.1), *with coefficients calculated from* (4.23.2), *converges to $f(x)$ at every continuity point of $f(x)$. At a discontinuity point, the series converges to*

$$\tfrac{1}{2}[f(x+0) + f(x-0)].$$

Theorem 3 can be proved by a method similar to that used in proving the corresponding theorem for Hermite polynomials (Theorem 2, p. 71).[47]

4.24. Examples of Expansions in Series of Laguerre Polynomials

In applying Theorem 3 to a given function $f(x)$, we have to evaluate the integrals in (4.23.2). In most cases this can be done by replacing $L_n^{\alpha}(\alpha)$ by its explicit expression (4.17.1) or by the integral representation (4.19.3). It is

[47] See J. V. Uspensky, *On the development of arbitrary functions in series of Hermite's and Laguerre's polynomials*, Annals of Math., (2), **28**, 593 (1927). For the case $\alpha > -\frac{1}{2}$, Uspensky imposes a less restrictive condition on the behavior of $f(x)$ near $x = 0$. For expansion theorems valid under other conditions on $f(x)$, see G. Szegö, *op. cit.*, and J. Korous, *On series of Laguerre polynomials* (in Czech), Rozpravy České Akademie, (2), **37**, no. 40 (1928).

sometimes helpful to make use of the generating function (4.17.3). The following examples serve to illustrate the technique of expanding functions in series of Laguerre polynomials:[48]

Example 1. The function

$$f(x) = x^{\nu}$$

satisfies the conditions of Theorem 3 if $\nu > -\frac{1}{2}(\alpha + 1)$, and we have

$$x^{\nu} = \sum_{n=0}^{\infty} c_n L_n^{\alpha}(x),$$

where

$$c_n = \frac{n!}{\Gamma(n + \alpha + 1)} \int_0^{\infty} e^{-x} x^{\nu+\alpha} L_n^{\alpha}(x)\, dx.$$

Substituting from (4.17.1) and integrating by parts n times, we find that

$$\begin{aligned} c_n &= \frac{1}{\Gamma(n + \alpha + 1)} \int_0^{\infty} x^{\nu} \frac{d^n}{dx^n} (e^{-x} x^{n+\alpha})\, dx \\ &= \frac{(-1)^n \nu(\nu - 1) \cdots (\nu - n + 1)}{\Gamma(n + \alpha + 1)} \int_0^{\infty} e^{-x} x^{\nu+\alpha}\, dx \\ &= (-1)^n \frac{\Gamma(\nu + \alpha + 1)\Gamma(\nu + 1)}{\Gamma(n + \alpha + 1)+\Gamma(\nu - n + 1)}, \end{aligned}$$

and hence

$$x^{\nu} = \Gamma(\nu + \alpha + 1)\Gamma(\nu + 1) \sum_{n=0}^{\infty} \frac{(-1)^n L_n^{\alpha}(x)}{\Gamma(n + \alpha + 1)\Gamma(\nu - n + 1)}, \qquad (4.24.1)$$

$$0 < x < \infty, \quad \alpha > -1.$$

In particular, if ν is a positive integer p, the series (4.24.1) terminates after a finite number of terms, and we have

$$x^p = \Gamma(p + \alpha + 1)p! \sum_{n=0}^{p} \frac{(-1)^n L_n^{\alpha}(x)}{\Gamma(n + \alpha + 1)(p - n)!}, \qquad (4.24.2)$$

$$0 < x < \infty, \quad \alpha > -1, \quad p = 0, 1, 2, \ldots$$

Example 2. The function

$$f(x) = e^{-ax}$$

satisfies the conditions of Theorem 3 if $a > -\frac{1}{2}$. In this case,

$$e^{-ax} = \sum_{n=0}^{\infty} c_n L_n^{\alpha}(x),$$

---

[48] It should be noted that the conditions imposed on the parameters in Examples 1–4 are sufficient, but the expansions may continue to hold in larger regions.

where

$$\begin{aligned} c_n &= \frac{n!}{\Gamma(n+\alpha+1)} \int_0^\infty e^{-(a+1)x} x^\alpha L_n^\alpha(x)\, dx \\ &= \frac{1}{\Gamma(n+\alpha+1)} \int_0^\infty e^{-ax} \frac{d^n}{dx^n} (e^{-x} x^{n+\alpha})\, dx \\ &= \frac{a^n}{\Gamma(n+\alpha+1)} \int_0^\infty e^{-(a+1)x} x^{n+\alpha}\, dx \\ &= \frac{a^n}{(a+1)^{n+\alpha+1}}, \qquad n = 0, 1, 2, \ldots \end{aligned}$$

With these values of c_n we have

$$e^{-ax} = (a+1)^{-\alpha-1} \sum_{n=0}^\infty \left(\frac{a}{a+1}\right)^n L_n^\alpha(x), \qquad 0 \leqslant x < \infty. \quad (4.24.3)$$

We get the same result by setting $t = a/(a+1)$ in the expansion (4.17.3) of the generating function.

Example 3. Consider the function

$$f(x) = (ax)^{-\alpha/2} J_\alpha(2\sqrt{ax}), \qquad x > 0, \quad a > 0, \quad \alpha > -1.$$

In this case, the desired expansion is

$$(ax)^{-\alpha/2} J_\alpha(2\sqrt{ax}) = \sum_{n=0}^\infty c_n L_n^\alpha(x),$$

where

$$c_n = \frac{n!}{\Gamma(n+\alpha+1)} \int_0^\infty e^{-x} \left(\frac{x}{a}\right)^{\alpha/2} J_\alpha(2\sqrt{ax}) L_n^\alpha(x)\, dx.$$

To evaluate the integral, we multiply the identity (4.17.3) by

$$e^{-x} \left(\frac{x}{a}\right)^{\alpha/2} J_\alpha(2\sqrt{ax})$$

and integrate with respect to x from 0 to ∞. Then, assuming that $|t|$ is sufficiently small, we obtain

$$\begin{aligned} (1-t)^{-\alpha-1} \int_0^\infty e^{-x/(1-t)} \left(\frac{x}{a}\right)^{\alpha/2} J_\alpha(2\sqrt{ax})\, dx &= e^{-a(1-t)} \\ = e^{-a} \sum_{n=0}^\infty \frac{a^n}{n!} t^n &= \sum_{n=0}^\infty t^n \int_0^\infty e^{-x} \left(\frac{x}{a}\right)^{\alpha/2} J_\alpha(2\sqrt{ax}) L_n^\alpha(x)\, dx, \end{aligned}$$

where we have used formula (5.15.2). Comparing coefficients of identical powers of t, we find that

$$c_n = \frac{e^{-a} a^n}{\Gamma(n+\alpha+1)},$$

and hence

$$(ax)^{-\alpha/2}J_\alpha(2\sqrt{ax}) = e^{-a}\sum_{n=0}^{\infty}\frac{a^n}{\Gamma(n+\alpha+1)}L_n^\alpha(x), \tag{4.24.4}$$
$$x > 0, \quad a > 0, \quad \alpha > -1,$$

Example 4. If we multiply (4.24.3) by $(a+1)^{\alpha-1}$ and integrate with respect to a from 0 to ∞, we obtain

$$\int_0^\infty e^{-ax}(a+1)^{\alpha-1}\,da = \sum_{n=0}^{\infty} L_n^\alpha(x)\int_0^\infty\left(\frac{a}{a+1}\right)^n\frac{da}{(a+1)^2} = \sum_{n=0}^{\infty}\frac{L_n^\alpha(x)}{n+1}.$$

The integral in the left-hand side can be expressed in terms of the complementary incomplete gamma function (see Problem 10, p. 15). This gives

$$e^x x^{-\alpha}\Gamma(\alpha, x) = \sum_{n=0}^{\infty}\frac{L_n^\alpha(x)}{n+1}, \qquad 0 < x < \infty, \quad \alpha > -1, \tag{4.24.5}$$

which for $\alpha = 0$ reduces to

$$-e^x Ei(-x) \equiv e^x\Gamma(0, x) = \sum_{n=0}^{\infty}\frac{L_n(x)}{n+1}, \qquad 0 < x < \infty. \tag{4.24.6}$$

Some other expansions in series of Laguerre polynomials are given in Problems 19–20, p. 96.

4.25. Application to the Theory of Propagation of Electromagnetic Waves. Reflection from the End of a Long Transmission Line Terminated by a Lumped Inductance

As a curious example of the application of Laguerre polynomials, we consider the problem of propagation of electromagnetic waves along a transmission line of length l. Suppose the line terminates at one end in a coil of inductance L_0, while at the other end a source of constant d-c voltage V_0 is suddenly switched on at time $t = 0$ (see Figure 12). Let the instantaneous values of the voltage and current be denoted by $V = V(x, t)$ and $I = I(x, t)$, and let the inductance and capacitance per unit length of the line be denoted by L and C. Then the problem reduces to the integration of the following system of linear differential equations,[49]

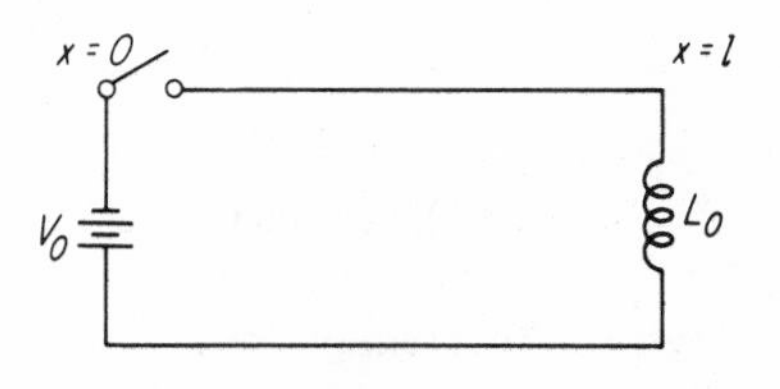

FIGURE 12

$$-\frac{\partial V}{\partial x} = L\frac{\partial I}{\partial t}, \qquad -\frac{\partial I}{\partial x} = C\frac{\partial V}{\partial t}, \tag{4.25.1}$$

---

[49] See S. Ramo and J. R. Whinnery, *Fields and Waves in Modern Radio*, second edition, John Wiley and Sons, New York (1953), p. 24.

subject to the initial conditions

$$V|_{t=0} = I|_{t=0} = 0 \tag{4.25.2}$$

and boundary conditions

$$V|_{x=0} = V_0, \qquad V|_{x=l} = L_0 \frac{\partial I}{\partial t}\bigg|_{x=l}. \tag{4.25.3}$$

To solve these equations, we use the method of the Laplace transform (see Secs. 2.6, 8), which converts (4.25.1) into a pair of ordinary differential equations. As usual, let $\bar{f}$ denote the Laplace transform of the function f:

$$\bar{f} = \int_0^\infty e^{-pt} f\, dt. \tag{4.25.4}$$

Then (4.25.1) goes into

$$-\frac{d\bar{V}}{dx} = Lp\bar{I}, \qquad -\frac{d\bar{I}}{dx} = Cp\bar{V},$$

and eliminating $\bar{I}$, we obtain a second-order differential equation

$$\frac{d^2\bar{V}}{dx^2} - LCp^2\bar{V} = 0, \tag{4.25.5}$$

subject to the boundary conditions

$$\bar{V}|_{x=0} = \frac{V_0}{p}, \qquad \frac{d\bar{V}}{dx} + \frac{L}{L_0}\bar{V}|_{x=l} = 0. \tag{4.25.6}$$

It follows from (4.25.5) and (4.25.6) that

$$\bar{V} = \frac{V_0}{p} \frac{\cosh\frac{p}{v}(l - x) + \frac{Z}{L_0 p}\sinh\frac{p}{v}(l - x)}{\cosh\frac{p}{v}l + \frac{Z}{L_0 p}\sinh\frac{p}{v}l}, \tag{4.25.7}$$

where $v = 1/\sqrt{LC}$ is the velocity of wave propagation along the line, and $Z = \sqrt{L/C}$ is the characteristic impedance.[50]

We now return to the original function V by using the Fourier-Mellin inversion theorem (cf. p. 25)

$$V = \frac{1}{2\pi i}\int_\Lambda e^{pt}\bar{V}\, dp, \tag{4.25.8}$$

where the integral is along a line Λ parallel to the imaginary axis and to the right of the origin. Being primarily interested in the voltage at the end of the line, we set $x = l$ in (4.25.7–8). Then

$$\frac{1}{V_0} V|_{x=l} = \frac{1}{2\pi i}\int_\Lambda \frac{e^{pt}}{p\cosh pt + \alpha \sinh pt}\, dp, \tag{4.25.9}$$

[50] S. Ramo and J. R. Whinnery, *op. cit.*, p. 27.

where $\alpha = Z/L_0$, and $T = l/v$ is the time it takes the wave to go from one end of the line to the other. To obtain the answer in a form which has a simple physical interpretation, we expand $\bar{V}|_{x=l}$ in powers of e^{-2pT} and integrate term by term. This gives

$$\frac{1}{2V_0} V|_{x=l} = \sum_{n=0}^{\infty} (-1)^n \frac{1}{2\pi i} \int_{\Lambda} \left(\frac{p-\alpha}{p+\alpha}\right)^n \frac{e^{p[t-(2n+1)T]}}{p+\alpha} \, dp,$$

or, if we introduce the new variable of integration $q = (p + \alpha)/2\alpha$,

$$\frac{1}{2V_0} V|_{x=l} = \sum_{n=0}^{\infty} (-1)^n e^{-\alpha[t-(2n+1)T]} \frac{1}{2\pi i} \int_{\Lambda'} \left(1 - \frac{1}{q}\right)^n e^{2q\alpha[t-(2n+1)T]} \frac{dq}{q}, \tag{4.25.10}$$

where Λ' is a line parallel to and to the right of Λ.

The evaluation of the integral in (4.25.10)

$$F(\tau) = \frac{1}{2\pi i} \int_{\Lambda'} \left(1 - \frac{1}{q}\right)^n \frac{e^{q\tau}}{q} \, dq \tag{4.25.11}$$

is accomplished by using residue theory applied to the closed contour consisting of Λ' and the arc of the circle $|q| = R$ (where R is arbitrarily large) lying to the left of Λ' if $\tau > 0$ or to the right of Λ' if $\tau < 0$. In the first case, we have

$$F(\tau) = \frac{1}{n!} \left[\frac{d^n}{dq^n} \{(q-1)e^{q\tau}\}\right]_{q=0} = \frac{e^{\tau}}{n!} \left[\frac{d^n}{dy^n} (y^n e^{-y})\right]_{y=\tau} = L_n(\tau), \tag{4.25.12}$$

where $L_n(\tau)$ is the nth Laguerre polynomial (see Sec. 4.17), while in the second case $F(\tau) = 0$. Substituting (4.25.12) into (4.5.10), we find that $V|_{x=l} = 0$ for $0 \leqslant t < T$, and

$$\frac{1}{2V_0} V|_{x=l} = \sum_{n=0}^{N-1} (-1)^n e^{-\alpha[t-(2n+1)T]} L_n\{2\alpha[t - (2n+1)T]\} \tag{4.25.13}$$

for

$$(2N - 1)T < t < (2n + 1)T, \qquad N = 1, 2, \ldots$$

Formula (4.25.13) represents the solution in closed form, and the appearance of new terms at intervals of $2T$ seconds corresponds to the arrival of additional reflected waves at the point $x = l$.

This method is applicable to transmission lines terminated by loads of other kinds, and in many other cases the answer can also be expressed in terms of Laguerre polynomials.

PROBLEMS

1. Show that all the roots of the equation $P_n(x) = 0$ are real and lie in the interval $(-1, 1)$.

Hint. Use Rolle's theorem.

2. Show that all the roots of the equation $H_n(x) = 0$ are real.

3. Prove the inequality[51]

$$(1 - x^2)^{1/4}|P_n(x)| < \left(\frac{2}{n\pi}\right)^{1/2}, \qquad -1 \leqslant x \leqslant 1, \quad n = 1, 2, \ldots$$

4. Using the expansions (4.9.2) and (4.17.2), prove *Uspensky's formula*

$$L_n^\alpha(x) = \frac{(-1)^n\Gamma(n + \alpha + 1)}{\sqrt{\pi}\,\Gamma(\alpha + \frac{1}{2})(2n)!}\int_{-1}^{1}(1 - t^2)^{\alpha - \frac{1}{2}}H_{2n}(\sqrt{x}t)\,dt, \qquad \alpha > -\tfrac{1}{2},$$

which expresses the Laguerre polynomials in terms of the Hermite polynomials.

5. Prove *Koshlyakov's formula*[52]

$$L_n^{\alpha+\beta}(x) = \frac{\Gamma(n + \alpha + \beta + 1)}{\Gamma(\beta)\Gamma(n + \alpha + 1)}\int_0^1 t^\alpha(1 - t)^{\beta-1}L_n^\alpha(xt)\,dt, \qquad \alpha > -1, \quad \beta > 0.$$

Hint. Replace the Laguerre polynomial $L_n^\alpha(xt)$ by its expansion (4.17.2), and integrate term by term.

Comment. For $\alpha = -\frac{1}{2}$, $\beta = \alpha + \frac{1}{2}$, Koshlyakov's formula reduces to Uspensky's formula.

6. In many cases, the evaluation of integrals of the form

$$\int_{-\infty}^{\infty} e^{-x^2}f(x)H_n^2(x)\,dx$$

can be accomplished by the following device: Multiply equation (4.9.6) by $f(x)$, integrate from $-\infty$ to ∞, and evaluate the integral in the left-hand side, calling the result $\varphi(t)$. Then expand $\varphi(t)$ in powers of t and equate coefficients of identical powers of t in both sides of the equation so obtained. Applying this method, show that

$$\int_{-\infty}^{\infty} e^{-x^2}H_n^2(x)\,dx = 2^n n!\sqrt{\pi},$$

$$\int_{-\infty}^{\infty} e^{-x^2}H_n^2(x)x^2\,dx = 2^n n!\sqrt{\pi}(n + \tfrac{1}{2}),$$

$$\int_{-\infty}^{\infty} e^{-2x^2}H_n^2(x)\,dx = 2^{n-\frac{1}{2}}\Gamma(n + \tfrac{1}{2}).$$

7. Prove that

$$\int_{-\infty}^{\infty} e^{-a^2x^2}H_{2n}(x)\,dx = \frac{(2n)!}{n!}\frac{\sqrt{\pi}}{a}\left(\frac{1 - a^2}{a^2}\right)^n, \qquad \operatorname{Re} a^2 > 0, \quad n = 0, 1, 2, \ldots,$$

$$\int_{-\infty}^{\infty} e^{-x^2}H_p^2(x)H_{2n}(x)\,dx = \frac{2^{p+n}(2n)!(p!)^2\sqrt{\pi}}{(p - n)!(n!)^2}, \qquad p = 0, 1, 2, \ldots, \quad n = 0, 1, 2, \ldots, p.$$

Hint. To derive the second formula, use the method of Problem 6.

---

[51] For a simple proof, see G. Szegö, *Orthogonal Polynomials*, revised edition, American Mathematical Society, New York (1959), Theorem 7.3.3, p. 163.

[52] N. S. Koshlyakov, *On Sonine's polynomials*, Messenger of Mathematics, **55**, 152 (1926).

8. Prove that

$$\frac{1}{2^n n!}\int_{-\infty}^{\infty}\frac{H_n^2(x)}{1+x^2}e^{-x^2}\,dx = \int_{-\infty}^{\infty}\left(\frac{1-x^2}{1+x^2}\right)^n\frac{e^{-x^2}}{1+x^2}\,dx, \qquad n = 0, 1, 2, \ldots$$

Hint. Use the method of Problem 6.

Comment. This formula was used in the proof of Theorem 2, p. 71.

9. Derive the integral representation

$$e^{-x/2}L_n(x) = \frac{1}{2^{n-1}n!\sqrt{\pi}}\int_0^{\infty}e^{-t^2}H_n^2(t)\cos(\sqrt{2x}\,t)\,dt.$$

Hint. To calculate the integral on the right, use the method of Problem 6.

10. Derive the formula

$$e^{-x^2}H_n^2(x) = \frac{2^n n!}{\sqrt{\pi}}\int_0^{\infty}e^{-s^2/4}L_n\left(\frac{s^2}{2}\right)\cos sx\,ds.$$

Hint. Use the result of Problem 9 and the Fourier integral theorem.[53]

11. Derive the following integral equation for the square of the Hermite polynomial of odd index:

$$\frac{e^{-x^2}H_{2n+1}^2(\sqrt{x})}{\sqrt{x}} = \int_0^{\infty}J_1(2\sqrt{xy})\,\frac{e^{-y}H_{2n+1}^2(\sqrt{y})}{\sqrt{y}}\,dy.$$

Hint. To calculate the integral on the right, use the method of Problem 6.

12. Derive the following integral equation for the square of the Laguerre polynomial:

$$e^{-x}x^{\alpha}[L_n^{\alpha}(x)]^2 = \int_0^{\infty}J_{2\alpha}(2\sqrt{xy})e^{-y}y^{\alpha}[L_n^{\alpha}(y)]^2\,dy, \qquad \alpha > -\tfrac{1}{2}.$$

Comment. The result of the preceding problem is a special case of this formula.

13. Prove the expansions

$$e^{t^2}\cos 2xt = \sum_{n=0}^{\infty}\frac{(-1)^n H_{2n}(x)}{(2n)!}\,t^{2n}, \qquad |t| < \infty,$$

$$e^{t^2}\sin 2xt = \sum_{n=0}^{\infty}\frac{(-1)^n H_{2n+1}(x)}{(2n+1)!}\,t^{2n+1}, \qquad |t| < \infty.$$

Comment. The expressions on the left in these formulas can be regarded as generating functions for the even and odd Hermite polynomials, respectively.

14. Verify the following expansions in Hermite polynomials (cf. Secs. 2.1.3):

$$e^{x^2}[1 - \Phi^2(x)] = \frac{4}{\pi}\sum_{n=0}^{\infty}\frac{(-1)^n}{2^{3n+\frac{1}{2}}n!}\,\frac{H_{2n}(x)}{2n+1},$$

$$F(x) = \sqrt{\pi}\sum_{n=0}^{\infty}\frac{(-1)^n}{2^{3n+3}\Gamma(n+\frac{3}{2})}\,H_{2n+1}(x).$$

---

[53] G. P. Tolstov, *op. cit.*, p. 190.

15. Derive the following expansion of the square of a Hermite polynomial in a series of Hermite polynomials:

$$H_p^2(x) = 2^p(p!)^2 \sum_{n=0}^{p} \frac{H_{2n}(x)}{2^n(n!)^2(p-n)!}, \qquad p = 0, 1, 2, \ldots$$

Hint. Use the result of Problem 7.

16. Derive the following expansion of a product of Hermite polynomials with different indices in a series of Hermite polynomials:

$$H_p(x)H_{p+r}(x) = 2^p p!(p+r)! \sum_{n=0}^{p} \frac{H_{2n+r}(x)}{2^n n!(n+r)!(p-n)!}, \qquad p, r = 0, 1, 2, \ldots$$

Hint. For $r = 1$ the required result is obtained by differentiating the formula found in the preceding problem. The general case can be obtained by using mathematical induction.

Comment. This expansion can be written in the symmetric form

$$H_p(x)H_q(x) = p!q! \sum_{n=0}^{\min(p,q)} \frac{2^n H_{p+q-2n}(x)}{n!(p-n)!(q-n)!}, \qquad p, q = 0, 1, 2, \ldots$$

17. Using the generating function (4.9.3), prove the following *addition theorem* for the Hermite polynomials:

$$H_p(x \cos \alpha + y \sin \alpha) = p! \sum_{n=0}^{p} \frac{H_n(x)H_{p-n}(y)}{n!(p-n)!} \cos^n \alpha \sin^{p-n} \alpha.$$

18. Prove the formula

$$L_p(x^2 + y^2) = \frac{(-1)^p}{2^{2p}} \sum_{n=0}^{p} \frac{H_{2n}(x)H_{2p-2n}(y)}{n!(p-n)!}, \qquad p = 0, 1, 2, \ldots$$

Hint. Use the expansion (4.17.3).

19. Derive the following expansion of the incomplete gamma function (see Problem 10, p. 15) in a series of Laguerre polynomials:

$$x^{-\alpha}\gamma(\alpha, x) = \sum_{n=0}^{\infty} \frac{L_n^\alpha(x)}{2^{n+\alpha}(n+\alpha)}, \qquad 0 < x < \infty, \quad \alpha > 0.$$

20. Derive the expansions

$$L_p^{\alpha+\beta+1}(x + y) = \sum_{n=0}^{p} L_n^\alpha(x)L_{p-n}^\beta(y), \qquad p = 0, 1, 2, \ldots,$$

$$L_p^\beta(x) = \sum_{n=0}^{p} \frac{\Gamma(\beta - \alpha + p - n)}{(p-n)!\Gamma(\beta - \alpha)} L_n^\alpha(x), \qquad p = 0, 1, 2, \ldots$$

Hint. Use the generating function (4.17.3).

21. The *Jacobi polynomials* $P_n^{(\alpha,\beta)}(x)$ are defined by the formula

$$P_n^{(\alpha,\beta)}(x) = \frac{(-1)^n}{2^n n!} (1-x)^{-\alpha}(1+x)^{-\beta} \frac{d^n}{dx^n} [(1-x)^{n+\alpha}(1+x)^{n+\beta}],$$
$$\alpha > -1, \quad \beta > -1, \quad n = 0, 1, 2, \ldots$$

Using the methods of this chapter, show that the Jacobi polynomials have the following properties:

(a) The function $u = P_n^{(\alpha,\beta)}(x)$ satisfies the differential equation

$$(1 - x^2)u'' + [\beta - \alpha - (\alpha + \beta + 2)x]u' + n(n + \alpha + \beta + 1)u = 0;$$

(b) The polynomials $P_n^{(\alpha,\beta)}(x)$ are orthogonal with weight

$$\rho(x) = (1 - x)^\alpha(1 + x)^\beta$$

on the interval $[-1, 1]$;

(c) The polynomials $P_n^{(\alpha,\beta)}(x)$ are the expansion coefficients of the generating function

$$w(x, t) = 2^{\alpha+\beta}R^{-1}(1 - t + R)^{-\alpha}(1 + t + R)^{-\beta} = \sum_{n=0}^{\infty} P_n^{(\alpha,\beta)}(x)t^n, \quad |t| < r,$$

where $R = (1 - 2xt + t^2)^{1/2}$, and r is given by formula (4.2.4).

22. The Chebyshev polynomials[54] are defined by the formula

$$T_n(x) = \cos(n \operatorname{arc\,cos} x), \qquad n = 0, 1, 2, \ldots$$

Show that the Chebyshev polynomials have the following properties:

(a) The function $u = T_n(x)$ satisfies the differential equation

$$(1 - x^2)u'' - xu' + n^2u = 0;$$

(b) The polynomials $T_n(x)$ are orthogonal with weight

$$\rho(x) = (1 - x^2)^{-1/2}$$

on the interval $[-1, 1]$;

(c) The polynomials $T_n(x)$ are the expansion coefficients of the generating function

$$w(x, t) = \frac{1 - t^2}{1 - 2xt + t^2} = T_0(x) + 2\sum_{n=1}^{\infty} T_n(x)t^n, \qquad |t| < r,$$

where r is again given by (4.2.4).

Comment. The Chebyshev polynomials play an important role in the theory of approximation.

---

[54] Sometimes transliterated as the "Tchebichef polynomials", as in G. Szegö, *op. cit.*, and in the Bateman Manuscript Project, *Higher Transcendental Functions*, *Vol. 2*, Chap. 10. We refer the reader to these sources for further information on the Jacobi and Chebyshev polynomials.

5

CYLINDER FUNCTIONS: THEORY

5.1. Introductory Remarks

By a *cylinder function* we mean a solution of the second-order linear differential equation

$$u'' + \frac{1}{z}u' + \left(1 - \frac{\nu^2}{z^2}\right)u = 0, \tag{5.1.1}$$

where z is a complex variable and ν is a parameter which can take arbitrary real or complex values. Equation (5.1.1), called *Bessel's equation of order* ν, is encountered in studying the boundary value problems of potential theory for cylindrical domains (see Sec. 6.3), which explains the origin of the term *cylinder function*. Certain special kinds of cylinder functions are known in the literature as *Bessel functions*, and this term is sometimes applied to the whole class of cylinder functions.

The cylinder functions, with their manifold applications, have been studied in great detail, and extensive tables of such functions are available. These functions are among the most important special functions, with very diverse applications to physics, engineering and mathematical analysis itself, ranging from abstract number theory and theoretical astronomy to concrete problems of physics and engineering. Some of these applications, mainly from the field of mathematical physics, will be considered in Chapter 6. The present chapter is devoted to a brief exposition of the elementary theory of cylinder functions. The reader who wishes to go further in his study of these functions should consult the special literature devoted to the subject (see the Bibliography on p. 300), notably the classic treatise by Watson,[1] to which we will make frequent reference.

[1] G. N. Watson, *A Treatise on the Theory of Bessel Functions*, second edition, Cambridge University Press, London (1962).

5.2. Bessel Functions of Nonnegative Integral Order

In many applied problems, one need only consider a special class of cylinder functions, corresponding to the case where the parameter ν in equation (5.1.1) is a nonnegative integer n. This case is much simpler than the case of arbitrary ν, and will serve to introduce the general theory.

We begin by showing that one of the solutions of Bessel's equation

$$u'' + \frac{1}{z}u' + \left(1 - \frac{n^2}{z^2}\right)u = 0, \qquad n = 0, 1, 2, \ldots \tag{5.2.1}$$

is the function $u_1 = J_n(z)$, known as the *Bessel function of the first kind of order n*, and defined for arbitrary z by the series

$$J_n(z) = \sum_{k=0}^{\infty} \frac{(-1)^k (z/2)^{n+2k}}{k!(n+k)!}, \qquad |z| < \infty. \tag{5.2.2}$$

Using the ratio test, we easily verify that this series converges in the whole complex plane, and hence represents an entire function of z. Suppose we denote the left-hand side of (5.2.1) by $l(u)$, and introduce the abbreviated notation

$$\alpha_k = \frac{(-1)^k}{2^{n+2k} k!(n+k)!}$$

for the coefficients of the series (5.2.2). Then we have

$$\begin{aligned}
l(u_1) &= \sum_{k=0}^{\infty} [(n+2k)(n+2k-1) + (n+2k) - n^2]\alpha_k z^{n+2k-2} + \sum_{k=0}^{\infty} \alpha_k z^{n+2k} \\
&= \sum_{k=1}^{\infty} 4\alpha_k k(n+k) z^{n+2k-2} + \sum_{k=0}^{\infty} \alpha_k z^{n+2k} \\
&= \sum_{k=0}^{\infty} [4\alpha_{k+1}(k+1)(n+k+1) + \alpha_k] z^{n+2k},
\end{aligned}$$

and therefore $l(u_1) \equiv 0$, since the expression in brackets vanishes. Thus $J_n(z)$ satisfies Bessel's equation (5.2.1), i.e., $J_n(z)$ is a cylinder function. The simplest functions of this kind are the Bessel functions of orders zero and one:

$$\begin{aligned}
J_0(z) &= 1 - \frac{(z/2)^2}{(1!)^2} + \frac{(z/2)^4}{(2!)^2} - \frac{(z/2)^6}{(3!)^2} + \cdots, \\
J_1(z) &= \frac{z}{2}\left[1 - \frac{(z/2)^2}{1!2!} + \frac{(z/2)^4}{2!3!} - \frac{(z/2)^6}{3!4!} + \cdots\right].
\end{aligned} \tag{5.2.3}$$

We now show that the Bessel functions of higher order can be expressed in terms of the two functions $J_0(z)$ and $J_1(z)$. Assuming that n is a positive integer, we multiply the series (5.2.2) by z^n and then differentiate with respect to z. This gives

$$\frac{d}{dz}[z^n J_n(z)] = \sum_{k=0}^{\infty} \frac{(-1)^k(2n+2k)}{2^{n+2k}k!(n+k)!} z^{2n+2k-1}$$

$$= z^n \sum_{k=0}^{\infty} \frac{(-1)^k}{k!(n-1+k)!}\left(\frac{z}{2}\right)^{n-1+2k} = z^n J_{n-1}(z),$$

or

$$\frac{d}{dz}[z^n J_n(z)] = z^n J_{n-1}(z), \qquad n = 1, 2, \ldots \tag{5.2.4}$$

Similarly, multiplying (5.2.2) by z^{-n}, we find that

$$\frac{d}{dz}[z^{-n} J_n(z)] = -z^{-n} J_{n+1}(z), \qquad n = 0, 1, 2, \ldots \tag{5.2.5}$$

Performing the differentiation in (5.2.4–5) and dividing by the factors $z^{\pm n}$, we arrive at the formulas

$$J_n'(z) + \frac{n}{z} J_n(z) = J_{n-1}(z), \qquad J_n'(z) - \frac{n}{z} J_n(z) = -J_{n+1}(z), \tag{5.2.6}$$

which immediately imply the following *recurrence relations* satisfied by the Bessel functions:

$$J_{n-1}(z) + J_{n+1}(z) = \frac{2n}{z} J_n(z), \qquad n = 1, 2, \ldots \tag{5.2.7}$$

$$J_{n-1}(z) - J_{n+1}(z) = 2J_n'(z), \qquad n = 1, 2, \ldots \tag{5.2.8}$$

Repeated application of (5.2.7) allows us to express a Bessel function of arbitrary order $\nu = n$ ($n = 0, 1, 2 \ldots$) in terms of $J_0(z)$ and $J_1(z)$, thereby greatly simplifying the effort needed to calculate tables of Bessel functions. Formula (5.2.8) allows us to express derivatives of Bessel functions in terms of other Bessel functions. For $n = 0$, (5.2.8) should be replaced by

$$J_0'(z) = -J_1(z) \tag{5.2.9}$$

[in keeping with (5.2.5)], which is an immediate consequence of the formulas (5.2.3).

The Bessel functions of the first kind $J_n(z)$ are simply related to the coefficients of the Laurent expansion of the function[2]

$$w(z, t) = e^{\frac{1}{2}z(t-t^{-1})} = \sum_{n=-\infty}^{\infty} c_n(z)t^n, \qquad 0 < |t| < \infty. \tag{5.2.10}$$

[2] Regarded as a function of t, $w(z, t)$ is analytic in the annulus $0 < \delta \leqslant t \leqslant A < \infty$, and therefore this expansion exists.

To calculate the coefficients $c_n(z)$, we multiply the power series

$$e^{zt/2} = 1 + \frac{(z/2)}{1!}t + \frac{(z/2)^2}{2!}t^2 + \cdots,$$

$$e^{-z/2t} = 1 - \frac{(z/2)}{1!}\frac{1}{t} + \frac{(z/2)^2}{2!}\frac{1}{t^2} + \cdots,$$

and then combine terms containing identical powers of t. As a result, we obtain

$$\begin{aligned} c_n(z) &= J_n(z), \qquad n = 0, 1, 2, \ldots, \\ c_n(z) &= (-1)^n J_{-n}(z), \qquad n = -1, -2, \ldots, \end{aligned} \tag{5.2.11}$$

which implies

$$w(z, t) = e^{\frac{1}{2}z(t-t^{-1})} = J_0(z) + \sum_{n=1}^{\infty} J_n(z)[t^n + (-1)^n t^{-n}], \quad 0 < |t| < \infty. \tag{5.2.12}$$

The function $w(z, t)$ is called the *generating function* of the Bessel functions of integral order, and formula (5.2.12) plays an important role in the theory of these functions.

To find a general solution of Bessel's equation (5.2.1), thereby obtaining an arbitrary cylinder function of integral order $\nu = n$ ($n = 0, 1, 2, \ldots$), we must construct a second solution of (5.2.1) which is linearly independent of $J_n(z)$. For such a solution we choose $u_2 = Y_n(z)$, called the *Bessel function of the second kind*, which will be defined in Sec. 5.4. It will be shown in Sec. 5.5 that this definition leads to the series expansion

$$\begin{aligned} Y_n(z) = \frac{2}{\pi}J_n(z)\log\frac{z}{2} &- \frac{1}{\pi}\sum_{k=0}^{n-1}\frac{(n-k-1)!}{k!}\left(\frac{z}{2}\right)^{2k-n} \\ &- \frac{1}{\pi}\sum_{k=0}^{\infty}\frac{(-1)^k (z/2)^{n+2k}}{k!(n+k)!}[\psi(k+1) + \psi(k+n+1)], \end{aligned} \tag{5.2.13}$$

where

$$\psi(m+1) = -\gamma + 1 + \frac{1}{2} + \ldots + \frac{1}{m}, \qquad \psi(1) = -\gamma,$$

γ is Euler's constant (see Sec. 1.3), and in the case $n = 0$, the first sum in (5.2.13) should be set equal to zero. The function $Y_n(z)$ is analytic in the complex plane cut along the segment $[-\infty, 0]$, and becomes infinite as $z \to 0$. Thus, the general expression for the cylinder function of order $\nu = n$ is a linear combination of Bessel functions of the first and second kinds, i.e.,

$$u = Z_n(z) = AJ_n(z) + BY_n(z), \qquad n = 0, 1, 2, \ldots, \tag{5.2.14}$$

where A and B are constants.

5.3. Bessel Functions of Arbitrary Order

The Bessel functions considered in the preceding section are a special case of the more general Bessel functions of the first kind of *arbitrary* order ν. To define these functions, consider the series

$$\sum_{k=0}^{\infty} \frac{(-1)^k (z/2)^{\nu+2k}}{\Gamma(k+1)\Gamma(k+\nu+1)}, \tag{5.3.1}$$

where z is a complex variable belonging to the plane cut along the segment $[-\infty, 0]$, and ν is a parameter which can take arbitrary real or complex values.[3] It is easily seen that (5.3.1) converges for all z and ν, and that the convergence is uniform in each variable in the region $|z| \leqslant R$, $|\nu| \leqslant N$ (where R and N are arbitrarily large). This follows from the fact that starting from some sufficiently large k, the ratio of the absolute value of the $(k+1)$th term to that of the kth term equals

$$\frac{|z|^2}{4(k+1)|k+1+\nu|} \leqslant \frac{R^2}{4(k+1)(k+1-N)},$$

where the right-hand side is positive, independent of z and ν, and approaches zero as $k \to \infty$.[4] Since the terms of (5.3.1) are analytic functions of z in the plane cut along $[-\infty, 0]$, the sum of the series is an analytic function of z in the same region. We call this function the *Bessel function of the first kind of order* ν, and denote it by $J_\nu(z)$, i.e.,

$$J_\nu(z) = \sum_{k=0}^{\infty} \frac{(-1)^k (z/2)^{\nu+2k}}{\Gamma(k+1)\Gamma(k+\nu+1)}, \qquad |z| < \infty, \quad |\arg z| < \pi. \tag{5.3.2}$$

To show that the function (5.3.2) satisfies Bessel's equation with parameter ν, we write

$$l(u) \equiv u'' + \frac{1}{z}u' + \left(1 - \frac{\nu^2}{z^2}\right)u = 0, \qquad u_1 = J_\nu(z),$$

and repeat the derivation given in Sec. 5.2,[5] obtaining

$$l(u_1) = \sum_{k=0}^{\infty} [4\alpha_{k+1}(k+1)(k+\nu+1) + \alpha_k] z^{\nu+2k},$$

---

[3] In general, the condition imposed on z is necessary for the function z^ν to be single-valued, but can be omitted if ν is an integer.

[4] A series of functions

$$\sum_{k=0}^{\infty} u_k(z)$$

converges uniformly in a domain D if

$$\left|\frac{u_{k+1}(z)}{u_k(z)}\right| \leqslant q < 1$$

for all z in D and $k \geqslant M$, where q is independent of z. See E. C. Titchmarsh, *op. cit.*, p. 4.

[5] Recall that a uniformly convergent series of analytic functions can be differentiated term by term.

where

$$\alpha_k = \frac{(-1)^k}{2^{\nu+2k}\Gamma(k+1)\Gamma(k+\nu+1)}.$$

Using (1.2.1), we see at once that $l(u_1) \equiv 0$.

Since for fixed z in the plane cut along the segment $[-\infty, 0]$, the terms of the series (5.3.2) are analytic functions of the variable ν (see Sec. 1.1), the fact that (5.3.2) is uniformly convergent implies that the Bessel function of the first kind is an entire function of *its order* ν. For integral $\nu = n$ ($n = 0, 1, 2, \ldots$), $\Gamma(k + \nu + 1) = (n + k)!$ and (5.3.2) reduces to (5.2.2). Therefore the functions defined in this section are the natural generalizations of those studied in the preceding section. For negative integral $\nu = -n$ ($n = 1, 2, \ldots.$), the first n terms of the series (5.3.2) vanish (see Sec. 1.2), and the series becomes

$$J_{-n}(z) = \sum_{k=n}^{\infty} \frac{(-1)^k (z/2)^{-n+2k}}{k!(k-n)!} = \sum_{s=0}^{\infty} \frac{(-1)^{n+s}(z/2)^{n+2s}}{(n+s)!s!},$$

and hence

$$J_{-n}(z) = (-1)^n J_n(z), \qquad n = 1, 2, \ldots \tag{5.3.3}$$

Thus, the Bessel functions of negative integral order differ only by sign from the corresponding functions of positive integral order. It follows that the expansion (5.2.12) can be written in the form

$$w(z, t) = e^{\frac{1}{2}z(t-t^{-1})} = \sum_{n=-\infty}^{\infty} J_n(z)t^n. \tag{5.3.4}$$

Many of the formulas derived earlier for Bessel functions of nonnegative integral order remain the same for Bessel functions of arbitrary order. For example,

$$\frac{d}{dz}[z^\nu J_\nu(z)] = z^\nu J_{\nu-1}(z), \qquad \frac{d}{dz}[z^{-\nu}J_\nu(z)] = -z^{-\nu}J_{\nu+1}(z), \tag{5.3.5}$$

$$J_{\nu-1}(z) + J_{\nu+1}(z) = \frac{2\nu}{z}J_\nu(z), \qquad J_{\nu-1}(z) - J_{\nu+1}(z) = 2J_\nu'(z), \tag{5.3.6}$$

generalize formulas (5.2.4–5, 7–8), and are proved in exactly the same way. We also have

$$\begin{aligned} \left(\frac{d}{z\,dz}\right)^m [z^\nu J_\nu(z)] &= z^{\nu-m}J_{\nu-m}(z), \\ \left(\frac{d}{z\,dz}\right)^m [z^{-\nu}J_\nu(z)] &= (-1)^m z^{-\nu-m}J_{\nu+m}(z), \end{aligned} \tag{5.3.7}$$

which are proved by repeated application of (5.3.6).

5.4. General Cylinder Functions. Bessel Functions of the Second Kind

By definition, a *cylinder function* is an arbitrary solution of the second-order linear differential equation

$$l(u) = u'' + \frac{1}{z}u' + \left(1 - \frac{\nu^2}{z^2}\right)u = 0, \tag{5.4.1}$$

and hence has the general form

$$u = Z_\nu(z) = C_1 u_1(z) + C_2 u_2(z), \tag{5.4.2}$$

where u_1 and u_2 are arbitrary linearly independent solutions of (5.4.1), and C_1, C_2 are constants which, in general, are arbitrary functions of the parameter ν. It is easy to obtain an expression for the general cylinder function in the case where ν is not an integer. In fact, choosing $u_1 = J_\nu(z)$, where $J_\nu(z)$ is the Bessel function defined in Sec. 5.3, we take the second function to be $u_2 = J_{-\nu}(z)$, which is also a solution of (5.4.1), since (5.4.1) does not change if ν is replaced by $-\nu$. For nonintegral ν, the asymptotic behavior of these solutions as $z \to 0$ is given by

$$u_1 \approx \frac{(z/2)^\nu}{\Gamma(1+\nu)}, \qquad u_2 \approx \frac{(z/2)^{-\nu}}{\Gamma(1-\nu)}, \tag{5.4.3}$$

and therefore these solutions are linearly independent.[6] Thus, the desired expression for the general cylinder function can be written as

$$u = Z_\nu(z) = C_1 J_\nu(z) + C_2 J_{-\nu}(z), \qquad \nu \neq 0, \pm 1, \pm 2, \ldots \tag{5.4.4}$$

If ν is an integer, then, because of (5.3.3), the particular solutions u_1 and u_2 are linearly dependent, and (5.4.4) is no longer a general solution of Bessel's equation (5.4.1). To obtain an expression for the general cylinder function which is suitable for arbitrary ν, we introduce the *Bessel functions of the second kind*, denoted by $Y_\nu(z)$ and defined by the formula

$$Y_\nu(z) = \frac{J_\nu(z)\cos\nu\pi - J_{-\nu}(z)}{\sin\nu\pi} \tag{5.4.5}$$

for arbitrary z belonging to the plane cut along the segment $[-\infty, 0]$.[7] For integral ν, the right-hand side of (5.4.5) becomes indeterminate [cf. (5.3.3)], and in this case we define $Y_n(z)$ as the limit

$$Y_n(z) = \lim_{\nu \to n} Y_\nu(z). \tag{5.4.6}$$

---

[6] This argument breaks down if ν is an integer (including zero).

[7] The function we denote by $Y_\nu(z)$ is sometimes denoted by $N_\nu(z)$ in the literature on Bessel functions.

Since both the numerator and denominator are entire functions of ν, and since

$$\frac{d}{d\nu} \sin \nu\pi = \pi \cos \nu\pi \neq 0 \qquad \text{if} \quad \nu = n,$$

this limit exists and can be calculated by L'Hospital's rule, application of which gives

$$Y_n(z) = \frac{1}{\pi} \left[\frac{\partial J_\nu(z)}{\partial \nu}\bigg|_{\nu=n} - (-1)^n \frac{\partial J_{-\nu}(z)}{\partial \nu}\bigg|_{\nu=n} \right]. \tag{5.4.7}$$

It follows from its definition that $Y_\nu(z)$ is an analytic function of z in the plane cut along $[-\infty, 0]$, and an entire function of the parameter ν for fixed z.

In view of (5.4.4), the fact that $Y_\nu(z)$ is a cylinder function, i.e., satisfies Bessel's equation (5.4.1), is obvious for nonintegral ν. To show that $Y_\nu(z)$ is a cylinder function for integral ν, we use the principle of analytic continuation, noting that since $l(Y_\nu)$ is an entire function of ν, $l(Y_\nu) \equiv 0$ for $\nu \neq n$ implies $l(Y_\nu)$ for all ν. The fact that the solutions $u_1 = J_\nu(z)$ and $u_2 = Y_\nu(z)$ are linearly independent follows from the linear independence of the solutions $J_\nu(z)$ and $J_{-\nu}(z)$ for nonintegral ν, and from a comparison of the behavior of u_1 and u_2 as $z \to 0$ [cf. (5.4.3) and (5.5.4), proved below] for integral ν. Thus, finally, the expression

$$u = Z_\nu(z) = C_1 J_\nu(z) + C_2 Y_\nu(z) \tag{5.4.8}$$

for the general cylinder function $Z_\nu(z)$ is suitable for arbitrary ν.

The Bessel functions of the second kind satisfy the same recurrence relations as the functions of the first kind, e.g.,

$$\begin{aligned} &\frac{d}{dz}[z^\nu Y_\nu(z)] = z^\nu Y_{\nu-1}(z), && \frac{d}{dz}[z^{-\nu} Y_\nu(z)] = -z^{-\nu} Y_{\nu+1}(z), \\ &Y_{\nu-1}(z) + Y_{\nu+1}(z) = \frac{2\nu}{z} Y_\nu(z), && Y_{\nu-1}(z) - Y_{\nu+1}(z) = 2Y'_\nu(z). \end{aligned} \tag{5.4.9}$$

For nonintegral ν, the validity of these formulas follows from the definition (5.4.5) and the corresponding formulas for $J_\nu(z)$. To obtain the same formulas for integral ν, we need only pass to the limit $\nu \to n$, observing that all the functions involved are continuous with respect to the index ν. We also note that (5.4.7) implies the relation

$$Y_{-n}(z) = (-1)^n Y_n(z), \qquad n = 0, 1, 2, \ldots, \tag{5.4.10}$$

which allows us to reduce the calculation of functions of negative integral order to that of functions of positive integral order.

By making changes of variables in Bessel's equation (5.4.1), we can easily obtain a number of other differential equations whose general solutions can be expressed in terms of cylinder functions. Of the various equations obtained in this way, those of greatest practical interest are

$$u'' + \frac{1-2\alpha}{z}u' + \left[(\beta\gamma z^{\gamma-1})^2 + \frac{\alpha^2 - \nu^2\gamma^2}{z^2}\right]u = 0,$$
$$u'' + \alpha z^{\gamma} u = 0, \tag{5.4.11}$$

with solutions

$$u = z^{\alpha} Z_{\nu}(\beta z^{\gamma}), \qquad u = z^{1/2} Z_{1/(\gamma+2)}\left(\frac{2\alpha^{1/2}}{\gamma+2} z^{1+(\gamma/2)}\right), \tag{5.4.12}$$

where $Z_{\nu}(z)$ denotes an arbitrary cylinder function.

5.5. Series Expansion of the Function $Y_n(z)$

To derive a series expansion of the function $Y_n(z)$, we use the expansion (5.3.2) to calculate the derivatives with respect to the index ν which appear in (5.4.7). Because of (5.4.10), we need only consider the case $\nu = n$ ($n = 0, 1, 2, \ldots$). Since, as already shown, the series (5.3.1) converges uniformly in ν, we can differentiate it term by term, obtaining[8]

$$\left.\frac{\partial J_{\nu}(z)}{\partial \nu}\right|_{\nu=n} = \sum_{k=0}^{\infty} \frac{(-1)^k (z/2)^{n+2k}}{k!(n+k)!}\left[\log\frac{z}{2} - \psi(k+n+1)\right],$$

where

$$\psi(z) = \frac{\Gamma'(z)}{\Gamma(z)}$$

is the logarithmic derivative of the gamma function (see Sec. 1.3). Similarly, we have

$$\frac{\partial J_{-\nu}(z)}{\partial \nu} = \sum_{k=0}^{\infty} \frac{(-1)^k (z/2)^{-\nu+2k}}{k!\Gamma(k-\nu+1)}\left[-\log\frac{z}{2} + \psi(k-\nu+1)\right].$$

For $k = 0, 1, 2, \ldots, n-1$,

$$\Gamma(k-\nu+1) \to \infty, \qquad \psi(k-\nu+1) \to \infty$$

as $\nu \to n$, so that the first n terms of the last series become indeterminate. However, using familiar formulas from the theory of the gamma function [see (1.2.2, 4) and (1.3.4)], we find that

$$\lim_{\nu\to n} \frac{\psi(k-\nu+1)}{\Gamma(k-\nu+1)} = \lim_{\nu\to n}\left[\Gamma(\nu-k)\sin\pi(\nu-k)\,\frac{\psi(\nu-k)+\pi\cot\pi(\nu-k)}{\pi}\right]$$
$$= (-1)^{n-k}(n-k-1)!, \qquad k = 0, 1, \ldots, n-1,$$

[8] The passage to the limit $\nu \to n$ behind the summation sign is legitimate, since a series obtained by term-by-term differentiation of a uniformly convergent series of analytic functions is itself uniformly convergent.

and therefore

$$\left.\frac{\partial J_{-\nu}(z)}{\partial \nu}\right|_{\nu=n} = (-1)^n \sum_{k=0}^{n-1} \frac{(n-k-1)!}{k!}\left(\frac{z}{2}\right)^{2k-n} + (-1)^n \sum_{p=0}^{\infty} \frac{(-1)^p}{(n+p)!p!}\left[-\log\frac{z}{2} + \psi(p+1)\right]\left(\frac{z}{2}\right)^{2p+n},$$

where we have introduced the new summation index $p = k - n$.

It now follows from (5.4.7) that the desired expansion of the function $Y_n(z)$ is

$$Y_n(z) = -\frac{1}{\pi}\sum_{k=0}^{n-1} \frac{(n-k-1)!}{k!}\left(\frac{z}{2}\right)^{2k-n} + \frac{1}{\pi}\sum_{k=0}^{\infty} \frac{(-1)^k(z/2)^{n+2k}}{k!(n+k)!}\left[2\log\frac{z}{2} - \psi(k+1) - \psi(k+n+1)\right]$$

$$|\arg z| < \pi, \quad n = 0, 1, 2, \ldots, \tag{5.5.1}$$

where the first sum should be set equal to zero if $n = 0$ [cf. (5.2.13)]. According to (1.3.6–7), the values of the logarithmic derivative of the gamma function are given by

$$\psi(1) = -\gamma, \qquad \psi(m+1) = -\gamma + 1 + \frac{1}{2} + \cdots + \frac{1}{m}, \qquad m = 1, 2, \ldots, \tag{5.5.2}$$

where $\gamma = 0.57721566\ldots$ is Euler's constant. Using (5.2.2), we can write the expansion (5.5.1) in a somewhat different form:

$$Y_n(z) = \frac{2}{\pi}J_n(z)\log\frac{z}{2} - \frac{1}{\pi}\sum_{k=0}^{n-1}\frac{(n-k-1)!}{k!}\left(\frac{z}{2}\right)^{2k-n} - \frac{1}{\pi}\sum_{k=0}^{\infty}\frac{(-1)^k(z/2)^{n+2k}}{k!(n+k)!}[\psi(k+1) + \psi(k+n+1)]. \tag{5.5.3}$$

Finally, we note that (5.5.1) implies the asymptotic representations

$$Y_0(z) \approx \frac{2}{\pi}\log\frac{z}{2}, \qquad z \to 0$$

$$Y_n(z) \approx -\frac{(n-1)!}{\pi}\left(\frac{z}{2}\right)^{-n}, \qquad z \to 0, \quad n = 1, 2, \ldots, \tag{5.5.4}$$

which show that $Y_n(z)$ becomes infinite as $z \to 0$.

5.6. Bessel Functions of the Third Kind

Next we discuss still another class of cylinder functions, i.e., the *Bessel functions of the third kind* or *Hankel functions*, denoted by $H_\nu^{(1)}(z)$ and $H_\nu^{(2)}(z)$.

These functions are defined in terms of the Bessel functions of the first and second kinds by the formulas

$$H_\nu^{(1)}(z) = J_\nu(z) + iY_\nu(z), \qquad H_\nu^{(2)}(z) = J_\nu(z) - iY_\nu(z), \tag{5.6.1}$$

where ν is arbitrary and z is any point of the plane cut along the segment $[-\infty, 0]$. The motivation for introducing the functions (5.6.1) is that these linear combinations of $J_\nu(z)$ and $Y_\nu(z)$ have very simple asymptotic expressions for large $|z|$ (see Sec. 5.11) and are frequently encountered in the applications.

It follows from (5.6.1) that the Hankel functions are entire functions of ν, and analytic functions of z in the plane cut along $[-\infty, 0]$. Clearly, the functions $H_\nu^{(1)}(z)$ and $H_\nu^{(2)}(z)$ are linearly independent of each other, and each is linearly independent of $J_\nu(z)$. Therefore we can write the general solution of Bessel's equation (5.4.1) in any of the forms

$$\begin{aligned} u = Z_\nu(z) &= A_1 J_\nu(z) + A_2 H_\nu^{(1)}(z) \\ &= B_1 J_\nu(z) + B_2 H_\nu^{(2)}(z) = D_1 H_\nu^{(1)}(z) + D_2 H_\nu^{(2)}(z), \end{aligned} \tag{5.6.2}$$

where $A_1, \ldots, D_2$ are arbitrary constants, as well as in the form (5.4.8).

Since the Hankel functions are linear combinations of the functions $J_\nu(z)$ and $Y_\nu(z)$, they satisfy the same recurrence relations as these functions, e.g.,

$$\frac{d}{dz}[z^\nu H_\nu^{(p)}(z)] = z^\nu H_{\nu-1}^{(p)}(z), \qquad \frac{d}{dz}[z^{-\nu} H_\nu^{(p)}(z)] = -z^{-\nu} H_{\nu+1}^{(p)}(z),$$

$$H_{\nu-1}^{(p)}(z) + H_{\nu+1}^{(p)}(z) = \frac{2\nu}{z} H_\nu^{(p)}(z), \quad H_{\nu-1}^{(p)}(z) - H_{\nu+1}^{(p)}(z) = 2\frac{dH_\nu^{(p)}(z)}{dz}, \tag{5.6.3}$$

where $p = 1, 2$. Using (5.4.5) to eliminate $Y_\nu(z)$ from (5.6.1), we obtain

$$H_\nu^{(1)}(z) = \frac{J_{-\nu}(z) - e^{-\nu\pi i}J_\nu(z)}{i \sin \nu\pi}, \quad H_\nu^{(2)}(z) = \frac{e^{\nu\pi i}J_\nu(z) - J_{-\nu}(z)}{i \sin \nu\pi}, \tag{5.6.4}$$

which imply the important formulas

$$H_{-\nu}^{(1)}(z) = e^{\nu\pi i} H_\nu^{(1)}(z), \qquad H_{-\nu}^{(2)}(z) = e^{-\nu\pi i} H_\nu^{(2)}(z). \tag{5.6.5}$$

5.7. Bessel Functions of Imaginary Argument

In the applications, one frequently encounters two functions $I_\nu(z)$ and $K_\nu(z)$, which are closely related to the Bessel functions. Let D be the complex plane cut along the negative real axis. Then, for all z in D, $I_\nu(z)$ and $K_\nu(z)$ are defined by the formulas

$$I_\nu(z) = \sum_{k=0}^{\infty} \frac{(z/2)^{\nu+2k}}{\Gamma(k+1)\Gamma(k+\nu+1)}, \qquad |z| < \infty, \quad |\arg z| < \pi, \tag{5.7.1}$$

$$K_\nu(z) = \frac{\pi}{2}\frac{I_{-\nu}(z) - I_\nu(z)}{\sin \nu\pi}, \qquad |\arg z| < \pi, \qquad \nu \neq 0, \pm 1, \pm 2, \ldots \tag{5.7.2}$$

where, for integral $\nu = n$,

$$K_n(z) = \lim_{\nu \to n} K_\nu(z), \qquad n = 0, \pm 1, \pm 2, \ldots \tag{5.7.3}$$

Repeating the considerations of Secs. 5.3–4, we find that $I_\nu(z)$ and $K_\nu(z)$ are analytic functions of z for all z in D, and entire functions of ν.

The functions $I_\nu(z)$ and $K_\nu(z)$ are simply related to the Bessel functions of argument $ze^{\pm \pi i/2}$. If

$$-\pi < \arg z < \frac{\pi}{2}, \quad \text{i.e.,} \quad -\frac{\pi}{2} < \arg (ze^{\pi i/2}) < \pi,$$

then (5.3.2) implies

$$J_\nu(ze^{\pi i/2}) = e^{\nu\pi i/2} \sum_{k=0}^{\infty} \frac{(z/2)^{\nu+2k}}{\Gamma(k+1)\Gamma(k+\nu+1)} = e^{\nu\pi i/2} I_\nu(z),$$

so that

$$I_\nu(z) = e^{-\nu\pi i/2} J_\nu(ze^{\pi i/2}), \qquad -\pi < \arg z < \frac{\pi}{2}. \tag{5.7.4}$$

Similarly, according to (5.6.4), for the same values of z we have

$$\begin{aligned} H_\nu^{(1)}(ze^{\pi i/2}) &= \frac{J_{-\nu}(ze^{\pi i/2}) - e^{-\nu\pi i} J_\nu(ze^{\pi i/2})}{i \sin \nu\pi} \\ &= \frac{e^{-\nu\pi i/2} I_{-\nu}(z) - e^{-\nu\pi i/2} I_\nu(z)}{i \sin \nu\pi} = \frac{2}{\pi i} e^{-\nu\pi i/2} K_\nu(z), \end{aligned}$$

and hence

$$K_\nu(z) = \frac{\pi i}{2} e^{\nu\pi i/2} H_\nu^{(1)}(ze^{\pi i/2}), \qquad -\pi < \arg z < \frac{\pi}{2}. \tag{5.7.5}$$

On the other hand, if

$$-\frac{\pi}{2} < \arg z < \pi, \qquad -\pi < \arg (ze^{-\pi i/2}) < \frac{\pi}{2},$$

then it is easily verified that

$$I_\nu(z) = e^{\nu\pi i/2} J_\nu(ze^{-\pi i/2}), \qquad K_\nu(z) = -\frac{\pi i}{2} e^{-\nu\pi i/2} H_\nu^{(2)}(ze^{-\pi i/2}). \tag{5.7.6}$$

Because of (5.7.4–6), $I_\nu(z)$ and $K_\nu(z)$ are often called *Bessel functions of imaginary argument*. However, this term is not too fortunate, and instead we will usually refer to $I_\nu(z)$ as the *modified Bessel function of the first kind* and to $K_\nu(z)$ as *Macdonald's function*.[9]

[9] $K_\nu(z)$ is called the *modified Bessel function of the third kind* in the Bateman Manuscript Project, *Higher Transcendental Functions, Vol. 2*, p. 5.

It is an immediate consequence of the formulas just derived that $I_\nu(z)$ and $K_\nu(z)$ are linearly independent solutions of the differential equation

$$u'' + \frac{1}{z}u' - \left(1 + \frac{\nu^2}{z^2}\right)u = 0, \tag{5.7.7}$$

which differs from Bessel's equation only by the sign of one term, and goes into Bessel's equation if we make the substitution $z = \pm it$. Equation (5.7.7) is often encountered in mathematical physics, and its general solution, for arbitrary ν, can be written in the form

$$u = C_1 I_\nu(z) + C_2 K_\nu(z). \tag{5.7.8}$$

The functions $I_\nu(z)$ and $K_\nu(z)$ satisfy simple recurrence relations, e.g.

$$\begin{aligned}
&\frac{d}{dz}[z^\nu I_\nu(z)] = z^\nu I_{\nu-1}(z), \qquad &&\frac{d}{dz}[z^{-\nu} I_\nu(z)] = z^{-\nu} I_{\nu+1}(z),\\
&\frac{d}{dz}[z^\nu K_\nu(z)] = -z^\nu K_{\nu-1}(z), \qquad &&\frac{d}{dz}[z^{-\nu} K_\nu(z)] = -z^{-\nu} K_{\nu+1}(z)\\
&I_{\nu-1}(z) + I_{\nu+1}(z) = 2I_\nu'(z), \qquad &&I_{\nu-1}(z) - I_{\nu+1}(z) = \frac{2\nu}{z} I_\nu(z),\\
&K_{\nu-1}(z) + K_{\nu+1}(z) = -2K_\nu'(z), \qquad &&K_{\nu-1}(z) - K_{\nu+1}(z) = -\frac{2\nu}{z} K_\nu(z).
\end{aligned} \tag{5.7.9}$$

The recurrence relations involving $I_\nu(z)$ are proved by substituting from (5.7.1). Then, using these formulas and (5.7.2), we derive the corresponding formulas involving $K_\nu(z)$ for nonintegral ν. Finally, we extend the results to the case of integral ν by using the continuity of $K_\nu(z)$ with respect to the index ν.

Two other useful formulas are

$$\begin{aligned}
I_{-n}(z) &= I_n(z), \qquad n = 0, \pm 1, \pm 2, \ldots,\\
K_{-\nu}(z) &= K_\nu(z),
\end{aligned} \tag{5.7.10}$$

where the first follows from (5.7.1) if we note that the first n terms of the expansion vanish if $\nu = -n$, while the second is an immediate consequence of the definition (5.7.2).

Using (5.7.3) and the method of Sec. 5.5, we can derive a series expansion of the function $K_n(z)$. The result of the calculations is

$$\begin{aligned}
K_n(z) &= \frac{1}{2}\sum_{k=0}^{n-1} \frac{(-1)^k(n-k-1)!}{k!}\left(\frac{z}{2}\right)^{2k-n}\\
&\quad + \frac{1}{2}(-1)^{n-1}\sum_{k=0}^{\infty}\frac{(z/2)^{2k+n}}{k!(k+n)!}\left[2\log\frac{z}{2} - \psi(k+1) - \psi(k+n+1)\right],\\
&\qquad |\arg z| < \pi, \quad n = 0, 1, 2, \ldots,
\end{aligned} \tag{5.7.11}$$

where $\psi(z)$ is the logarithmic derivative of the gamma function [whose values can be found from (5.5.2)], and the first sum should be set equal to zero if $n = 0$. We note that (5.7.11) implies the asymptotic representations

$$\begin{aligned} K_0(z) &\approx \log\frac{2}{z}, \qquad z \to 0, \\ K_n(z) &\approx \frac{1}{2}(n-1)!\left(\frac{z}{2}\right)^{-n}, \qquad z \to 0, \quad n = 1, 2, \ldots, \end{aligned} \tag{5.7.12}$$

which show that $K_n(z)$ becomes infinite as $z \to 0$.

5.8. Cylinder Functions of Half-Integral Order

We now consider the special class of cylinder functions of order $n + \frac{1}{2}$ $(n = 0, \pm 1, \pm 2, \ldots)$. In this case, the cylinder functions can be expressed in terms of elementary functions. To see this, we first find the values of the functions $J_{\pm 1/2}(z)$. Setting $\nu = \pm\frac{1}{2}$ in (5.3.1) and using the duplication formula (1.2.3) for the gamma function, we obtain

$$\begin{aligned} J_{1/2}(z) &= \sum_{k=0}^{\infty} \frac{(-1)^k (z/2)^{2k+\frac{1}{2}}}{\Gamma(k+1)\Gamma(k+\frac{3}{2})} \\ &= \left(\frac{2z}{\pi}\right)^{1/2} \sum_{k=0}^{\infty} \frac{(-1)^k z^{2k}}{\Gamma(2k+2)} = \left(\frac{2}{\pi z}\right)^{1/2} \sin z \end{aligned} \tag{5.8.1}$$

and similarly,

$$J_{-1/2}(z) = \left(\frac{2}{\pi z}\right)^{1/2} \cos z. \tag{5.8.2}$$

The fact that any Bessel function of the first kind of half-integral order can be expressed in terms of elementary functions now follows from the recurrence relation

$$J_{\nu-1}(z) + J_{\nu+1}(z) = \frac{2\nu}{z} J_\nu(z)$$

[see (5.3.6)], repeated application of which gives

$$J_{3/2}(z) = \frac{1}{z} J_{1/2}(z) - J_{-1/2}(z) = \left(\frac{2}{\pi z}\right)^{1/2} \left[\frac{\sin z}{z} - \cos z\right],$$

$$J_{-3/2}(z) = -\left(\frac{2}{\pi z}\right)^{1/2} \left[\sin z + \frac{\cos z}{z}\right],$$

and so on. Using (5.3.7), we can write the general expression for $J_{n+\frac{1}{2}}(z)$ in terms of elementary functions. For example, setting $\nu = \frac{1}{2}$ in the second of the formulas (5.3.7) and taking account of (5.8.1), we find that

$$J_{n+\frac{1}{2}}(z) = (-1)^n \left(\frac{2}{\pi}\right)^{1/2} z^{n+\frac{1}{2}} \left(\frac{d}{z\,dz}\right)^n \frac{\sin z}{z}, \qquad n = 0, 1, 2, \ldots \tag{5.8.3}$$

To derive the corresponding formulas for Bessel functions of the second and third kinds, we start from the expressions (5.4.5) and (5.6.4) of these functions in terms of Bessel functions of the first kind, and use (5.8.1–2). For example,

$$Y_{1/2}(z) = -J_{-1/2}(z) = -\left(\frac{2}{\pi z}\right)^{1/2} \cos z,$$
$$H_{1/2}^{(1)}(z) = -i\left(\frac{2}{\pi z}\right)^{1/2} e^{iz}, \qquad H_{1/2}^{(2)}(z) = i\left(\frac{2}{\pi z}\right)^{1/2} e^{-iz}, \tag{5.8.4}$$

and so on.

Finally, we note that

$$I_{1/2}(z) = \left(\frac{2}{\pi z}\right)^{1/2} \sinh z, \quad I_{-1/2}(z) = \left(\frac{2}{\pi z}\right)^{1/2} \cosh z, \quad K_{1/2}(z) = \left(\frac{\pi}{2z}\right)^{1/2} e^{-z}, \tag{5.8.5}$$

where the formulas for general index $n + \frac{1}{2}$ are obtained from (5.8.5) and the recurrence relations (5.7.9). It has been shown by Liouville that the case of half-integral order is the only case where the cylinder functions reduce to elementary functions.

5.9. Wronskians of Pairs of Solutions of Bessel's Equation

By the *Wronskian* of a pair $u_1(z)$, $u_2(z)$ of solutions of a linear homogeneous second-order differential equation is meant the determinant

$$W\{u_1(z), u_2(z)\} = \begin{vmatrix} u_1(z) & u_2(z) \\ u_1'(z) & u_2'(z) \end{vmatrix},$$

where the prime denotes differentiation with respect to the independent variable z. The solutions u_1 and u_2 are linearly independent if and only if the Wronskian does not vanish identically.[10] We now calculate the Wronskians of various pairs of solutions of Bessel's equation

$$u'' + \frac{1}{z}u' + \left(1 - \frac{\nu^2}{z^2}\right)u = 0,$$

thereby obtaining a number of formulas which are useful in the applications. In particular, these formulas show that the solutions in question are linearly independent, a fact proved earlier by other means.

To calculate the Wronskian, we write the equations for u_1 and u_2 in the form

$$\frac{d}{dz}(zu_1') + \left(z - \frac{\nu^2}{z}\right)u_1 = 0, \qquad \frac{d}{dz}(zu_2') + \left(z - \frac{\nu^2}{z}\right)u_2 = 0,$$

[10] E. A. Coddington, *op. cit.*, Theorem 6, p. 111.

and then subtract the first equation multiplied by u_2 from the second equation multiplied by u_1. The result is

$$\frac{d}{dz}[zW\{u_1(z), u_2(z)\}] = 0,$$

which implies

$$W\{u_1(z), u_2(z)\} = \frac{C}{z},$$

where C is a constant, independent of z, whose value can be determined, for example, from the relation

$$C = \lim_{z \to 0} zW\{u_1(z), u_2(z)\}.$$

In particular, choosing $u_1 = J_\nu(z)$, $u_2 = J_{-\nu}(z)$, where ν is not an integer, and using the expansion (5.3.2) and formulas (1.2.1–2) from the theory of the gamma function, we find that

$$C = \lim_{z \to 0} \frac{-2\nu}{\Gamma(1 + \nu)\Gamma(1 - \nu)} [1 + O(z)^2] = -\frac{2 \sin \nu\pi}{\pi},$$

which implies

$$W\{J_\nu(z), J_{-\nu}(z)\} = -\frac{2 \sin \nu\pi}{\pi z}. \tag{5.9.1}$$

The validity of (5.9.1) for integral ν follows by continuity, and we have $W \equiv 0$, as must be expected. The Wronskians of other pairs of solutions of Bessel's equation can be found in the same way, or else they can be deduced from (5.9.1) and the relations (5.4.5), (5.6.4). We always begin by considering the case of nonintegral ν, and then use continuity to extend the result to arbitrary values of ν. In this way, we find that

$$W\{J_\nu(z), Y_\nu(z)\} = \frac{2}{\pi z}, \tag{5.9.2}$$

$$W\{J_\nu(z), H_\nu^{(2)}(z)\} = -\frac{2i}{\pi z}, \tag{5.9.3}$$

$$W\{H_\nu^{(1)}(z), H_\nu^{(2)}(z)\} = -\frac{4i}{\pi z}, \tag{5.9.4}$$

and so on. For the Bessel functions of imaginary argument we have

$$W\{I_\nu(z), K_\nu(z)\} = -\frac{1}{z}. \tag{5.9.5}$$

5.10. Integral Representations of the Cylinder Functions

The cylinder functions have simple integral representations in terms of definite integrals and contour integrals containing z as a parameter. The representations by contour integrals have greater generality, and are usually valid in larger regions of values of the argument z and parameter ν than the representations by definite integrals, but the latter are more frequently encountered in the applications. Therefore we will be primarily concerned with representations by definite integrals.[11]

One of the simplest integral representations of the Bessel functions is due to Poisson. Consider the identity

$$\frac{1}{\Gamma(k+\nu+1)} = \frac{1}{\Gamma(k+\frac{1}{2})\,\Gamma(\nu+\frac{1}{2})}\int_{-1}^{1} t^{2k}(1-t^2)^{\nu-\frac{1}{2}}\,dt, \qquad \operatorname{Re}\nu > -\tfrac{1}{2}, \tag{5.10.1}$$

implied by (1.5.6). Substituting (5.10.1) into the expansion (5.3.2) and reversing the order of integration and summation,[12] we obtain

$$\begin{aligned} J_\nu(z) &= \sum_{k=0}^{\infty}\frac{(-1)^k(z/2)^{\nu+2k}}{\Gamma(k+1)}\,\frac{1}{\Gamma(k+\frac{1}{2})\Gamma(\nu+\frac{1}{2})}\int_{-1}^{1} t^{2k}(1-t^2)^{\nu-\frac{1}{2}}\,dt \\ &= \frac{(z/2)^\nu}{\Gamma(\nu+\frac{1}{2})}\int_{-1}^{1}(1-t^2)^{\nu-\frac{1}{2}}\,dt\sum_{k=0}^{\infty}\frac{(-1)^k(zt)^{2k}}{2^{2k}\Gamma(k+1)\Gamma(k+\frac{1}{2})} \\ &= \frac{(z/2)^\nu}{\Gamma(\frac{1}{2})\Gamma(\nu+\frac{1}{2})}\int_{-1}^{1}(1-t^2)^{\nu-\frac{1}{2}}\cos zt\,dt, \end{aligned} \tag{5.10.2}$$

where we have used the duplication formula (1.2.3) for the gamma function:

$$2^{2k}\Gamma(k+1)\Gamma(k+\tfrac{1}{2}) = \Gamma(\tfrac{1}{2})\Gamma(2k+1) = \Gamma(\tfrac{1}{2})(2k)!.$$

Thus

$$J_\nu(z) = \frac{(z/2)^\nu}{\Gamma(\frac{1}{2})\Gamma(\nu+\frac{1}{2})}\int_{-1}^{1}(1-t^2)^{\nu-\frac{1}{2}}\cos zt\,dt, \qquad \operatorname{Re}\nu > -\tfrac{1}{2}, \quad |\arg z| < \pi, \tag{5.10.3}$$

or equivalently,

$$J_\nu(z) = \frac{(z/2)^\nu}{\Gamma(\frac{1}{2})\Gamma(\nu+\frac{1}{2})}\int_0^{\pi}\cos(z\cos\theta)\sin^{2\nu}\theta\,d\theta, \quad \operatorname{Re}\nu > -\tfrac{1}{2}, \quad |\arg z| < \pi, \tag{5.10.4}$$

where we have made the substitution $t = \cos\theta$.

---

[11] The reader with a special interest in integral representations of cylinder functions should consult G. N. Watson, *op. cit.*, Chap. 6.

[12] To justify reversing the order of integration and summation, we note that

$$\sum_{k=0}^{\infty}\frac{|z/2|^{\nu+2k}}{\Gamma(k+1)}\,\frac{1}{\Gamma(k+\frac{1}{2})\Gamma(\nu+\frac{1}{2})}\int_{-1}^{1} t^{2k}(1-t^2)^{\nu-\frac{1}{2}}\,dt = \sum_{k=0}^{\infty}\frac{|z/2|^{\nu+2k}}{\Gamma(k+1)\Gamma(k+\nu+1)} \equiv I_\nu(|z|) < \infty,$$

if $\operatorname{Re}\nu > -\frac{1}{2}$.

To obtain another important integral representation of $J_\nu(z)$, we start from the formula

$$\frac{1}{\Gamma(k+\nu+1)} = \frac{1}{2\pi i}\int_C e^s s^{-(k+\nu+1)}\,ds, \qquad (5.10.5)$$

proved in Problem 9, p. 15, where C is the contour shown in Fig. 13. Substituting (5.10.5) into (5.3.2), we find that

$$\begin{aligned} J_\nu(z) &= \sum_{k=0}^{\infty}\frac{(-1)^k(z/2)^{\nu+2k}}{\Gamma(k+1)}\frac{1}{2\pi i}\int_C e^s s^{-(k+\nu+1)}\,ds \\ &= \left(\frac{z}{2}\right)^\nu \frac{1}{2\pi i}\int_C e^s s^{-\nu-1}\,ds\sum_{k=0}^{\infty}\frac{(-1)^k(z^2/4s)^k}{\Gamma(k+1)} \\ &= \left(\frac{z}{2}\right)^\nu \frac{1}{2\pi i}\int_C e^{s-(z^2/4s)} s^{-\nu-1}\,ds, \end{aligned} \qquad (5.10.6)$$

where reversing the order of integration and summation is again easily justified by an absolute convergence argument. Assuming temporarily that z is a positive real number and setting $s = zt/2$, we can write (5.10.6) in the form

$$J_\nu(z) = \frac{1}{2\pi i}\int_{C'} e^{\frac{1}{2}z(t-t^{-1})} t^{-\nu-1}\,dt, \qquad (5.10.7)$$

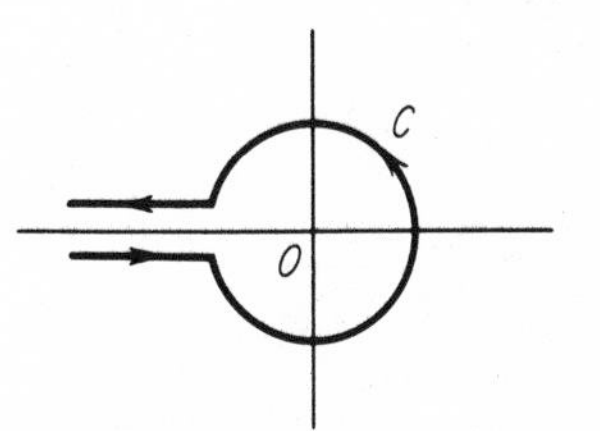

FIGURE 13

where C' is a contour resembling C. By the principle of analytic continuation, this result is valid in the whole region $|\arg z| < \pi/2$. Writing $t = \rho e^{i\theta}$ and choosing the radius of the circular part of C' to be 1, we have

$$J_\nu(z) = \frac{1}{\pi}\int_0^\pi \cos(z\sin\theta - \nu\theta)\,d\theta - \frac{\sin\nu\pi}{\pi}\int_1^\infty e^{-\frac{1}{2}z(\rho-\rho^{-1})}\rho^{-\nu-1}\,d\rho,$$

which, after the substitution $\rho = e^\alpha$, becomes

$$J_\nu(z) = \frac{1}{\pi}\int_0^\pi \cos(z\sin\theta - \nu\theta)\,d\theta - \frac{\sin\nu\pi}{\pi}\int_0^\infty e^{-z\sinh\alpha-\nu\alpha}\,d\alpha, \qquad \operatorname{Re} z > 0, \qquad (5.10.8)$$

where ν is arbitrary. In the case $\nu = n$ ($n = 0, \pm 1, \pm 2, \ldots$), the second term on the right vanishes, and (5.10.8) takes a simpler form.

In many cases, one can derive integral representations of Bessel functions of the second and third kinds from the corresponding integral representations of Bessel functions of the first kind, by using formulas (5.4.5) and (5.6.4).

For example, if $\operatorname{Re} z > 0$ and ν is nonintegral, it follows from (5.4.5) and (5.10.8) that

$$Y_\nu(z) = \frac{\cot \nu\pi}{\pi}\int_0^\pi \cos(z\sin\theta - \nu\theta)\,d\theta - \frac{\cos\nu\pi}{\pi}\int_0^\infty e^{-z\sinh\alpha - \nu\alpha}\,d\alpha$$
$$- \frac{\csc\nu\pi}{\pi}\int_0^\pi \cos(z\sin\theta + \nu\theta)\,d\theta - \frac{1}{\pi}\int_0^\infty e^{-z\sinh\alpha + \nu\alpha}\,d\alpha.$$

Replacing θ by $\pi - \theta$ in the third integral on the right, we find after some simple calculations that

$$Y_\nu(z) = \frac{1}{\pi}\int_0^\pi \sin(z\sin\theta - \nu\theta)\,d\theta - \frac{1}{\pi}\int_0^\infty e^{-z\sinh\alpha}(e^{\nu\alpha} + e^{-\nu\alpha}\cos\nu\pi)\,d\alpha. \tag{5.10.9}$$

In proving (5.10.9), it was assumed that ν is nonintegral, but the formula holds for arbitrary ν by the principle of analytic continuation, since both sides are entire functions of ν.

Integral representations of the Hankel functions can be obtained by using (5.10.8–9) and the definitions (5.6.1). For example, if $\operatorname{Re} z > 0$,

$$H_\nu^{(1)}(z) = J_\nu(z) + iY_\nu(z) = \frac{1}{\pi}\int_0^\pi e^{i(z\sin\theta - \nu\theta)}\,d\theta$$
$$+ \frac{1}{\pi i}\int_0^\infty e^{-z\sinh\alpha}[e^{\nu\alpha} + e^{-\nu(\alpha+\pi i)}]\,d\alpha$$
$$= \frac{1}{\pi i}\int_{-\infty}^0 e^{z\sinh\alpha - \nu\alpha}\,d\alpha + \frac{1}{\pi i}\int_{\theta=0}^\pi e^{z\sinh i\theta - \nu i\theta}\,d(i\theta)$$
$$+ \frac{1}{\pi i}\int_{\alpha=0}^\infty e^{z\sinh(\alpha+\pi i) - \nu(\alpha+\pi i)}\,d(\alpha + \pi i),$$

which, after the substitution $t = \alpha + i\theta$, reduces to

$$H_\nu^{(1)}(z) = \frac{1}{\pi i}\int_{C_1} e^{z\sinh t - \nu t}\,dt, \qquad \operatorname{Re} z > 0, \tag{5.10.10}$$

where C_1 is the contour shown in Fig. 14(a). Similarly,

$$H_\nu^{(2)}(z) = -\frac{1}{\pi i}\int_{C_2} e^{z\sinh t - \nu t}\,dt, \qquad \operatorname{Re} z > 0 \tag{5.10.11}$$

where C_2 is the contour shown in Fig. 14(b). Thus (5.10.10) and (5.10.11) are the same, except for the choice of the contour of integration. Substituting $t = u \pm \frac{1}{2}\pi i$ into (5.10.10–11), we find that

$$H_\nu^{(1)}(z) = \frac{e^{-\nu\pi i/2}}{\pi i}\int_{D_1} e^{iz\cosh u - \nu u}\,du, \qquad \operatorname{Re} z > 0, \tag{5.10.12}$$

$$H_\nu^{(2)}(z) = -\frac{e^{\nu\pi i/2}}{\pi i}\int_{D_2} e^{-iz\cosh u - \nu u}\,du, \qquad \operatorname{Re} z > 0, \tag{5.10.13}$$

where the paths of integration D_1 and D_2 are shown in Figure 15.

To further transform these integrals, we assume temporarily that z is a positive real number and that the parameter ν is confined to the strip $-1 < \operatorname{Re} \nu < 1$. Then, according to Cauchy's integral theorem, the integral

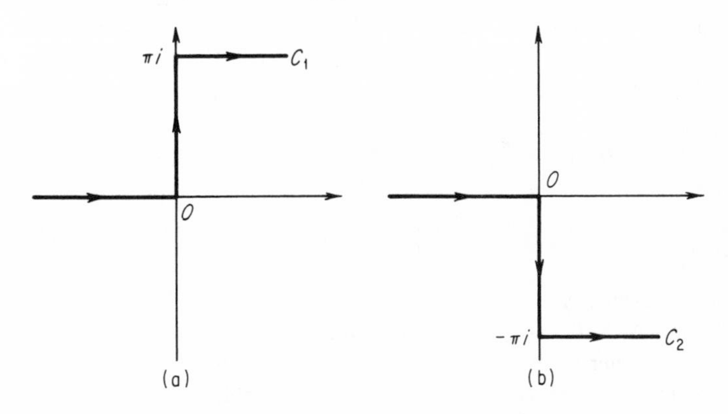

FIGURE 14

along the left-hand part of the broken line D_1 (or D_2), up to the point $u = 0$, can be replaced by an integral along the negative real axis, and the integral

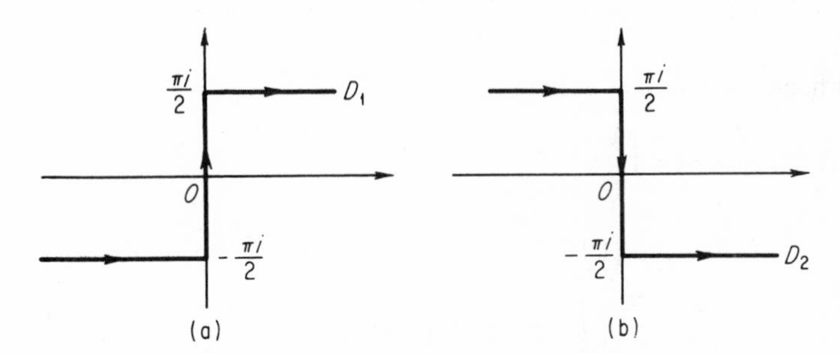

FIGURE 15

along the right-hand part of the broken line can be replaced by an integral along the positive real axis.[13] Thus formulas (5.10.12–13) become

$$H_\nu^{(1)}(z) = \frac{e^{-\nu\pi i/2}}{\pi i} \int_{-\infty}^{\infty} e^{iz \cosh u - \nu u}\, du, \tag{5.10.14}$$

$$H_\nu^{(2)}(z) = -\frac{e^{\nu\pi i/2}}{\pi i} \int_{-\infty}^{\infty} e^{-iz \cosh u - \nu u}\, du, \tag{5.10.15}$$

[13] It is easily verified that the integral along the vertical segment needed to complete each contour to which we apply Cauchy's integral theorem approaches zero as the segment is moved indefinitely far to the left (or to the right) of the imaginary axis. To show that the condition $-1 < \operatorname{Re} \nu < 1$ guarantees the convergence of (5.10, 14–15), consider the substitution $y = e^u$.

where $z > 0$, $-1 < \operatorname{Re} \nu < 1$. Using the principle of analytic continuation, we easily see that (5.10.14) remains valid for $0 \leqslant \arg z < \pi$, while (5.10.15) remains valid for $-\pi < \arg z \leqslant 0$, since in each case both sides of (5.10.14–15) are analytic functions of z in the indicated region. Moreover, the condition $-1 < \operatorname{Re} \nu < 1$ can be dropped if $\operatorname{Im} z > 0$ in (5.10.14), or if $\operatorname{Im} z < 0$ in (5.10.15). Finally, therefore, we have the integral representations

$$H_\nu^{(1)}(z) = \frac{e^{-\nu\pi i/2}}{\pi i} \int_{-\infty}^{\infty} e^{iz \cosh u - \nu u}\, du, \qquad \operatorname{Im} z > 0, \quad (5.10.16)$$

$$H_\nu^{(2)}(z) = -\frac{e^{\nu\pi i/2}}{\pi i} \int_{-\infty}^{\infty} e^{-iz \cosh u - \nu u}\, du, \qquad \operatorname{Im} z < 0, \quad (5.10.17)$$

where ν is arbitrary.

Formulas (5.10.16–17) are the basic integral representations of the Hankel functions. Other integral representations of the Hankel functions, useful in the applications, can be derived by making suitable transformations of the integrals in (5.10.16–17). For example, consider formula (5.10.16), let $\operatorname{Re} \nu > -\frac{1}{2}$, and for the time being assume that $\arg z = \pi/2$, so that $-iz$ is positive. According to (1.5.1),

$$y^{-\nu-\frac{1}{2}} = \frac{1}{\Gamma(\nu + \frac{1}{2})} \int_0^{\infty} e^{-xy} x^{\nu-\frac{1}{2}}\, dx, \qquad \operatorname{Re} \nu > -\tfrac{1}{2}, \quad (5.10.18)$$

and hence, setting $y = e^u$ in (5.10.16), we have

$$\begin{aligned} H_\nu^{(1)}(z) &= \frac{e^{-\nu\pi i/2}}{\pi i} \int_0^{\infty} e^{\frac{1}{2}iz(y + y^{-1})} y^{-\nu-1}\, dy \\ &= \frac{e^{-\nu\pi i/2}}{\pi i \Gamma(\nu + \frac{1}{2})} \int_0^{\infty} e^{\frac{1}{2}iz(y + y^{-1})} y^{-1/2}\, dy \int_0^{\infty} e^{-xy} x^{\nu-\frac{1}{2}}\, dx \\ &= \frac{e^{-\nu\pi i/2}}{\pi i \Gamma(\nu + \frac{1}{2})} \int_0^{\infty} x^{\nu-\frac{1}{2}}\, dx \int_0^{\infty} \exp\left[-y\left(x - \frac{iz}{2}\right) + \frac{iz}{2y}\right] y^{-1/2}\, dy, \end{aligned}$$

where the reversal of the order of integration is easily justified by proving the absolute convergence of the double integral. To calculate the inner integral, we use the formula[14]

$$\int_0^{\infty} e^{-av^2 - (b/v^2)}\, dv = \frac{\sqrt{\pi}}{2\sqrt{a}} e^{-2\sqrt{ab}}, \qquad a > 0, \quad b > 0. \quad (5.10.19)$$

This gives

$$H_\nu^{(1)}(z) = \frac{e^{-\nu\pi i/2}}{i\sqrt{\pi}\Gamma(\nu + \frac{1}{2})} \int_0^{\infty} \frac{e^{-2\sqrt{-iz/2}\sqrt{x-(iz/2)}}}{\sqrt{x - (iz/2)}} x^{\nu-\frac{1}{2}}\, dx,$$

---

[14] After making the transformation $\tau = v^2$, the integral (5.10.19) becomes the Laplace transform of the function $\frac{1}{2}\tau^{-1/2} e^{-b/\tau}$, evaluated at $p = a$.

or

$$H_\nu^{(1)}(z) = \frac{2e^{-\nu\pi i}}{i\sqrt{\pi}\,\Gamma(\nu + \frac{1}{2})}\left(\frac{z}{2}\right)^\nu \int_1^\infty e^{izt}(t^2 - 1)^{\nu - \frac{1}{2}}\,dt, \qquad \operatorname{Re}\nu > -\tfrac{1}{2}, \tag{5.10.20}$$

where we introduce the new variable of integration

$$t = \frac{\sqrt{x - (iz/2)}}{\sqrt{-iz/2}}.$$

By the principle of analytic continuation, this formula, proved under the assumption that $-iz > 0$, remains valid for arbitrary complex z belonging to the sector $0 < \arg z < \pi$. In just the same way, we have the formula

$$H_\nu^{(2)}(z) = \frac{-2e^{\nu\pi i}}{i\sqrt{\pi}\,\Gamma(\nu + \frac{1}{2})}\left(\frac{z}{2}\right)^\nu \int_1^\infty e^{-izt}(t^2 - 1)^{\nu - \frac{1}{2}}\,dt, \tag{5.10.21}$$

$$\operatorname{Re}\nu > -\tfrac{1}{2}, \quad -\pi < \arg z < 0$$

for the second Hankel function. The integral representations (5.10.20–21) play an important role in the derivation of asymptotic representations of the cylinder functions as $|z| \to \infty$.

Integral representations for the Bessel functions of imaginary argument can either be obtained directly by a slight modification of the considerations of this section, or else deduced from (5.7.4–6) and the corresponding integral representations of the Bessel functions and Hankel functions. Thus, it follows from (5.10.3) that

$$I_\nu(z) = \frac{(z/2)^\nu}{\sqrt{\pi}\,\Gamma(\nu + \frac{1}{2})}\int_{-1}^1 (1 - t^2)^{\nu - \frac{1}{2}} \cosh zt\,dt, \tag{5.10.22}$$

$$|\arg z| < \pi, \quad \operatorname{Re}\nu > -\tfrac{1}{2},$$

and from (5.10.16, 20) that

$$K_\nu(z) = \frac{1}{2}\int_{-\infty}^\infty e^{-z\cosh u - \nu u}\,du = \int_0^\infty e^{-z\cosh u}\cosh \nu u\,du, \tag{5.10.23}$$

$$\operatorname{Re} z > 0, \quad \nu \text{ arbitrary},$$

$$K_\nu(z) = \frac{\sqrt{\pi}}{\Gamma(\nu + \frac{1}{2})}\left(\frac{z}{2}\right)^\nu \int_1^\infty e^{-zt}(t^2 - 1)^{\nu - \frac{1}{2}}\,dt, \tag{5.10.24}$$

$$\operatorname{Re} z > 0, \quad \operatorname{Re}\nu > -\tfrac{1}{2}.$$

We also call attention to another integral representation

$$K_\nu(z) = \frac{1}{2}\left(\frac{z}{2}\right)^\nu \int_0^\infty e^{-t - (z^2/4t)}t^{-\nu - 1}\,dt, \qquad |\arg z| < \frac{\pi}{4}, \tag{5.10.25}$$

which is useful in the applications, and is obtained from (5.10.23) by changing the variable of integration.

Some other useful integral representations of the cylinder functions and their products are given in Problems 1–9, p. 139.

5.11. Asymptotic Representations of the Cylinder Functions for Large $|z|$

There are simple asymptotic formulas which allow us to approximate the cylinder functions for large $|z|$ and fixed ν. The leading terms of these asymptotic expansions can be derived starting from the differential equations satisfied by the cylinder functions, but to obtain more exact expressions, it is preferable to use the integral representations found in the preceding section.

Asymptotic representations of the cylinder functions for large $|\nu|$ and fixed z can be obtained rather simply from formulas (5.3.2), (5.4.5), (5.6.4) and (5.7.1.–2) by using Stirling's formula (1.4.22). The problem of approximating the cylinder functions when both $|z|$ and $|\nu|$ are large is one of the most difficult problems of the theory. Some basic results along these lines can be found in Chapter 8 of Watson's treatise, and new formulas of this type have been obtained in recent years by Langer[15] and Cherry.[16]

Of all the cylinder functions, the Hankel functions have the simplest asymptotic representations. We now derive an asymptotic representation of the function $H_\nu^{(1)}(z)$, starting from formula (5.10.20). Making the substitution $t = 1 + 2s$, we find that

$$H_\nu^{(1)}(z) = \frac{2^{\nu+1}e^{i(z-\nu\pi)}z^\nu}{i\sqrt{\pi}\,\Gamma(\nu+\frac{1}{2})}\int_0^\infty e^{2zis}s^{\nu-\frac{1}{2}}(1+s)^{\nu-\frac{1}{2}}\,ds, \tag{5.11.1}$$

$$\operatorname{Re}\nu > -\tfrac{1}{2}, \quad 0 > \arg z < \pi.$$

Replacing $(1+s)^{\nu-\frac{1}{2}}$ by its binomial expansion

$$(1+s)^{\nu-\frac{1}{2}} = \sum_{k=0}^{n}\frac{(-1)^k(\frac{1}{2}-\nu)_k}{k!}s^k + \frac{(-1)^{n+1}(\frac{1}{2}-\nu)_{n+1}}{n!}s^{n+1}\int_0^1(1-t)^n(1+st)^{\nu-n-\frac{3}{2}}\,dt \tag{5.11.2}$$

---

[15] R. E. Langer, *On the asymptotic solutions of ordinary differential equations, with an application to the Bessel functions of large order*, Trans. Amer. Math. Soc., **33**, 23 (1931); *On the asymptotic solutions of differential equations, with an application to the Bessel functions of large complex order*, ibid., **34**, 447 (1942).

[16] T. M. Cherry, *Uniform asymptotic expansions*, J. Lond. Math. Soc., **24**, 121 (1949). *On expansion in eigenfunctions, particularly in Bessel functions*, Proc. Lond. Math. Soc., **51**, 14 (1949); *Uniform asymptotic formulae for functions with transition points*, Trans. Amer. Math. Soc., **68**, 224 (1950).

with remainder,[17] and integrating term by term, we obtain

$$H_\nu^{(1)}(z) = \left(\frac{2}{\pi z}\right)^{1/2} e^{i(z - \frac{1}{2}\nu\pi - \frac{1}{4}\pi)}\left[\sum_{k=0}^{n} \frac{(\frac{1}{2} - \nu)_k(\frac{1}{2} + \nu)_k}{k!} (2zi)^{-k} + r_n(z)\right].$$

Here

$$r_n(z) = \frac{(-1)^{n+1}(\frac{1}{2} - \nu)_{n+1}(-2zi)^{\nu + \frac{1}{2}}}{n!\Gamma(\nu + \frac{1}{2})}$$
$$\times \int_0^\infty e^{2zis} s^{\nu + n + \frac{1}{2}}\, ds \int_0^1 (1 - t)^n(1 + st)^{\nu - n - \frac{3}{2}}\, dt,$$

and we have used the formula

$$\int_0^\infty e^{2zis} s^{k + \nu - \frac{1}{2}}\, ds = \Gamma(\nu + \tfrac{1}{2})(\nu + \tfrac{1}{2})_k(-2zi)^{-(k + \nu + \frac{1}{2})},$$
$$\operatorname{Re} \nu > -\tfrac{1}{2}, \quad 0 < \arg z < \pi, \quad k = 0, 1, 2, \ldots,$$

implied by (1.5.1).

Now suppose that $\delta \leqslant \arg z \leqslant \pi - \delta$, where δ is an arbitrarily small positive number, and for the time being, assume that $\operatorname{Re} \nu - n - \frac{3}{2} \leqslant 0$. Then, estimating $|r_n(z)|$, we find that[18]

$$|r_n(z)| \leqslant \frac{|(\frac{1}{2} - \nu)_{n+1}|(2|z|)^{\operatorname{Re} \nu + \frac{1}{2}} e^{\pi |\operatorname{Im} \nu|}}{n!|\Gamma(\nu + \frac{1}{2})|}$$
$$\times \int_0^\infty e^{-2|z|s \sin \delta} s^{\operatorname{Re} \nu + n + \frac{1}{2}}\, ds \int_0^1 (1 - t)^n\, dt$$
$$= \frac{|(\frac{1}{2} - \nu)_{n+1}|(2|z|)^{\operatorname{Re} \nu + \frac{1}{2}} e^{\pi|\operatorname{Im} \nu|}\Gamma(\operatorname{Re} \nu + n + \frac{3}{2})}{(n + 1)!|\Gamma(\nu + \frac{1}{2})|(2|z| \sin \delta)^{\operatorname{Re} \nu + n + \frac{3}{2}}} = O(|z|^{-n-1})$$

for fixed ν. Therefore

$$H_\nu^{(1)}(z) = \left(\frac{2}{\pi z}\right)^{1/2} e^{i(z - \frac{1}{2}\nu\pi - \frac{1}{4}\pi)}\left[\sum_{k=0}^{n} \frac{(\frac{1}{2} - \nu)_k(\frac{1}{2} + \nu)_k}{k!} (2zi)^{-k} + O(|z|^{-n-1})\right],$$
$$\operatorname{Re} \nu > -\tfrac{1}{2}, \quad \delta \leqslant \arg z \leqslant \pi - \delta, \quad n \geqslant \operatorname{Re} \nu - \tfrac{3}{2} \quad (5.11.3)$$

for large $|z|$. Actually, the condition imposed on n can be dropped, since if

$$\operatorname{Re} \nu - n - \tfrac{3}{2} > 0$$

---

[17] Note that

$$(1 + \zeta)^\mu = \sum_{k=0}^{n} (-1)^k \frac{(-\mu)_k}{k!} \zeta^k + (-1)^{n+1} \frac{(-\mu)_{n+1}}{n!} \zeta^{n+1} \int_0^1 (1 - t)^n(1 + \zeta t)^{\mu - n - 1}\, dt,$$

where

$$|\arg (1 + \zeta)| < \pi, \qquad (\lambda)_0 = 1, \qquad (\lambda)_k = \frac{\Gamma(\lambda + k)}{\Gamma(\lambda)} = \lambda(\lambda + 1)\cdots(\lambda + k - 1).$$

[18] For complex a and b we have

$$|a^b| = |a|^{\operatorname{Re} b}\, e^{-\operatorname{Im} b \cdot \arg a}.$$

we can always find an integer $m > n$ such that

$$\operatorname{Re} \nu - m - \tfrac{3}{2} \leqslant 0.$$

Then, representing $H_\nu^{(1)}(z)$ by (5.11.3) with n replaced by m, and noting that

$$\begin{aligned}\sum_{k=0}^{m} \cdots + O(|z|^{-m-1}) &= \sum_{k=0}^{n} \cdots + \sum_{k=n+1}^{m} \cdots + O(|z|^{-m-1}) \\ &= \sum_{k=0}^{n} \cdots + O(|z|^{-n-1}),\end{aligned}$$

we again arrive at (5.11.3). Moreover, the relation

$$H_\nu^{(1)}(z) = e^{-\nu\pi i} H_{-\nu}^{(1)}(z)$$

[cf. (5.6.5)] allows us to eliminate the condition imposed on the parameter ν, and in fact, by using an integral representation of a somewhat more general type than (5.10.20), it can be shown that the asymptotic formula (5.11.3) remains valid in the larger sector $|\arg z| \leqslant \pi - \delta$.[19] Finally, therefore, we have

$$H_\nu^{(1)}(z) = \left(\frac{2}{\pi z}\right)^{1/2} e^{i(z - \frac{1}{2}\nu\pi - \frac{1}{4}\pi)} \left[\sum_{k=0}^{n} (-1)^k (\nu, k)(2iz)^{-k} + O(|z|^{-n-1})\right],$$

$$|\arg z| \leqslant \pi - \delta \quad (5.11.4)$$

for large $|z|$, where we introduce the notation

$$(\nu, k) = \frac{(-1)^k}{k!} (\tfrac{1}{2} - \nu)_k (\tfrac{1}{2} + \nu)_k = \frac{(4\nu^2 - 1)(4\nu^2 - 3^2) \cdots (4\nu^2 - (2k-1)^2)}{2^{2k} k!},$$

$$(\nu, 0) = 1.$$

An asymptotic representation of the function $H_\nu^{(2)}(z)$ can be obtained in the same way, starting from formula (5.10.21). The result is

$$H_\nu^{(2)}(z) = \left(\frac{2}{\pi z}\right)^{1/2} e^{-i(z - \frac{1}{2}\nu\pi - \frac{1}{4}\pi)} \left[\sum_{k=0}^{n} (\nu, k)(2iz)^{-k} + O(|z|^{-n-1})\right],$$

$$|\arg z| \leqslant \pi - \delta, \quad (5.11.5)$$

which differs from (5.11.4) only by the sign of i.

Asymptotic representations for the Bessel functions of the first and second kinds can be deduced from formulas (5.11.4–5) and the relations (5.6.1). Thus we find that[20]

$$\begin{aligned}J_\nu(z) = {} & \left(\frac{2}{\pi z}\right)^{1/2} \cos (z - \tfrac{1}{2}\nu\pi - \tfrac{1}{4}\pi)\left[\sum_{k=0}^{n} (-1)^k (\nu, 2k)(2z)^{-2k} + O(|z|^{-2n-2})\right] \\ & - \left(\frac{2}{\pi z}\right)^{1/2} \sin (z - \tfrac{1}{2}\nu\pi - \tfrac{1}{4}\pi) \\ & \times \left[\sum_{k=0}^{n} (-1)^k (\nu, 2k+1)(2z)^{-2k-1} + O(|z|^{-2n-3})\right],\end{aligned}$$

$$|\arg z| \leqslant \pi - \delta, \quad (5.11.6)$$

[19] G. N. Watson, *op. cit.*, p. 196.

[20] In (5.11.6–8) the integer n need not be the same in both sums.

and

$$Y_\nu(z) = \left(\frac{2}{\pi z}\right)^{1/2} \cos\left(z - \tfrac{1}{2}\nu\pi - \tfrac{1}{4}\pi\right) \times \left[\sum_{k=0}^{n} (-1)^k(\nu, 2k+1)(2z)^{-2k-1} + O(|z|^{-2n-3})\right]$$
$$+ \left(\frac{2}{\pi z}\right)^{1/2} \sin\left(z - \tfrac{1}{2}\nu\pi - \tfrac{1}{4}\pi\right)\left[\sum_{k=0}^{n} (-1)^k(\nu, 2k)(2z)^{-2k} + O(|z|^{-2n-2})\right],$$
$$|\arg z| \leqslant \pi - \delta. \quad (5.11.7)$$

Similarly, asymptotic formulas for the Bessel functions of imaginary argument can be derived from the integral representations (5.10.22, 24), or else by using the relations given in Sec. 5.7, in conjunction with formulas (5.11.4–5). In this way, we find that

$$I_\nu(z) = e^z(2\pi z)^{-1/2}\left[\sum_{k=0}^{n} (-1)^k(\nu, k)(2z)^{-k} + O(|z|^{-n-1})\right]$$
$$+ e^{-z\pm\pi i(\nu+\frac{1}{2})}(2\pi z)^{-1/2}\left[\sum_{k=0}^{n} (\nu, k)(2z)^{-k} + O(|z|^{-n-1})\right],$$
$$|\arg z| \leqslant \pi - \delta, \qquad (5.11.8)$$

and

$$K_\nu(z) = \left(\frac{\pi}{2z}\right)^{1/2} e^{-z}\left[\sum_{k=0}^{n} (\nu, k)(2z)^{-k} + O(|z|)^{-n-1}\right], \qquad |\arg z| \leqslant \pi - \delta, \qquad (5.11.9)$$

where in (5.11.8) we choose the plus sign if $\operatorname{Im} z > 0$ and the minus sign if $\operatorname{Im} z < 0$. The second term in (5.11.8) will be small if $|\arg z| \leqslant \frac{1}{2}\pi - \delta$, and then

$$I_\nu(z) = e^z(2\pi z)^{-1/2}\left[\sum_{k=0}^{n} (-1)^k(\nu, k)(2z)^{-k} + O(|z|^{-n-1})\right], \quad |\arg z| \leqslant \frac{\pi}{2} - \delta. \qquad (5.11.10)$$

The divergent series obtained by formally setting $n = \infty$ in each of the formulas (5.11.4–10) is the asymptotic series (see Sec. 1.4) of the function appearing in the left-hand side.

The method used here to derive asymptotic expansions gives only the order of magnitude of the remainder term $r_n(z)$, and does not furnish more exact information about the size of $|r_n(z)|$. With suitable assumptions concerning z and ν, the considerations given above can be modified to yield much more exact results. For example, it can be shown[21] that if z and ν are

[21] G. N. Watson, *op. cit.*, p. 206.

positive real numbers, and if n is so large that $2n \geqslant \nu - \frac{1}{2}$, then the remainder in the asymptotic expansion of $J_\nu(z)$ or $Y_\nu(z)$ is smaller in absolute value than the first neglected term, while the same is true of the asymptotic expansion of $K_\nu(z)$ if $n \geqslant \nu - \frac{3}{2}$.

5.12. Addition Theorems for the Cylinder Functions

Given an arbitrary triangle with sides r_1, r_2 and R, let θ and ψ be the angles opposite the sides R and r_1, respectively (see Figure 16), so that

$$R = \sqrt{r_1^2 + r_2^2 - 2r_1r_2 \cos \theta}, \qquad \sin \psi = \frac{r_1}{R} \sin \theta.$$

By an *addition theorem* for cylinder functions we mean an identity of the form

$$Z_\nu(\lambda R) = f_\nu(r_1, r_2, \theta) \sum_{(m)} \Phi_\nu^{(m)}(\lambda r_1)\Psi_\nu^{(m)}(\lambda r_2)\Theta_\nu^{(m)}(\theta), \tag{5.12.1}$$

where λ is an arbitrary complex number with $|\arg \lambda| < \pi$ (for integral ν, this condition can be dropped), and m ranges over some set of indices. Formula (5.12.1) is an expansion of the general cylinder function $Z_\nu(\lambda R)$ in a series whose terms are obtained by multiplying some function $f_\nu(r_1, r_2, \theta)$, which is independent of the summation index m, by three factors, each of which depends on only one of the variables r_1, r_2, θ.

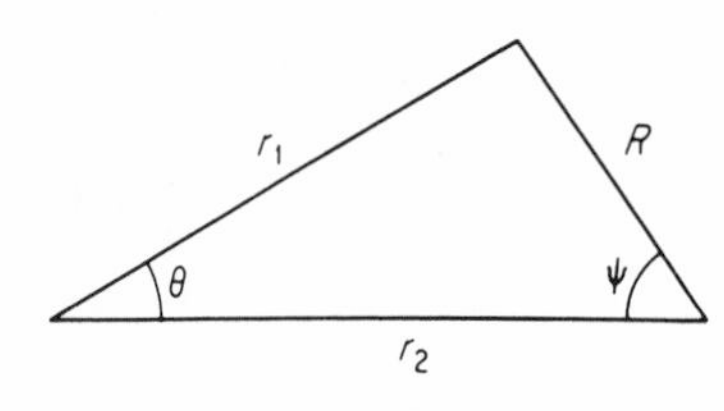

FIGURE 16

Formulas of this kind play an important role in the applications, especially in mathematical physics. The simplest such formula is the following addition theorem for the Bessel function of the first kind of order zero:

$$\begin{aligned} J_0(\lambda R) &= \sum_{m=-\infty}^{\infty} J_m(\lambda r_1)J_m(\lambda r_2)e^{im\theta} \\ &= J_0(\lambda r_1)J_0(\lambda r_2) + 2\sum_{m=1}^{\infty} J_m(\lambda r_1)J_m(\lambda r_2)\cos m\theta. \end{aligned} \tag{5.12.2}$$

To prove (5.12.2), we first note that

$$J_n(z) = \frac{1}{2\pi i}\int_C e^{\frac{1}{2}z(t-t^{-1})}t^{-n-1}\,dt, \qquad n = 0, \pm 1, \pm 2, \ldots \tag{5.12.3}$$

where C is an arbitrary closed contour surrounding the point $t = 0$.[22] Introducing a new variable of integration u by writing

$$t = \frac{r_1e^{i\theta} - r_2}{R}u,$$

[22] Formula (5.12.3) is a special case of (5.10.7) and can be proved immediately by using residues, after recalling the expansion (5.3.4).

and using the fact that

$$R^2 = (r_1 e^{i\theta} - r_2)(r_1 e^{-i\theta} - r_2),$$

we have

$$J_0(\lambda R) = \frac{1}{2\pi i}\int_{C'} \exp\left[\frac{\lambda r_1}{2}\left(ue^{i\theta} - \frac{1}{ue^{i\theta}}\right) - \frac{\lambda r_2}{2}\left(u - \frac{1}{u}\right)\right]\frac{du}{u},$$

where the integration is along a contour C' resembling C. Moreover, according to (5.3.4),

$$\exp\left[\frac{\lambda r_1}{2}\left(ue^{i\theta} - \frac{1}{ue^{i\theta}}\right)\right] = \sum_{m=-\infty}^{\infty} J_m(\lambda r_1)e^{im\theta}u^m, \tag{5.12.4}$$

where the convergence is uniform in u on the contour C'. Therefore, substituting (5.12.4) into (5.12.3) and integrating term by term, we find that

$$\begin{aligned} J_0(\lambda R) &= \sum_{m=-\infty}^{\infty} J_m(\lambda r_1)e^{im\theta}\frac{1}{2\pi i}\int_{C'} \exp\left[-\frac{\lambda r_2}{2}\left(u - \frac{1}{u}\right)\right] u^{m-1}\,du \\ &= \sum_{m=-\infty}^{\infty} J_m(\lambda r_1)J_{-m}(-\lambda r_2)e^{im\theta} = \sum_{m=-\infty}^{\infty} J_m(\lambda r_1)J_m(\lambda r_2)e^{im\theta}, \end{aligned}$$

which proves (5.12.2).

We now give two generalizations of formula (5.12.2) to the case of Bessel functions of arbitrary order ν, referring the reader elsewhere for proofs.[23] The first generalization is of the form[24]

$$J_\nu(\lambda R)\,\frac{\cos \nu\psi}{\sin \nu\psi} = \sum_{m=-\infty}^{\infty} J_{\nu+m}(\lambda r_2)J_m(\lambda r_1)\,\frac{\cos m\theta}{\sin m\theta}, \tag{5.12.5}$$

where ψ is shown in Figure 16, and $r_2 > r_1$ if ν is nonintegral (for integral ν, this restriction can be dropped, i.e., r_1 and r_2 can be interchanged). The second generalization of (5.12.2) is given by the formula

$$\frac{J_\nu(\lambda R)}{(\lambda R)^\nu} = 2^\nu\Gamma(\nu)\sum_{m=0}^{\infty}(\nu + m)\frac{J_{\nu+m}(\lambda r_1)J_{\nu+m}(\lambda r_2)}{(\lambda r_1)^\nu(\lambda r_2)^\nu}\,C_m^\nu(\cos\theta),$$

$$\nu \neq 0, -1, -2, \ldots, \tag{5.12.6}$$

where r_1 and r_2 are arbitrary. Here the functions $C_m^\nu(x)$, $m = 0, 1, 2, \ldots$, known as the *Gegenbauer polynomials*, are defined as the coefficients in the expansion

$$(1 - 2tx + t^2)^{-\nu} = \sum_{m=0}^{\infty} C_m^\nu(x)t^m, \tag{5.12.7}$$

[so that the function on the left is the generating function of the polynomials $C_m^\nu(x)$], and have the following explicit expressions:

$$C_m^\nu(x) = \sum_{k=0}^{[m/2]} (-1)^k 2^{m-2k}\frac{\Gamma(\nu + m - k)}{\Gamma(\nu)k!(m - 2k)!}\,x^{m-2k} \tag{5.12.8}$$

---

[23] G. N. Watson, *op. cit.*, Chap. 11.

[24] Formula (5.12.5) is an abbreviated way of writing two formulas, one involving cosines in both sides, the other sines.

$[C_0^\nu(x) = 1]$. For $\nu = \frac{1}{2}$ the expansion (5.12.7) reduces to formula (4.2.3), and then the Gegenbauer polynomials coincide with the Legendre polynomials:

$$C_m^{1/2}(x) = P_m(x). \tag{5.12.9}$$

For $\nu = 0$ we have

$$C_m^0(x) \equiv 0, \qquad m = 1, 2, \ldots,$$

but the product $\Gamma(\nu)C_m^\nu(x)$ approaches a finite limit as $\nu \to 0$:

$$\lim_{\nu \to 0} \Gamma(\nu)(\nu + m)C_m^\nu(x) = 2 \cos(m \arccos x), \qquad m = 1, 2, \ldots \tag{5.12.10}$$

Therefore both formulas (5.12.5–6) reduce to (5.12.2) in the limit $\nu \to 0$.

For cylinder functions of other kinds, we have similar addition theorems, among which we cite the following:

$$Z_\nu(\lambda R) \frac{\cos \nu\psi}{\sin \nu\psi} = \sum_{m=-\infty}^{\infty} Z_{\nu+m}(\lambda r_2) J_m(\lambda r_1) \frac{\cos m\theta}{\sin m\theta}, \tag{5.12.11}$$

$$\frac{Z_\nu(\lambda R)}{(\lambda R)^\nu} = 2^\nu \Gamma(\nu) \sum_{m=0}^{\infty} (\nu + m) \frac{Z_{\nu+m}(\lambda r_2) J_{\nu+m}(\lambda r_1)}{(\lambda r_2)^\nu (\lambda r_1)^\nu} C_m^\nu(\cos\theta), \tag{5.12.12}$$

$$I_\nu(\lambda R) \frac{\cos \nu\psi}{\sin \nu\psi} = \sum_{m=-\infty}^{\infty} (-1)^m I_{\nu+m}(\lambda r_2) I_m(\lambda r_1) \frac{\cos m\theta}{\sin m\theta}, \tag{5.12.13}$$

$$\frac{I_\nu(\lambda R)}{(\lambda R)^\nu} = 2^\nu \Gamma(\nu) \sum_{m=0}^{\infty} (-1)^m (\nu + m) \frac{I_{\nu+m}(\lambda r_2) I_{\nu+m}(\lambda r_1)}{(\lambda r_2)^\nu (\lambda r_1)^\nu} C_m^\nu(\cos\theta), \tag{5.12.14}$$

$$K_\nu(\lambda R) \frac{\cos \nu\psi}{\sin \nu\psi} = \sum_{m=-\infty}^{\infty} K_{\nu+m}(\lambda r_2) I_m(\lambda r_1) \frac{\cos m\theta}{\sin m\theta}, \tag{5.12.15}$$

$$\frac{K_\nu(\lambda R)}{(\lambda R)^\nu} = 2^\nu \Gamma(\nu) \sum_{m=0}^{\infty} (\nu + m) \frac{K_{\nu+m}(\lambda r_2) I_{\nu+m}(\lambda r_1)}{(\lambda r_2)^\nu (\lambda r_1)^\nu} C_m^\nu(\cos\theta). \tag{5.12.16}$$

In formulas (5.12.11–13, 15–16), it is assumed that $r_2 > r_1$ unless ν is an integer or $Z_{\nu+m} = J_{\nu+m}$ in (5.12.12).

An important special case of these addition theorems, encountered in mathematical physics, occurs when $\nu = \frac{1}{2}$. The formulas corresponding to this case are easily obtained by using (5.12.9), together with the results of Sec. 5.8.[25]

5.13. Zeros of the Cylinder Functions

In solving many applied problems, one needs information about the location of the zeros of cylinder functions in the complex plane, and in particular,

---

[25] G. N. Watson, *op. cit.*, p. 368.

one must be able to make approximate calculations of the values of these zeros. Here we cite without proof some important results along these lines.[26] We begin by considering the distribution of zeros of the Bessel functions of the first kind, i.e., roots of the equation

$$J_\nu(z) = 0. \tag{5.13.1}$$

Theorem 1 deals with the case of nonnegative integral ν, and Theorem 2 with the case of arbitrary real ν:

THEOREM 1. *The function $J_n(z)$, $n = 0, 1, 2, \ldots$ has no complex zeros, and has an infinite number of real zeros symmetrically located with respect to the point $z = 0$, which is itself a zero if $n > 0$. All the zeros of $J_n(z)$ are simple, except the point $z = 0$, which is a zero of order n if $n > 0$.*

THEOREM 2. *Let ν be an arbitrary real number, and suppose that $|\arg z| < \pi$. Then the function $J_\nu(z)$ has an infinite number of positive real zeros, and a finite number $2N(\nu)$ of conjugate complex zeros, where*

1. *$N(\nu) = 0$ if $\nu > -1$ or $\nu = -1, -2, \ldots;$*
2. *$N(\nu) = m$ if $-(m + 1) < \nu < -m, \quad m = 1, 2, \ldots$*

(In the second case, if $[-\nu]$ is odd, there is a pair of purely imaginary zeros among the conjugate complex zeros.) Moreover, all the zeros are simple, except possibly the zero at the point $z = 0$.

The following generalization of equation (5.13.1) is often encountered in mathematical physics (A and B are real):

$$AJ_\nu(z) + BzJ'_\nu(z) = 0, \qquad \nu > -1, \qquad |\arg z| < \pi. \tag{5.13.2}$$

It can be shown that this equation has infinitely many positive real roots and no complex roots, unless

$$\frac{A}{B} + \nu < 0,$$

in which case (5.13.2) also has two purely imaginary roots.[27]

The distribution of zeros of the function $I_\nu(z)$ can be deduced from Theorem 2 and the relations of Sec. 5.7. In particular, it should be noted that all the zeros of $I_\nu(z)$ are purely imaginary if $\nu > -1$. If ν is real, Macdonald's function $K_\nu(z)$ has no zeros in the region $|\arg z| \leqslant \pi/2$. In the rest of the z-plane cut along the segment $[-\infty, 0]$, $K_\nu(z)$ has a finite number of zeros.[28]

---

[26] The problem of the distribution of the zeros of cylinder functions is also of considerable theoretical interest, but lies outside the scope of this book. We again refer the reader interested in details to the specialized literature, e.g., Chap. 15 of Watson's treatise. It should be noted that some of the results on zeros of cylinder functions can be derived by arguments of a completely elementary character.

[27] G. N. Watson, *op. cit.*, p. 482.

[28] *Ibid.*, p. 511.

To make approximate calculations of the roots of equations involving cylinder functions, one can use the method of successive approximations, where in many cases a good first approximation is given by the roots of the equations obtained when the cylinder functions are replaced by their asymptotic representations.

5.14. Expansions in Series and Integrals Involving Cylinder Functions

In mathematical physics, it is often necessary to expand a given function in terms of cylinder functions, where the form of the expansion depends on the specific nature of the problem (see Secs. 6.3–6.7). We now consider the most important of these expansions, whose role in various problems involving cylinder functions resembles that of Fourier series and Fourier integrals in problems involving trigonometric functions. Foremost among such expansions are series of the form

$$f(r) = \sum_{m=1}^{\infty} c_m J_\nu\left(x_{\nu m} \frac{r}{a}\right), \qquad 0 < r < a, \quad \nu \geq -\tfrac{1}{2}, \qquad (5.14.1)$$

where $f(r)$ is a given real function defined in the interval $(0, a)$, $J_\nu(x)$ is a Bessel function of the first kind of real order $\nu \geq -\frac{1}{2}$, and

$$0 < x_{\nu 1} < \cdots < x_{\nu m} < \cdots$$

are the positive roots of the equation $J_\nu(x) = 0$. The expansion coefficients c_m can be determined by using an orthogonality property of the system of functions

$$J_\nu\left(x_{\nu m} \frac{r}{a}\right), \qquad m = 1, 2, \ldots, \qquad (5.14.2)$$

which is proved as follows: Let α and β be distinct nonzero real numbers, and let

$$u_\alpha'' + \frac{1}{r} u_\alpha' + \left(\alpha^2 - \frac{\nu^2}{r^2}\right) u_\alpha = 0, \qquad u_\beta'' + \frac{1}{r} u_\beta' + \left(\beta^2 - \frac{\nu^2}{r^2}\right) u_\beta = 0$$

be the equations satisfied by the functions $u_\alpha = J_\nu(\alpha r)$ and $u_\beta = J_\nu(\beta r)$. Subtracting the second equation multiplied by $r u_\alpha$ from the first equation multiplied by $r u_\beta$, and integrating the result from 0 to a, we find that

$$(\alpha^2 - \beta^2) \int_0^a r u_\alpha u_\beta \, dr = r(u_\alpha u_\beta' - u_\beta u_\alpha')\Big|_0^a,$$

which implies

$$\int_0^a r J_\nu(\alpha r) J_\nu(\beta r) \, dr = \frac{a\beta J_\nu(\alpha a) J_\nu'(\beta a) - a\alpha J_\nu(\beta a) J_\nu'(\alpha a)}{\alpha^2 - \beta^2} \qquad (5.14.3)$$

if $\nu > -1$. Setting $\alpha = x_{\nu m}/a$, $\beta = x_{\nu n}/a$ in (5.14.3), we obtain the formula

$$\int_0^a rJ_\nu\left(x_{\nu m}\frac{r}{a}\right)J_\nu\left(x_{\nu n}\frac{r}{a}\right)dr = 0 \quad \text{if} \quad m \neq n, \tag{5.14.4}$$

which shows that the system (5.14.2) is orthogonal with weight r on the interval $[0, a]$ (see Sec. 4.1).

Taking the limit of (5.14.3) as $\beta \to \alpha$, with the aid of L'Hospital's rule, and using Bessel's equation to eliminate J_ν'', we find that[29]

$$\int_0^a rJ_\nu^2(\alpha r)\,dr = \frac{a^2}{2}\left[J_\nu'^2(\alpha a) + \left(1 - \frac{\nu^2}{\alpha^2 a^2}\right)J_\nu^2(\alpha a)\right], \tag{5.14.5}$$

or, using the relations (5.3.5),

$$\int_0^a rJ_\nu^2\left(x_{\nu n}\frac{r}{a}\right)dr = \frac{a^2}{2}J_\nu'^2(x_{\nu n}) = \frac{a^2}{2}J_{\nu+1}^2(x_{\nu n}). \tag{5.14.6}$$

Then, assuming that an expansion of the form (5.14.1) is possible, multiplying by $rJ_\nu(x_{\nu n}\,r/a)$ and integrating term by term from 0 to a, we obtain the following formal values of the coefficients c_m:

$$c_m = \frac{2}{a^2J_{\nu+1}^2(x_{\nu m})}\int_0^a rf(r)J_\nu\left(x_{\nu m}\frac{r}{a}\right)dr, \qquad m = 1, 2, \ldots \tag{5.14.7}$$

The series (5.14.1), with coefficients calculated from (5.14.7), is called the *Fourier-Bessel series* of the function $f(r)$.

We now cite a theorem which gives conditions under which the Fourier-Bessel series of the function $f(r)$ actually converges and has the sum $f(r)$:

THEOREM 3.[30] *Suppose the real function* $f(r)$ *is piecewise continuous in* $(0, a)$ *and of bounded variation in every subinterval* $[r_1, r_2]$,[31] *where* $0 < r_1 < r_2 < a$. *Then, if the integral*

$$\int_0^a \sqrt{r}\,|f(r)|\,dr$$

is finite, the Fourier-Bessel series (5.14.1) *converges to* $f(r)$ *at every continuity point* of $f(r)$, *and to*

$$\tfrac{1}{2}[f(r+0) + f(r-0)]$$

at every discontinuity point of $f(r)$.

Next, we consider an important generalization of the concept of a Fourier-Bessel series. Suppose the function $f(r)$ is expanded in a series of the form (5.14.1), where this time the numbers

$$0 < x_{\nu 1} < \cdots < x_{\nu m} < \cdots$$

---

[29] The details are given in G. P. Tolstov, *op. cit.*, p. 218.

[30] For the proof, see G. N. Watson, *op. cit.*, p. 591.

[31] Concerning *functions of bounded variation*, see E. C. Titchmarsh, *op. cit.*, p. 355.

are the roots of the equation

$$AJ_\nu(x) + BxJ_\nu'(x) = 0, \tag{5.14.8}$$

instead of the equation $J_\nu(x) = 0$. Then it is an immediate consequence of formulas (5.14.3, 5, 8) that

$$\int_0^a rJ_\nu\left(x_{\nu m}\frac{r}{a}\right)J_\nu\left(x_{\nu n}\frac{r}{a}\right)dr = \begin{cases} 0 & \text{if} \quad m \neq n, \\ \dfrac{a^2}{2}\left[J_\nu'^2(x_{\nu n}) + \left(1 - \dfrac{\nu^2}{x_{\nu n}^2}\right)J_\nu^2(x_{\nu n})\right] & \text{if} \quad m = n, \end{cases} \tag{5.14.9}$$

and therefore the coefficients c_m are now given by

$$c_m = \frac{2}{a^2\{J_\nu'^2(x_{\nu m}) + [1 - (\nu^2/x_{\nu m}^2)]J_\nu^2(x_{\nu m})\}}\int_0^a rf(r)J_\nu\left(x_{\nu m}\frac{r}{a}\right)dr. \tag{5.14.10}$$

The series (5.14.1), with coefficients calculated from (5.14.10), is called the *Dini series*[32] of the function $f(r)$. If $f(r)$ satisfies the conditions of Theorem 3, and if $AB^{-1} + \nu > 0$, then the Dini series of $f(r)$ actually converges to $f(r)$ at every continuity point.[33] Both Fourier-Bessel series and Dini series play an important role in problems of mathematical physics, and examples of such expansions will be given in Secs. 6.3 and 6.7.

We now turn to expansions of a function $f(r)$ defined in the infinite interval $(0, \infty)$, in terms of *integrals* involving Bessel functions. Among such expansions, the one of greatest practical importance is the *Fourier-Bessel integral*, defined by

$$f(r) = \int_0^\infty \lambda J_\nu(\lambda r)\,d\lambda \int_0^\infty \rho J_\nu(\lambda\rho)f(\rho)\,d\rho, \qquad 0 < r < \infty, \quad \nu > -\tfrac{1}{2}. \tag{5.14.11}$$

Formula (5.14.11) is sometimes called *Hankel's integral theorem*, and is valid at every continuity point of $f(r)$ provided that

1. The function $f(r)$, defined in the infinite interval $(0, \infty)$, is piecewise continuous and of bounded variation in every finite subinterval $[r_1, r_2]$, where $0 < r_1 < r_2 < \infty$;
2. The integral
$$\int_0^\infty \sqrt{r}\,|f(r)|\,dr$$
is finite.[34]

---

[32] Called a *Fourier-Bessel series of the second type* in G. P. Tolstov, *op. cit.*, p. 237.

[33] For the proof, see G. N. Watson, *op. cit.*, p. 596 ff., where one will also find the modifications that must be made in the Dini series if $AB^{-1} + \nu \leqslant 0$.

[34] For the proof, see G. N. Watson, *op. cit.*, p. 456 ff. At discontinuity points, the integral in the right-hand side of (5.4.11) equals

$$\tfrac{1}{2}[f(r + 0) + f(r - 0)].$$

As examples of Fourier-Bessel integrals, consider the expansions

$$\frac{1}{\sqrt{z^2 + r^2}} = \int_0^\infty e^{-\lambda|z|} J_0(\lambda r)\, d\lambda, \tag{5.14.12}$$

$$\frac{e^{-k\sqrt{z^2+r^2}}}{\sqrt{z^2 + r^2}} = \int_0^\infty e^{-|z|\sqrt{\lambda^2+k^2}} \frac{\lambda J_0(\lambda r)}{\sqrt{\lambda^2 + k^2}}\, d\lambda \tag{5.14.13}$$

(with real z and r), implied by formulas (5.15.1, 7) below.

The author has studied another integral expansion of a completely different type, involving integration with respect to the *order* of the cylinder function.[35] This expansion, which turns out to be very useful in solving certain problems of mathematical physics (see Secs. 6.5–6) is of the form

$$f(x) = \frac{2}{\pi^2} \int_0^\infty \tau \sinh \pi\tau \frac{K_{i\tau}(x)}{\sqrt{x}}\, d\tau \int_0^\infty f(\xi) \frac{K_{i\tau}(\xi)}{\sqrt{\xi}}\, d\xi, \qquad x > 0, \tag{5.14.14}$$

where $K_\nu(x)$ is Macdonald's function of imaginary order $\nu = i\tau$. Formula (5.14.14) is valid at every continuity point of $f(x)$ provided that

1. The function $f(x)$, defined in the infinite interval $(0, \infty)$, is piecewise continuous and of bounded variation in every finite subinterval $[x_1,x_2]$, where $0 < x_1 < x_2 < \infty$;
2. The integrals

$$\int_0^{1/2} |f(x)| x^{-1/2} \log \frac{1}{x}\, dx, \qquad \int_{1/2}^\infty |f(x)|\, dx \tag{5.14.15}$$

are finite.

Example. An expansion of this type is[36]

$$f(x) = \sqrt{x} e^{-x \cos \alpha} = \frac{2}{\pi} \int_0^\infty \frac{\tau \sinh \alpha\tau}{\sin \alpha} \frac{K_{i\tau}(x)}{\sqrt{x}}\, d\tau. \tag{5.14.16}$$

5.15. Definite Integrals Involving Cylinder Functions

In the applications, it is often necessary to evaluate integrals involving cylinder functions in combination with various elementary functions or special

---

[35] N. N. Lebedev, *Sur une formule d'inversion*, Dokl. Akad. Nauk SSSR, **52**, 655 (1946); *Expansion of an arbitrary function in an integral with respect to cylinder functions of imaginary order and argument* (in Russian), Prikl. Mat. Mekh., **13**, 465 (1949); *Some Integral Transformations of Mathematical Physics* (in Russian), Dissertation, Izd. Leningrad. Gos. Univ. (1951). At discontinuity points, the integral in the right-hand side of (5.14.14) equals

$$\tfrac{1}{2}[f(x + 0) + f(x - 0)].$$

[36] To derive (5.14.16), use (5.14.14) and the Bateman Manuscript Project, *Tables of Integral Transforms, Vol. 1*, formula (24), p. 197.

functions of other kinds. Such integrals are usually evaluated by replacing the cylinder function by a series or by a suitable integral representation, and then reversing the order in which the operations are carried out. Since an extremely detailed treatment of this whole topic is available in the literature,[37] we confine ourselves here to a few examples which illustrate the method and lead to some results needed later in the book.

Example 1. *Evaluate the integral*

$$\int_0^\infty e^{-ax}J_0(bx)\,dx, \qquad a > 0, \quad b > 0.$$

Replacing $J_0(bx)$ by its integral representation (5.10.8), we find that

$$\begin{aligned}\int_0^\infty e^{-ax}J_0(bx)\,dx &= \int_0^\infty e^{-ax}\,dx\,\frac{2}{\pi}\int_0^{\pi/2}\cos(bx\sin\varphi)\,d\varphi \\ &= \frac{2}{\pi}\int_0^{\pi/2}d\varphi\int_0^\infty e^{-ax}\cos(bx\sin\varphi)\,dx \\ &= \frac{2}{\pi}\int_0^{\pi/2}\frac{a\,d\varphi}{a^2+b^2\sin^2\varphi},\end{aligned}$$

where the absolute convergence of the double integral justifies reversing the order of integration. Evaluating the last integral, we have

$$\int_0^\infty e^{-ax}J_0(bx)\,dx = \frac{1}{\sqrt{a^2+b^2}}, \qquad a > 0, \quad b > 0. \tag{5.15.1}$$

Example 2. *Evaluate Weber's integral*

$$\int_0^\infty e^{-a^2x^2}J_\nu(bx)x^{\nu+1}\,dx, \qquad a > 0, \quad b > 0, \quad \operatorname{Re}\nu > -1.$$

Replacing $J_\nu(bx)$ by its series expansion (5.3.2) and integrating term by term, we find that

$$\begin{aligned}\int_0^\infty e^{-a^2x^2}J_\nu(bx)x^{\nu+1}\,dx &= \int_0^\infty e^{-a^2x^2}x^{\nu+1}\,dx\sum_{k=0}^\infty\frac{(-1)^k(bx/2)^{\nu+2k}}{k!\Gamma(k+\nu+1)} \\ &= \sum_{k=0}^\infty\frac{(-1)^k}{k!\Gamma(k+\nu+1)}\left(\frac{b}{2}\right)^{\nu+2k}\int_0^\infty e^{-a^2x^2}x^{2\nu+2k+1}\,dx \\ &= \sum_{k=0}^\infty\frac{(-1)^k}{k!\Gamma(k+\nu+1)}\left(\frac{b}{2}\right)^{\nu+2k}\frac{1}{2a^{2\nu+2k+2}}\int_0^\infty e^{-t}t^{\nu+k}\,dt \\ &= \frac{b^\nu}{(2a^2)^{\nu+1}}\sum_{k=0}^\infty\frac{(-b^2/4a^2)^k}{k!},\end{aligned}$$

---

[37] G. N. Watson, *op. cit.*, Chaps. 12–13, the Bateman Manuscript Project, *Higher Transcendental Functions, Vol. 2*, Chap. 7, and ibid., *Tables of Integral Transforms, Vols. 1, 2*. See also F. Oberhettinger, *Tabellen zur Fourier Transformation*, Springer-Verlag, Berlin (1957).

where reversing the order of integration and summation is again justified by an absolute convergence argument. Summing the last series, we have

$$\int_0^\infty e^{-a^2x^2} J_\nu(bx) x^{\nu+1}\, dx = \frac{b^\nu}{(2a^2)^{\nu+1}} e^{-b^2/4a^2}, \tag{5.15.2}$$
$$a > 0,\ b > 0, \quad \operatorname{Re}\nu > -1.$$

Example 3. *Evaluate the integral*

$$\int_0^\infty \frac{x^{\nu+1}J_\nu(bx)}{(x^2+a^2)^{\mu+1}}\, dx, \qquad a > 0, \quad b > 0, \quad -1 < \operatorname{Re}\nu < 2\operatorname{Re}\mu + \tfrac{3}{2},$$

often encountered in the applications. First we replace the function $(x^2 + a^2)^{-\mu-1}$ by an integral of the type (1.5.1), i.e.,

$$\frac{1}{(x^2+a^2)^{\mu+1}} = \frac{1}{\Gamma(\mu+1)} \int_0^\infty e^{-(x^2+a^2)t} t^\mu\, dt, \ \operatorname{Re}\mu > -1, \tag{5.15.3}$$

assuming temporarily that $-1 < \operatorname{Re}\nu < 2\operatorname{Re}\mu + \frac{1}{2}$ (this guarantees absolute convergence of the relevant double integral). Then, using (5.15.2) and the integral representation (5.10.25) of Macdonald's function, we find that

$$\begin{aligned}
\int_0^\infty \frac{x^{\nu+1}J_\nu(bx)}{(x^2+a^2)^{\mu+1}}\, dx &= \frac{1}{\Gamma(\mu+1)} \int_0^\infty e^{-a^2t} t^\mu\, dt \int_0^\infty e^{-x^2t} J_\nu(bx) x^{\nu+1}\, dx \\
&= \frac{b^\nu}{2^{\nu+1}\Gamma(\mu+1)} \int_0^\infty e^{-a^2t-(b^2/4t)} \frac{dt}{t^{\nu+1-\mu}} \\
&= \frac{b^\nu a^{2\nu-2\mu}}{2^{\nu+1}\Gamma(\mu+1)} \int_0^\infty e^{-u-[(ab)^2/4u]} \frac{du}{u^{\nu-\mu+1}} \\
&= \frac{a^{\nu-\mu}b^\mu}{2^\mu\Gamma(\mu+1)} K_{\nu-\mu}(ab).
\end{aligned}$$

The extension of this result to values of the parameter μ satisfying the weaker condition $-1 < \operatorname{Re}\nu < 2\operatorname{Re}\mu + \frac{3}{2}$ is accomplished by using the principle of analytic continuation. Thus we have

$$\int_0^\infty \frac{x^{\nu+1}J_\nu(bx)}{(x^2+a^2)^{\mu+1}}\, dx = \frac{a^{\nu-\mu}b^\mu}{2^\mu\Gamma(\mu+1)} K_{\nu-\mu}(ab), \tag{5.15.4}$$
$$a > 0,\ b > 0, \quad -1 < \operatorname{Re}\nu < 2\operatorname{Re}\mu + \tfrac{3}{2}.$$

In particular, setting $\mu = -\frac{1}{2}$, $\nu = 0$ and using (5.8.5), we obtain the integral

$$\int_0^\infty \frac{xJ_0(bx)}{\sqrt{x^2+a^2}}\, dx = \frac{e^{-ab}}{b}, \qquad a \geqslant 0, \quad b > 0. \tag{5.15.5}$$

Example 4. *Evaluate the integral*

$$\int_0^\infty \frac{K_\mu(a\sqrt{x^2+y^2})}{(x^2+y^2)^{\mu/2}} J_\nu(bx) x^{\nu+1}\, dx,$$
$$a > 0, \quad b > 0, \quad y > 0, \quad \operatorname{Re}\nu > -1,$$

which also has numerous applications to mathematical physics. Using the integral representation (5.10.25) and formula (5.15.2), we find that

$$\int_0^\infty \frac{K_\mu(a\sqrt{x^2+y^2})}{(x^2+y^2)^{\mu/2}} J_\nu(bx)x^{\nu+1}\,dx$$

$$= \frac{a^\mu}{2^{\mu+1}} \int_0^\infty J_\nu(bx)x^{\nu+1}\,dx \int_0^\infty e^{-t-[a^2(x^2+y^2)/4t]} \frac{dt}{t^{\mu+1}}$$

$$= \frac{a^\mu}{2^{\mu+1}} \int_0^\infty e^{-t-(a^2y^2/4t)} \frac{dt}{t^{\mu+1}} \int_0^\infty e^{-a^2x^2/4t} J_\nu(bx)x^{\nu+1}\,dx$$

$$= 2^{\nu-\mu}a^{\mu-2\nu-2}b^\nu \int_0^\infty e^{-t(1+b^2/a^2)-(a^2y^2/4t)} \frac{dt}{t^{\mu-\nu}}$$

$$= \frac{2^{\nu-\mu}b^\nu}{a^\mu}(a^2+b^2)^{\mu-\nu-1} \int_0^\infty e^{-u-[y^2(a^2+b^2)/4u]} \frac{du}{u^{\mu-\nu}}$$

$$= \frac{b^\nu}{a^\mu}\left(\frac{\sqrt{a^2+b^2}}{y}\right)^{\mu-\nu-1} K_{\mu-\nu-1}(y\sqrt{a^2+b^2}).$$

By choosing various values of the parameters in the identity

$$\int_0^\infty \frac{K_\mu(a\sqrt{x^2+y^2})}{(x^2+y^2)^{\mu/2}} J_\nu(bx)x^{\nu+1}\,dx = \frac{b^\nu}{a^\mu}\left(\frac{\sqrt{a^2+b^2}}{y}\right)^{\mu-\nu-1} K_{\mu-\nu-1}(y\sqrt{a^2+b^2}),$$

$$a > 0, \quad b > 0, \quad y > 0, \quad \operatorname{Re}\nu > -1, \tag{5.15.6}$$

we can derive a number of useful formulas encountered in the applications. For example, setting $\mu = \frac{1}{2}$, $\nu = 0$, we have

$$\int_0^\infty \frac{e^{-a\sqrt{x^2+y^2}}}{\sqrt{x^2+y^2}} J_0(bx)x\,dx = \frac{e^{-y\sqrt{a^2+b^2}}}{\sqrt{a^2+b^2}}. \tag{5.15.7}$$

5.16. Cylinder Functions of Nonnegative Argument and Order

We now collect some elementary and easily verified results pertaining to the very important case of cylinder functions where both the argument x and the order ν are nonnegative real numbers:

1. *Bessel functions of the first kind.* For $x \geqslant 0$ and $\nu \geqslant 0$, the function $J_\nu(x)$ is real and bounded, and has an oscillatory character. Its behavior for small and large values of x is described by the asymptotic formulas

$$J_\nu(x) \approx \frac{x^\nu}{2^\nu\Gamma(1+\nu)}, \qquad x \to 0,$$
$$J_\nu(x) \approx \sqrt{\frac{2}{\pi x}}\cos\left(x - \tfrac{1}{2}\nu\pi - \tfrac{1}{4}\pi\right), \qquad x \to \infty. \tag{5.16.1}$$

$J_\nu(x)$ has infinitely many zeros, including the point $x = 0$ if $\nu > 0$. The graphs of $J_0(x)$ and $J_1(x)$ are shown in Figure 17.

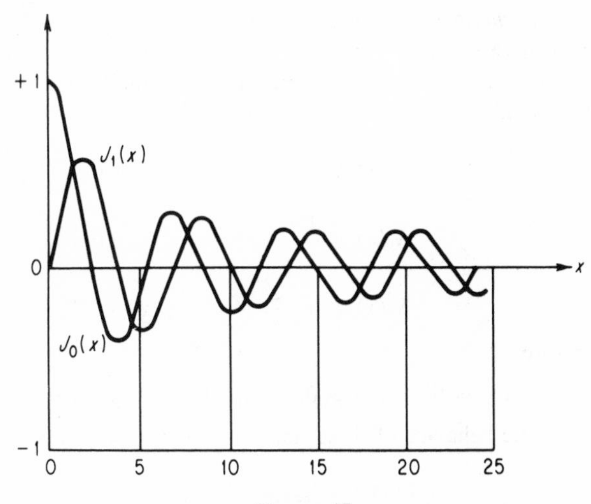

FIGURE 17

2. *Bessel functions of the second kind.* For $x > 0$ and $\nu \geqslant 0$, the function $Y_\nu(x)$ is an oscillatory real function, which is bounded at infinity. Its behavior for small and large values of x is described by the asymptotic formulas

$$Y_\nu(x) \approx -\frac{2^\nu \Gamma(\nu)}{\pi x^\nu} \qquad x \to 0, \quad \nu > 0,$$

$$Y_\nu(x) \approx \sqrt{\frac{2}{\pi x}} \sin(x - \tfrac{1}{2}\nu\pi - \tfrac{1}{4}\pi), \qquad x \to \infty, \tag{5.16.2}$$

$$Y_0(x) \approx -\frac{2}{\pi} \log \frac{2}{x}, \qquad x \to 0,$$

which show, in particular, that $Y_\nu(x) \to -\infty$ as $x \to 0$.

3. *Bessel functions of the third kind.* For $x > 0$ and $\nu \geqslant 0$, the Hankel functions $H_\nu^{(1)}(x)$ and $H_\nu^{(2)}(x)$ are conjugate complex functions, which are bounded at infinity. Their behavior for small and large values of x is described by the asymptotic formulas

$$H_\nu^{(p)}(x) \approx \mp i \left(\frac{2}{x}\right)^\nu \frac{\Gamma(\nu)}{\pi}, \qquad x \to 0, \quad \nu > 0,$$

$$H_\nu^{(p)}(x) \approx \sqrt{\frac{2}{\pi x}}\, e^{\pm i(x - \frac{1}{2}\nu\pi - \frac{1}{4}\pi)}, \qquad x \to \infty, \tag{5.16.3}$$

$$H_0^{(p)}(x) \approx \mp i \frac{2}{\pi} \log \frac{2}{x}, \qquad x \to 0,$$

where the upper sign corresponds to the case $p = 1$, and the lower sign to the case $p = 2$. Obviously, $H_\nu^{(p)}(x) \to \infty$ as $x \to 0$.

4. *Bessel functions of imaginary argument.* For $x > 0$ and $\nu \geqslant 0$, $I_\nu(x)$ is a positive function which increases monotonically as $x \to \infty$, while $K_\nu(x)$ is a positive function which decreases monotonically as $x \to \infty$.[38] For small x we have the asymptotic formulas

$$\begin{aligned} I_\nu(x) &\approx \frac{x^\nu}{2^\nu \Gamma(1+\nu)}, & x &\to 0, \\ K_\nu(x) &\approx \frac{2^{\nu-1}\Gamma(\nu)}{x^\nu}, & x &\to 0, \\ K_0(x) &\approx \log \frac{2}{x}, & x &\to 0, \end{aligned} \tag{5.16.4}$$

and therefore

$$I_\nu(0) = 0 \quad \text{if } \nu > 0, \qquad I_0(0) = 1, \quad K_\nu(0) = \infty.$$

The asymptotic behavior of these functions as $x \to \infty$ is given by

$$\begin{aligned} I_\nu(x) &\approx \frac{e^x}{\sqrt{2\pi x}}, & x &\to \infty, \\ K_\nu(x) &\approx \sqrt{\frac{\pi}{2x}}\, e^{-x}, & x &\to \infty. \end{aligned} \tag{5.16.5}$$

Clearly, neither function has any zeros for $x > 0$.

5.17. Airy Functions

The solutions of the second-order linear differential equation

$$u'' - zu = 0 \tag{5.17.1}$$

are called *Airy functions*. These functions are closely related to the cylinder functions, and play an important role in the theory of asymptotic representations of various special functions arising as solutions of linear differential equations.[39] In particular, the Airy functions turn out to be useful in deriving asymptotic representations of the cylinder functions for large values of $|z|$ and $|\nu|$, valid in an extended region of values of z and ν. The Airy functions also have a variety of applications to mathematical physics, e.g., the theory of diffraction of radio waves around the earth's surface.[40]

---

[38] This fact about $K_\nu(x)$ follows from the integral representation (5.10.23).

[39] See R. E. Langer, *op. cit.*, T. M. Cherry, *op. cit.*, and V. A. Fock, *Tables of the Airy Fuctions* (in Russian), Izd. Inform. Otdel. Nauchno-Issled. Inst., Moscow (1946).

[40] See V. A. Fock, *Diffraction of Radio Waves Around the Earth's Surface* (in Russian), Izd. Akad. Nauk SSSR, Moscow (1946).

We now present the rudiments of the theory of Airy functions. Choosing $\alpha = -1, \gamma = 1$ in the second of the equations (5.4.11–12), and using the results of Sec. 5.7, we find that the general solution of (5.17.1) can be expressed in terms of Bessel functions of imaginary argument of order $\nu = \pm\frac{1}{3}$. In particular, two linearly independent solutions of (5.17.1) are

$$u = u_1 = \operatorname{Ai}(z) = \frac{z^{1/2}}{3}\left[I_{-1/3}\left(\frac{2z^{3/2}}{3}\right) - I_{1/3}\left(\frac{2z^{3/2}}{3}\right)\right]$$

$$\equiv \frac{1}{\pi}\left(\frac{z}{3}\right)^{1/2} K_{1/3}\left(\frac{2z^{3/2}}{3}\right), \qquad |\arg z| < \frac{2\pi}{3},$$

$$u = u_2 = \operatorname{Bi}(z) = \left(\frac{z}{3}\right)^{1/2}\left[I_{-1/3}\left(\frac{2z^{3/2}}{3}\right) + I_{1/3}\left(\frac{2z^{3/2}}{3}\right)\right], \qquad |\arg z| < \frac{2\pi}{3}, \tag{5.17.2}$$

called the *Airy functions of the first and second kind*, respectively. Replacing $I_{\pm 1/3}$ by the series expansion (5.7.1), we obtain the expansions

$$\operatorname{Ai}(z) = \sum_{k=0}^{\infty} \frac{z^{3k}}{3^{2k+2/3}k!\Gamma(k+\frac{2}{3})} - \sum_{k=0}^{\infty} \frac{z^{3k+1}}{3^{2k+4/3}k!\Gamma(k+\frac{4}{3})}, \qquad |z| < \infty,$$

$$\operatorname{Bi}(z) = 3^{1/2}\left[\sum_{k=0}^{\infty} \frac{z^{3k}}{3^{2k+2/3}k!\Gamma(k+\frac{2}{3})} + \sum_{k=0}^{\infty} \frac{z^{3k+1}}{3^{2k+4/3}k!\Gamma(k+\frac{4}{3})}\right], \qquad |z| < \infty, \tag{5.17.3}$$

which show that the Airy functions are entire functions of z.

We can also write (5.17.3) in another, somewhat more concise form. For example, the first expansion is equivalent to

$$\operatorname{Ai}(z) = \frac{2}{3^{7/6}} \sum_{k=0}^{\infty} \frac{\sin\left[\frac{2\pi}{3}(k+1)\right]}{\Gamma\left(\frac{k+2}{3}\right)\Gamma\left(\frac{k+3}{3}\right)}\left(\frac{z}{3^{2/3}}\right)^k, \qquad |z| < \infty. \tag{5.17.4}$$

Using the "triplication formula" for the gamma function [Problem 4, formula (i), p. 14] we can transform (5.17.4) into

$$\operatorname{Ai}(z) = \frac{3^{-2/3}}{\pi} \sum_{k=0}^{\infty} \frac{\Gamma\left(\frac{k+1}{3}\right)\sin\frac{2\pi}{3}(k+1)}{k!}(3^{1/3}z)^k, \qquad |z| < \infty. \tag{5.17.5}$$

It follows from (5.17.3) that the Airy functions $\operatorname{Ai}(z)$ and $\operatorname{Bi}(z)$ can be defined as the solutions of equation (5.17.1) satisfying the initial conditions

$$\begin{aligned} u_1(0) &= \operatorname{Ai}(0) = \frac{3^{-2/3}}{\Gamma(\frac{2}{3})}, & u_1'(0) &= \operatorname{Ai}'(0) = -\frac{3^{-4/3}}{\Gamma(\frac{4}{3})}, \\ u_2(0) &= \operatorname{Bi}(0) = \frac{3^{-1/6}}{\Gamma(\frac{2}{3})}, & u_2'(0) &= \operatorname{Bi}'(0) = \frac{3^{-5/6}}{\Gamma(\frac{4}{3})}. \end{aligned} \tag{5.17.6}$$

The Wronskian of this pair of solutions is

$$W\{\mathrm{Ai}(z), \mathrm{Bi}(z)\} = W\{\mathrm{Ai}(z), \mathrm{Bi}(z)\}_{z=0} = \frac{1}{\pi}, \tag{5.17.7}$$

where we again use the triplication formula for the gamma function.[41] We can also calculate (5.17.7) directly from (5.17.2) and (5.9.5).

Asymptotic representations of the Airy functions for large $|z|$ can be deduced from the corresponding results of Sec. 5.11. In particular, we have

$$\mathrm{Ai}(z) = \frac{\pi^{-1/2}}{2} z^{-1/4} e^{-\frac{2}{3}z^{3/2}}[1 + O(|z|^{-3/2})], \qquad |\arg z| \leqslant \frac{2\pi}{3} - \delta, \tag{5.17.8}$$

$$\mathrm{Bi}(z) = \pi^{-1/2} z^{-1/4} e^{\frac{2}{3}z^{3/2}}[1 + O(|z|^{-3/2})], \qquad |\arg z| \leqslant \frac{\pi}{3} - \delta. \tag{5.17.9}$$

It follows at once from (5.17.3), (5.7.1) and (5.3.2) that the Airy functions of argument $-z$ can be expressed in terms of Bessel functions of the first kind of order $\nu = \pm\frac{1}{3}$:

$$\begin{aligned} \mathrm{Ai}(-z) &= \frac{z^{1/2}}{3}[J_{-1/3}(\tfrac{2}{3}z^{3/2}) + J_{1/3}(\tfrac{2}{3}z^{3/2})], \qquad |\arg z| < \frac{2\pi}{3}. \\ \mathrm{Bi}(-z) &= \left(\frac{z}{3}\right)^{1/2}[J_{-1/3}(\tfrac{2}{3}z^{3/2}) - J_{1/3}(\tfrac{2}{3}z^{3/2})], \qquad |\arg z| < \frac{2\pi}{3}. \end{aligned} \tag{5.17.10}$$

Then, using the asymptotic representation (5.11.6), we find that

$$\begin{aligned} \mathrm{Ai}(-x) &\approx \pi^{-1/2} x^{-1/4} \cos\left(\frac{2}{3}x^{3/2} - \frac{\pi}{4}\right), \qquad x \to \infty, \\ \mathrm{Bi}(-x) &\approx -\pi^{-1/2} x^{-1/4} \sin\left(\frac{2}{3}x^{3/2} - \frac{\pi}{4}\right), \qquad x \to \infty, \end{aligned} \tag{5.17.11}$$

which shows that the Airy functions have an oscillatory character for large negative values of the argument.

Finally, we note that the definition of $\mathrm{Ai}(x)$ and the integral representation of Macdonald's function given in Problem 6, formula (ii), p. 140, imply

$$\mathrm{Ai}(x) = \frac{2x^{1/2}}{3\pi}\int_0^\infty \cos\left(\frac{2x^{3/2}}{3}\sinh y\right)\cosh\frac{y}{3}\,dy, \qquad x > 0.$$

After making the substitution

$$\sinh\frac{y}{3} = \frac{1}{2}x^{-1/2}t,$$

this gives the following integral representation of $\mathrm{Ai}(x)$:

$$\mathrm{Ai}(x) = \frac{1}{\pi}\int_0^\infty \cos(\tfrac{1}{3}t^3 + xt)\,dt, \qquad x \geqslant 0. \tag{5.17.12}$$

---

[41] For a proof of the first equality in (5.17.7), cf. E. A. Coddington, *op. cit.*, Theorem 8, p. 113.

A somewhat more complicated argument gives the following integral representation of $\mathrm{Bi}(x)$:[42]

$$\mathrm{Bi}(x) = \frac{1}{\pi}\int_0^\infty [e^{-\frac{1}{3}t^3 + xt} + \sin(\tfrac{1}{3}t^3 + xt)]\,dt, \qquad x \geqslant 0.$$

For an integral representation of $[\mathrm{Ai}(x)]^2$, see Problem 22, p. 142.

PROBLEMS

1. Derive the integral representation[43]

$$J_n^2(z) = \frac{2}{\pi}\int_0^{\pi/2} J_{2n}(2z\cos\theta)\,d\theta = (-1)^n\frac{2}{\pi}\int_0^{\pi/2} J_0(2z\cos\theta)\cos 2n\theta\,d\theta,$$
$$n = 0, 1, 2, \ldots$$

2. Derive the following formula involving products of Bessel functions:[44]

$$J_\mu(z)J_\nu(z) = \frac{2}{\pi}\int_0^{\pi/2} J_{\mu+\nu}(2z\cos\theta)\cos(\mu-\nu)\theta\,d\theta, \qquad \mathrm{Re}\,(\mu+\nu) > -1.$$

3. Prove that

$$J_n(z)J_n(z') = \frac{1}{\pi}\int_0^{\pi} J_0(\sqrt{z^2 + z'^2 - 2zz'\cos\theta})\cos n\theta\,d\theta, \qquad n = 0, 1, 2, \ldots$$

Hint. Use the addition theorem (5.12.2).

4. Derive the integral representations

$$J_\nu(x) = \frac{2}{\pi}\int_0^\infty \sin\left(x\cosh t - \frac{\nu\pi}{2}\right)\cosh\nu t\,dt, \qquad -1 < \mathrm{Re}\,\nu < 1, \quad x > 0,$$

$$Y_\nu(x) = -\frac{2}{\pi}\int_0^\infty \cos\left(x\cosh t - \frac{\nu\pi}{2}\right)\cosh\nu t\,dt, \quad -1 < \mathrm{Re}\,\nu < 1, \quad x > 0.$$

Hint. Use formulas (5.10.14, 15).

5. Derive the formulas

$$H_\nu^{(1)}(z) = \left(\frac{2}{\pi z}\right)^{1/2}\frac{e^{i(z - \frac{1}{2}\nu\pi - \frac{1}{4}\pi)}}{\Gamma(\nu + \frac{1}{2})}\int_0^\infty e^{-s}s^{\nu-\frac{1}{2}}\left(1 - \frac{s}{2iz}\right)^{\nu-\frac{1}{2}}ds,$$
$$\mathrm{Re}\,\nu > -\tfrac{1}{2}, \quad -\frac{\pi}{2} < \arg z < \pi,$$

$$H_\nu^{(2)}(z) = \left(\frac{2}{\pi z}\right)^{1/2}\frac{e^{-i(z - \frac{1}{2}\nu\pi - \frac{1}{4}\pi)}}{\Gamma(\nu + \frac{1}{2})}\int_0^\infty e^{-s}s^{\nu-\frac{1}{2}}\left(1 + \frac{s}{2iz}\right)^{\nu-\frac{1}{2}}ds,$$
$$\mathrm{Re}\,\nu > -\tfrac{1}{2}, \quad -\pi < \arg z < \frac{\pi}{2}.$$

---

[42] H. Jeffreys and B. S. Jeffreys, *op. cit.*, p. 510.
[43] G. N. Watson, *op. cit.*, p. 32.
[44] *Ibid.*, p. 150.

6. Prove the following integral representations of Macdonald's function:[45]

$$K_\nu(z) = \frac{\sqrt{\pi}\,z^\nu}{2^\nu\Gamma(\nu+\frac{1}{2})}\int_0^\infty e^{-z\cosh t}\sinh^{2\nu} t\,dt, \qquad \operatorname{Re} z > 0, \quad \operatorname{Re}\nu > -\tfrac{1}{2},$$

$$K_\nu(x) = \frac{2^\nu\Gamma(\nu+\frac{1}{2})}{x^\nu\sqrt{\pi}}\int_0^\infty \frac{\cos xt}{(1+t^2)^{\nu+\frac{1}{2}}}\,dt, \qquad x > 0, \quad \operatorname{Re}\nu > -\tfrac{1}{2}, \tag{i}$$

$$K_\nu(x) = \frac{1}{\cos\frac{\nu\pi}{2}}\int_0^\infty \cos(x\sinh t)\cosh\nu t\,dt, \qquad x > 0, \quad |\operatorname{Re}\nu| < 1, \tag{ii}$$

$$K_\nu(z) = \left(\frac{\pi}{2z}\right)^{1/2}\frac{e^{-z}}{\Gamma(\nu+\frac{1}{2})}\int_0^\infty e^{-s}s^{\nu-\frac{1}{2}}\left(1+\frac{s}{2z}\right)^{\nu-\frac{1}{2}} ds,$$

$$|\arg z| < \pi, \quad \operatorname{Re}\nu > -\tfrac{1}{2}.$$

7. Prove the following formulas involving products of Macdonald functions:[46]

$$K_\nu(x)K_\nu(y) = \frac{1}{2}\int_0^\infty e^{-\frac{1}{2}[t+(x^2+y^2)/t]}K_\nu\left(\frac{xy}{t}\right)\frac{dt}{t}$$

$$= \int_0^\infty K_0(\sqrt{x^2+y^2+2xy\cosh t})\cosh\nu t\,dt, \qquad x > 0, \quad y > 0,$$

$$K_\nu(x)K_\nu(y) = \frac{\pi}{2\sin\nu\pi}\int_{\log(y/x)}^\infty J_0(\sqrt{2xy\cosh t - x^2 - y^2})\sinh\nu t\,dt,$$

$$x > 0, \quad y > 0, \quad |\operatorname{Re}\nu| < \tfrac{1}{4}. \tag{iii}$$

8. Derive the integral representation

$$I_\nu(x)K_\nu(y) = \frac{1}{2}\int_{\log(y/x)}^\infty J_0(\sqrt{2xy\cosh t - x^2 - y^2})\,e^{-\nu t}\,dt,$$

$$x > 0, \quad y > 0, \quad \operatorname{Re}\nu > -\tfrac{1}{4}.$$

9. Derive the integral representation

$$K_\mu(x)K_\nu(x) = \int_0^\infty K_{\mu-\nu}\left(2x\cosh\frac{t}{2}\right)\cosh\frac{\mu+\nu}{2}t\,dt, \qquad x > 0, \quad y > 0.$$

10. Derive the following asymptotic representations for large values of the order $|\nu|$:

$$J_\nu(z) \approx \frac{1}{\sqrt{2\pi}}e^{\nu+\nu\log(z/2)-(\nu+\frac{1}{2})\log\nu}, \qquad |\nu| \to \infty, \quad |\arg\nu| \leqslant \pi - \delta,$$

$$K_{i\tau}(x) \approx \left(\frac{2\pi}{\tau}\right)^{1/2} e^{-\pi\tau/2}\sin\left(\frac{\pi}{4} + \tau\log\tau - \tau - \tau\log\frac{x}{2}\right), \qquad \tau \to \infty.$$

(In the second formula, x is a fixed positive number.)

---

[45] G. N. Watson, *op. cit.*, 172, 183.

[46] Concerning Problems 7–9, see *ibid.*, p. 439. The most detailed investigation of various integral representations of products of cylinder functions is due to A. L. Dixon and W. L. Ferrar, *Integrals for the product of two Bessel functions*, Quart. J. Math. Oxford Ser., **4**, 193 (1933); *Part II*, ibid., **4**, 297 (1933).

11. Prove the formulas

$$J_\nu(-x + i0) - J_\nu(-x - i0) = 2i \sin \nu\pi \, J_\nu(x),$$
$$Y_\nu(-x + i0) - Y_\nu(-x - i0) = 2i[J_\nu(x) \cos \nu\pi + J_{-\nu}(x)],$$
$$H_\nu^{(1)}(-x + i0) - H_\nu^{(1)}(-x - i0) = -2[J_{-\nu}(x) + e^{-\nu\pi i} J_\nu(x)],$$
$$H_\nu^{(2)}(-x + i0) - H_\nu^{(2)}(-x - i0) = 2[J_\nu(x) + e^{\nu\pi i} J_\nu(x)],$$

where $x > 0$, characterizing the behavior of the cylinder functions on the cut $[-\infty, 0]$.

12. Verify that

$$I_\nu(-x + i0) - I_\nu(-x - i0) = 2i \sin \nu\pi \, I_\nu(x),$$
$$K_\nu(-x + i0) - K_\nu(-x - i0) = -\pi i[I_{-\nu}(x) + I_\nu(x)],$$

where $x > 0$.

Comment. The formulas given in Problems 11–12 take a particularly simple form if $\nu = n$ $(n = 0, \pm 1, \pm 2, \ldots)$.

13. Verify the expansion

$$\int_0^z J_\nu(t)\, dt = 2 \sum_{k=0}^{\infty} J_{\nu+2k+1}(z), \qquad \operatorname{Re} \nu > -1.$$

Hint. Use the recurrence relation (5.3.6) to show that both sides have the same derivative.

14. Derive the recurrence relation

$$\int_0^z t^\mu J_\nu(t)\, dt = z^\mu J_{\nu+1}(z) - (\mu - \nu - 1) \int_0^z t^{\mu-1} J_{\nu+1}(t)\, dt, \quad \operatorname{Re}(\mu + \nu) > -1.$$

Hint. Apply (5.3.5) in the form

$$t^{\nu+1} J_\nu(t) = \frac{d}{dt}[t^{\nu+1} J_{\nu+1}(t)],$$

and then integrate by parts.

15. Using the result of Problem 14, show that the evaluation of integrals of the form

$$\int_0^z t^m J_\nu(t)\, dt, \qquad \operatorname{Re} \nu > -1, \quad m = 0, 1, 2, \ldots$$

reduces to the evaluation of the integral

$$\int_0^z J_{\nu+m}(t)\, dt,$$

whose value was found in Problem 13.

Comment. If $\nu = \pm(m - 1), \pm(m - 3), \pm(m - 5), \ldots,$ then the coefficient of the last integral vanishes, and the original integral can be expressed in closed form in terms of Bessel functions.

16. Verify the formula[47]

$$\int_0^\infty \frac{J_\nu(x)}{x^\mu}\,dx = \frac{\Gamma\left(\dfrac{\nu+1-\mu}{2}\right)}{2^\mu\Gamma\left(\dfrac{\nu+1+\mu}{2}\right)}, \qquad \operatorname{Re}\mu > \tfrac{1}{2}, \quad \operatorname{Re}(\nu-\mu) > -1.$$

17. Verify the formula

$$\int_0^\infty e^{-ax}J_\nu(bx)\,dx = \frac{[\sqrt{a^2+b^2}-a]^\nu}{b^\nu\sqrt{a^2+b^2}}, \qquad \operatorname{Re}\nu > -1, \quad a > 0, \quad b > 0.$$

18. Show that the Bessel function $J_0(x)$ satisfies the following integral equal tion:

$$J_0(x) = \frac{2}{\pi}\int_0^\infty \frac{\sin(x+y)}{x+y}J_0(y)\,dy, \qquad 0 \leqslant x < \infty.$$

19. The *integral Bessel function of order* ν is defined by the formula

$$\operatorname{Ji}_\nu(z) = \int_\infty^z \frac{J_\nu(t)}{t}\,dt, \qquad |\arg z| < \pi.$$

Show that $\operatorname{Ji}_\nu(z)$ is an entire function of ν and an analytic function of z in the plane cut along the segment $[-\infty, 0]$ (in fact, an entire function of z for $\nu = \pm 1, \pm 2\ldots$). Verify the formulas

$$\nu\operatorname{Ji}_\nu(z) = \nu\int_0^z \frac{J_\nu(t)}{t}\,dt - 1, \tag{iv}$$

$$\nu\operatorname{Ji}_\nu(z) = \int_0^z J_{\nu-1}(t)\,dt - J_\nu(t) - 1, \qquad \operatorname{Re}\nu > 0, \quad |\arg z| < \pi.$$

Hint. Use the results of Problems 14 and 16.

20. Prove the following expansions of the integral Bessel functions:

$$\operatorname{Ji}_0(z) = \log\frac{z}{2} + \gamma + \sum_{k=1}^\infty \frac{(-1)^k(z/2)^{2k}}{(2k)(k!)^2}, \qquad |z| < \infty, \quad |\arg z| < \pi$$

$$\operatorname{Ji}_n(z) = -\frac{1}{n} + \sum_{k=0}^\infty \frac{(-1)^k(z/2)^{2k+n}}{(2k+n)k!(n+k)!}, \qquad |z| < \infty, \quad n = 1, 2, \ldots$$

Hint. Substitute (5.3.2) into Problem 19, formula (iv).

21. Derive the asymptotic formula

$$\operatorname{Ji}_\nu(x) \approx \left(\frac{2}{\pi x}\right)^{1/2}\frac{\sin(x - \frac{1}{2}\nu\pi - \frac{1}{4}\pi)}{x}.$$

22. Prove the integral representation

$$[\operatorname{Ai}(x)]^2 = \frac{1}{4\pi\sqrt{3}}\int_0^\infty J_0\left(\frac{1}{12}t^3 + xt\right)t\,dt, \qquad x \geqslant 0$$

for the square of the Airy function of the first kind.

Hint. Use Problem 7, formula (iii).

---

[47] G. N. Watson, *op. cit.*, p. 391.

6

CYLINDER FUNCTIONS: APPLICATIONS

6.1. Introductory Remarks

As already noted in Sec. 5.1, the cylinder functions have a very wide range of applications to physics and engineering, which cannot even be touched upon in a book of this size. Instead, we confine ourselves to a discussion of a few selected problems of mathematical physics involving cylinder functions,[1] where the selection has been made with the aim of illustrating the application of the theory of Chapter 6. We are mainly concerned with the solution of boundary value problems for various special domains. In addition to several examples of an elementary character, we include some that are more complicated, e.g., the Dirichlet problem for a wedge (see Sec. 6.5).

6.2. Separation of Variables in Cylindrical Coordinates

Consider the partial differential equation

$$\nabla^2 u = \frac{1}{a^2}\frac{\partial^2 u}{\partial t^2} + b\frac{\partial u}{\partial t} + cu, \tag{6.2.1}$$

where ∇^2 is the Laplacian (operator), t is the time, and a, b, c are given constants. A variety of important differential equations occurring in mathematical physics (e.g., in electrodynamics, the theory of vibrations, the theory of heat conduction) are special cases of (6.2.1). The boundary conditions imposed on the function u often require the use of a system of cylindrical

[1] We assume that the reader has already encountered the simplest problems of this type in a first course on mathematical physics.

coordinates r, φ, z, related to the rectangular coordinates x, y, z by the formulas

$$x = r \cos \varphi, \quad y = r \sin \varphi, \quad z = z,$$

where

$$0 \leqslant r < \infty, \quad -\pi < \varphi \leqslant \pi, \quad -\infty < z < \infty.$$

In cylindrical coordinates, equation (6.2.1) becomes

$$\frac{1}{r}\frac{\partial}{\partial r}\left(r\frac{\partial u}{\partial r}\right) + \frac{1}{r^2}\frac{\partial^2 u}{\partial \varphi^2} + \frac{\partial^2 u}{\partial z^2} = \frac{1}{a^2}\frac{\partial^2 u}{\partial t^2} + b\frac{\partial u}{\partial t} + cu, \tag{6.2.2}$$

and has infinitely many solutions of the form

$$u = R(r)Z(z)\Phi(\varphi)T(t), \tag{6.2.3}$$

where each of the functions on the right depends on only one variable. Substituting (6.2.3) into (6.2.2) and dividing by $RZ\Phi T$, we obtain

$$\frac{1}{Rr}\frac{d}{dr}\left(r\frac{dR}{dr}\right) + \frac{1}{r^2\Phi}\frac{d^2\Phi}{d\varphi^2} + \frac{1}{Z}\frac{d^2 Z}{dz^2} - c = \frac{1}{T}\left(\frac{1}{a^2}\frac{d^2T}{dt^2} + b\frac{dT}{dt}\right). \tag{6.2.4}$$

Since the variables r, φ, z and t are independent, both sides of (6.2.4) must equal a constant, which we denote by $-\varkappa^2$. This leads to two equations

$$\frac{1}{a^2}\frac{d^2T}{dt^2} + b\frac{dT}{dt} + \varkappa^2 T = 0 \tag{6.2.5}$$

and

$$\frac{1}{Rr}\frac{d}{dr}\left(r\frac{dR}{dr}\right) + \varkappa^2 + \frac{1}{r^2\Phi}\frac{d^2\Phi}{d\varphi^2} = c - \frac{1}{Z}\frac{d^2Z}{dz^2}.$$

The same reasoning shows that both sides of the last equation must equal a constant, which this time we denote by $-\lambda^2$, obtaining the equations

$$\frac{d^2Z}{dz^2} - (\lambda^2 + c)Z = 0 \tag{6.2.6}$$

and

$$r^2\left[\frac{1}{Rr}\frac{d}{dr}\left(r\frac{dR}{dr}\right) + (\lambda^2 + \varkappa^2)\right] = -\frac{1}{\Phi}\frac{d^2\Phi}{d\varphi^2}.$$

Again, both sides of the last equation must equal a constant, denoted by μ^2, which implies

$$\frac{d^2\Phi}{d\varphi^2} + \mu^2\Phi = 0 \tag{6.2.7}$$

and

$$\frac{1}{r}\frac{d}{dr}\left(r\frac{dR}{dr}\right) + \left(\lambda^2 + \varkappa^2 - \frac{\mu^2}{r^2}\right)R = 0. \tag{6.2.8}$$

The process just described is called *separation of variables*, and leads to infinitely many solutions of the form (6.2.3), depending on the parameters $\varkappa$, λ, μ, which can take real or complex values.[2]

Thus, determining the factors in the product (6.2.3) reduces to the relatively simple problem of solving the ordinary differential equations (6.2.5–8). The first three of these equations can be solved in terms of elementary functions, but if we introduce a new variable proportional to r, the fourth equation becomes Bessel's equation, whose solutions involve cylinder functions. The required solution of the given physical problem is obtained by superposition of the particular solutions (6.2.3), where the specific conditions of the problem dictate the choice of the parameters $\varkappa$, λ, μ and the corresponding solutions of (6.2.5–8).

Finally, we call attention to two important special cases of equation (6.2.1), obtained by making certain choices of the constants a, b and c:

1. *Laplace's equation* $\nabla^2 u = 0$ (corresponding to the choice $a \to \infty$, $b = c = 0$). This equation has particular solutions of the form

$$u = R(r)Z(z)\Phi(\varphi), \tag{6.2.9}$$

where

$$\frac{1}{r}\frac{d}{dr}\left(r\frac{dR}{dr}\right) + \left(\lambda^2 - \frac{\mu^2}{r^2}\right)R = 0,$$
$$\frac{d^2Z}{dz^2} - \lambda^2 Z = 0, \qquad \frac{d^2\Phi}{d\varphi^2} + \mu^2\Phi = 0. \tag{6.2.10}$$

In the special case where the conditions of the problem are such that u is independent of the angular coordinate φ, we have

$$u = R(r)Z(z) \tag{6.2.11}$$

where

$$\frac{1}{r}\frac{d}{dr}\left(r\frac{dR}{dr}\right) + \lambda^2 R = 0, \qquad \frac{d^2Z}{dz^2} - \lambda^2 Z = 0. \tag{6.2.12}$$

2. *Helmholtz's equation* $\nabla^2 u + k^2 u = 0$ (corresponding to the choice $a \to \infty$, $b = 0$, $c = -k^2$). In this case, application of the method of separation of variables leads to particular solutions of the form

$$u = R(r)Z(z)\Phi(\varphi), \tag{6.2.13}$$

where

$$\frac{1}{r}\frac{d}{dr}\left(r\frac{dR}{dr}\right) + \left(\lambda^2 - \frac{\mu^2}{r^2}\right)R = 0,$$
$$\frac{d^2Z}{dz^2} - (\lambda^2 - k^2)Z = 0, \qquad \frac{d^2\Phi}{d\varphi^2} + \mu^2\Phi = 0. \tag{6.2.14}$$

---

[2] Without loss of generality, we can assume that each of the parameters $\varkappa$, λ, μ belongs to an arbitrarily chosen half-plane, since changing the sign of $\varkappa$, λ, μ does not affect the "separation constants" $-\varkappa^2$, $-\lambda^2$, μ^2.

6.3. The Boundary Value Problems of Potential Theory. The Dirichlet Problem for a Cylinder

A function $u = u(x, y, z)$ is said to be *harmonic* in a domain τ if u and its first and second partial derivatives with respect to x, y and z are continuous and satisfy Laplace's equation $\nabla^2 u = 0$ in τ. Consider the problem of finding a function u which is harmonic in τ and satisfies one of the three boundary conditions

$$u|_\sigma = f, \tag{6.3.1a}$$

$$\left.\frac{\partial u}{\partial n}\right|_\sigma = f, \tag{6.3.1b}$$

$$\left.\left(\frac{\partial u}{\partial n} + hu\right)\right|_\sigma = f, \qquad h > 0, \tag{6.3.1c}$$

where σ is the boundary of τ, f is a given function of a variable point of σ,[3] and $\partial/\partial n$ denotes the derivative with respect to the exterior normal to σ. This problem is called the *first boundary value problem of potential theory* or the *Dirichlet problem* if the boundary condition is of the form (6.3.1a), the *second boundary value problem of potential theory* or the *Neumann problem* if it is of the form (6.3.1b), and the *third* or *mixed boundary value problem of potential theory* if it is of the form (6.3.1c). These problems play a very important role in mathematical physics.[4] We now consider the Dirichlet problem for the case where τ is a cylinder of length l and radius a.

Let r, φ, z be a cylindrical coordinate system, with z-axis along the axis of the cylinder and origin in one face of the cylinder (see Figure 18). To satisfy the boundary condition (6.3.1a), we first solve two simpler problems corresponding to the boundary conditions

$$u|_{r=a} = 0, \qquad u|_{z=0} = f_0, \qquad u|_{z=l} = f_l, \tag{6.3.2a}$$

$$u|_{r=a} = F, \qquad u|_{z=0} = u|_{z=l} = 0. \tag{6.3.2b}$$

(In the first case, f vanishes on the lateral surface of the cylinder, and in the second case, f vanishes on the ends of the cylinder.) Obviously, the sum of the solutions satisfying the boundary conditions (6.3.2a) and (6.3.2b) will then satisfy the more general boundary condition (6.3.1a).[5]

---

[3] If $f \equiv 0$, the boundary condition is said to be *homogeneous*, and otherwise *inhomogeneous*. Here it is assumed that u is continuous in the closed domain $\tau + \sigma$ (cf. Sec. 8.1).

[4] For a more detailed formulation of boundary value problems, and for conditions guaranteeing the existence and uniqueness of solutions under various assumptions concerning the domain τ and the boundary function f, see the books by Frank and von Mises, Tikhonov and Samarski, Courant and Hilbert, and Smirnov (Vol. IV), cited in the Bibliography on p. 300.

[5] It should be noted that in many problems involving inhomogeneous boundary conditions, repeated use of the superposition method leads to solutions of excessively complicated form. This can often be avoided by using another method, due to G. A. Grinberg, *Selected Topics in the Mathematical Theory of Electric and Magnetic Phenomena* (in Russian), Izd. Akad. Nauk SSSR, Moscow (1948).

For simplicity, we temporarily assume that the boundary conditions are independent of the angular coordinate φ, so that

$$f_0 = f_0(r), \qquad f_l = f_l(r), \qquad F = F(z).$$

Then the solution u will also be independent of φ, and therefore, according to (6.2.11, 12) the particular solutions of Laplace's equation take the form $u = R(r)Z(z)$, where $R(r)$ and $Z(z)$ satisfy the differential equations

$$\frac{1}{r}\frac{d}{dr}\left(r\frac{dR}{dr}\right) + \lambda^2 R = 0, \qquad \frac{d^2Z}{dz^2} - \lambda^2 Z = 0. \tag{6.3.3}$$

Solving these equations, we find that

$$R = AJ_0(\lambda r) + BY_0(\lambda r), \qquad Z = C\cosh\lambda z + D\sinh\lambda z, \tag{6.3.4}$$

where $J_0(x)$ and $Y_0(x)$ are Bessel functions of order zero, of the first and second kinds, respectively.

First we consider the boundary conditions (6.3.2a). Since $J_0(\lambda r) \to 1$, $Y_0(\lambda r) \to \infty$ as $r \to 0$, and since the solution R must satisfy the physical requirement of being bounded on the axis of the cylinder, the constant B must equal zero. Then the homogeneous boundary condition becomes

$$AJ_0(\lambda a) = 0,$$

and hence the admissible values of the parameter λ are $\lambda_n = x_n/a$, where the x_n are the positive zeros of the Bessel function $J_0(x)$ [see Sec. 5.13]. Thus we obtain the following set of particular solutions of Laplace's equation:

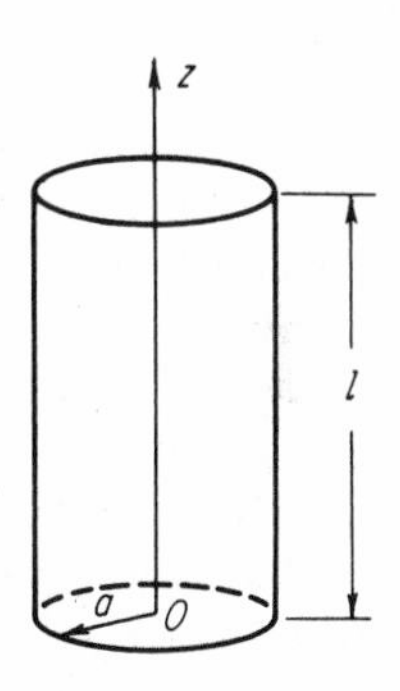

FIGURE 18

$$u = u_n = \left[M_n\cosh\left(x_n\frac{z}{a}\right) + N_n\sinh\left(x_n\frac{z}{a}\right)\right]J_0\left(x_n\frac{r}{a}\right), \qquad n = 1, 2, \ldots \tag{6.3.5}$$

By superposition of these solutions, we can construct a solution of our problem. In fact, suppose each of the functions $f_0(r)$ and $f_l(r)$ can be expanded in a Fourier-Bessel series (see Sec. 5.14), i.e.,

$$f_0(r) = \sum_{n=1}^{\infty} f_{0,n}J_0\left(x_n\frac{r}{a}\right), \qquad f_l(r) = \sum_{n=1}^{\infty} f_{l,n}J_0\left(x_n\frac{r}{a}\right), \tag{6.3.6}$$

where

$$f_{p,n} = \frac{2}{a^2J_1^2(x_n)}\int_0^a rf_p(r)J_0\left(x_n\frac{r}{a}\right)dr, \qquad p = 0, l. \tag{6.3.7}$$

Then the series

$$u = \sum_{n=1}^{\infty}\left[f_{0,n}\frac{\sinh\left(x_n\dfrac{l-z}{a}\right)}{\sinh\left(x_n\dfrac{l}{a}\right)} + f_{l,n}\frac{\sinh\left(x_n\dfrac{z}{a}\right)}{\sinh\left(x_n\dfrac{l}{a}\right)}\right]J_0\left(x_n\frac{r}{a}\right), \tag{6.3.8}$$

whose terms are of the form (6.3.5), clearly satisfies both Laplace's equation and the boundary conditions (6.3.2a).[6]

Next we consider the boundary conditions (6.3.2b). In this case, we must set $C = 0$ and choose

$$\lambda = \frac{n\pi i}{l}, \qquad n = 1, 2, \ldots$$

if the homogeneous boundary conditions are to be satisfied. Then the solutions of (6.3.3) take the form

$$\begin{aligned} R &= AI_0\left(\frac{n\pi r}{l}\right) + BK_0\left(\frac{n\pi r}{l}\right), \\ Z &= D \sin\left(\frac{n\pi z}{l}\right), \end{aligned} \tag{6.3.9}$$

where $I_0(x)$ and $K_0(x)$ are Bessel functions of imaginary argument (see Sec. 5.7). Since $K_0(n\pi r/l) \to \infty$ as $r \to 0$, we must also set $B = 0$. Therefore the particular solutions of Laplace's equation are now

$$u = u_n = M_n I_0\left(\frac{n\pi r}{l}\right) \sin\left(\frac{n\pi z}{l}\right), \qquad n = 1, 2, \ldots \tag{6.3.10}$$

Applying the *superposition method* just described,[7] we find that the solution of Laplace's equation satisfying the boundary conditions (6.3.2b) is given by the series

$$u = \sum_{n=1}^{\infty} F_n \frac{I_0\left(\dfrac{n\pi r}{l}\right)}{I_0\left(\dfrac{n\pi a}{l}\right)} \sin\frac{n\pi z}{l}, \tag{6.3.11}$$

where the F_n are the Fourier coefficients of $F(z)$ in a series expansion with respect to the functions $\sin(n\pi z/l)$:

$$F_n = \frac{2}{l}\int_0^l F(z) \sin\frac{n\pi z}{l}\, dz. \tag{6.3.12}$$

Remark 1. The solution of the Neumann problem and the mixed problem, involving the boundary conditions (6.3.1b) and (6.3.1c), is obtained in the same way, but now we must use *Dini series* (see Sec. 5.14) instead of Fourier-Bessel series.

Remark 2. To generalize our results to the case of boundary conditions involving the angular coordinate φ, we construct particular solutions of the

---

[6] Here we have in mind formal solutions, whose validity needs subsequent verification. A somewhat more rigorous point of view is adopted in Chap. 8 (cf. p. 208).

[7] Often called the *Fourier method*, or the *eigenfunction method.*

more general form (6.2.9), satisfying the equations (6.2.10). The values of the parameter μ are now determined by imposing the continuity conditions

$$u|_{\varphi=-\pi} = u|_{\varphi=\pi}, \qquad \left.\frac{\partial u}{\partial \varphi}\right|_{\varphi=-\pi} = \left.\frac{\partial u}{\partial \varphi}\right|_{\varphi=\pi}. \tag{6.3.13}$$

This is equivalent to the physical requirement that the solutions be periodic in φ, and gives $\mu = m$ $(m = 0, 1, 2, \ldots)$. The rest of the analysis differs only slightly from that just given, and leads to the following particular solutions of Laplace's equation

$$u = u_{mn} = \left[M_{mn} \cosh\left(x_{mn}\frac{z}{a}\right) + N_{mn} \sinh\left(x_{mn}\frac{z}{a}\right)\right] J_m\left(x_{mn}\frac{r}{a}\right) \begin{matrix}\cos m\varphi \\ \sin m\varphi\end{matrix}, \tag{6.3.14}$$

$$u = u_{mn} = M_{mn} I_m\left(\frac{n\pi r}{l}\right) \sin\frac{n\pi z}{l} \begin{matrix}\cos m\varphi \\ \sin m\varphi\end{matrix}, \tag{6.3.15}$$

corresponding to (6.3.2a) and (6.3.2b), respectively, where the numbers x_{mn} $(m = 0, 1, 2, \ldots; n = 1, 2, \ldots)$ denote the positive zeros of the Bessel function $J_m(x)$. Then the boundary value problems are solved by superpositions of these solutions in the form of double series, with coefficients obtained by expanding the functions

$$f_0 = f_0(r, \varphi), \qquad f_l = f_l(r, \varphi), \qquad F = F(z, \varphi)$$

in appropriate double series.

Example. *Find the stationary distribution of temperature u in a cylinder of length l and radius a, with one end held at temperature u_0, while the rest of the surface is held at temperature zero.*

The desired solution is found at once from (6.3.8) by setting $f_0 = u_0$, $f_l = 0$, and using (5.3.5) to evaluate the integral (6.3.7):

$$u = 2u_0 \sum_{n=1}^{\infty} \frac{\sinh\left(x_n \dfrac{l-z}{a}\right)}{\sinh\left(x_n \dfrac{l}{a}\right)} \frac{J_0\left(x_n \dfrac{r}{a}\right)}{x_n J_1(x_n)}. \tag{6.3.16}$$

6.4 The Dirichlet Problem for a Domain Bounded by Two Parallel Planes

Using the superposition method, we can also solve the boundary value problems of potential theory for the domain consisting of the layer between two parallel planes (see Figure 19). Let the boundary conditions be of the form (6.3.1a), and consider the case of rotational symmetry, where the functions f_0 and f_l appearing in the conditions

$$u|_{z=0} = f_0, \qquad u|_{z=l} = f_l \tag{6.4.1}$$

depend only on the variable r. A function which is harmonic in the domain $0 < z < l$ and satisfies the conditions (6.4.1) can be found by integration with respect to λ of the following particular solutions of Laplace's equation:

$$u = u_\lambda = [M_\lambda \cosh \lambda z + N_\lambda \sinh \lambda z] J_0(\lambda r), \qquad \lambda \geqslant 0. \tag{6.4.2}$$

In fact, assuming that each of the functions f_0 and f_l can be represented as a Fourier-Bessel integral (5.14.11), we find that the formal solution of the problem is given by

$$u = \int_0^\infty \lambda J_0(\lambda r) \left[f_{0,\lambda} \frac{\sinh \lambda(l - z)}{\sinh \lambda l} + f_{l,\lambda} \frac{\sinh \lambda z}{\sinh \lambda l} \right] d\lambda, \tag{6.4.3}$$

where

$$f_{p,\lambda} = \int_0^\infty r f_p(r) J_0(\lambda r)\, dr, \qquad p = 0, l. \tag{6.4.4}$$

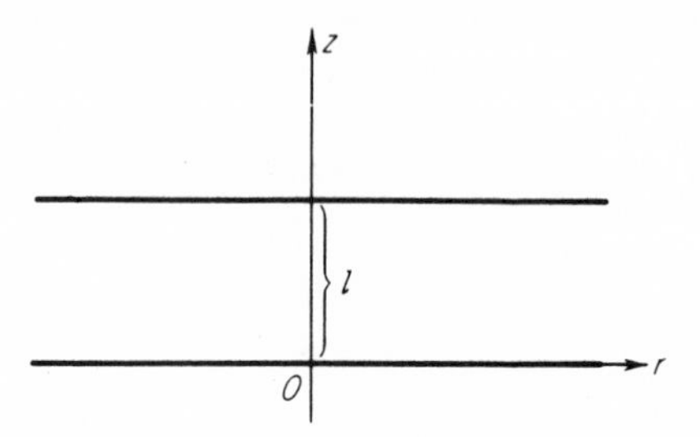

FIGURE 19

The boundary value problem for the half-space $z > 0$ can be solved in the same way. In fact, the solution turns out to be

$$u = \int_0^\infty \lambda J_0(\lambda r) f_\lambda e^{-\lambda z}\, d\lambda,$$

where

$$f_\lambda = \int_0^\infty r f(r) J_0(\lambda r)\, dr,$$

if the boundary condition is of the form

$$u|_{z=0} = f(r).$$

6.5. The Dirichlet Problem for a Wedge

In the case of a wedge-shaped domain, bounded by two intersecting planes (see Figure 20), the boundary value problems of potential theory can also be solved by the superposition method, with the help of cylinder functions. To obtain a suitable set of particular solutions of Laplace's equation $\nabla^2 u = 0$, we introduce a cylindrical coordinate system whose z-axis coincides with the line in which the two planes intersect, and we set

$$\lambda = i\sigma, \qquad 0 \leqslant \sigma < \infty,$$
$$\mu = i\tau, \qquad 0 \leqslant \tau < \infty$$

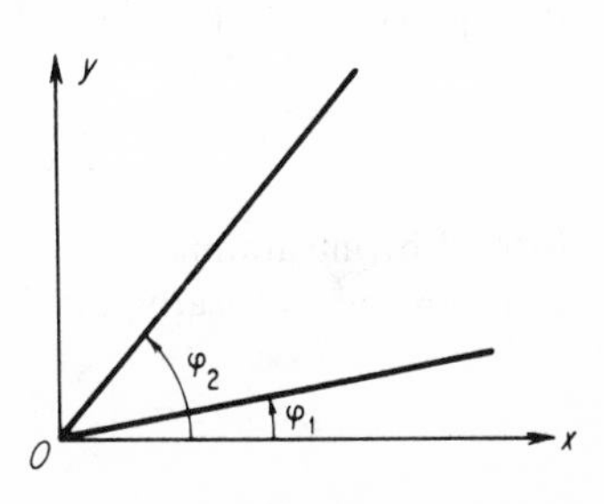

FIGURE 20

in the differential equations (6.2.10). Then, according to Sec. 5.7, the solutions of these equations become

$$R = AI_{i\tau}(\sigma r) + BK_{i\tau}(\sigma r),$$
$$\Phi = C \cosh \tau\varphi + D \sinh \tau\varphi,$$
$$Z = E \cos \sigma z + F \sin \sigma z,$$

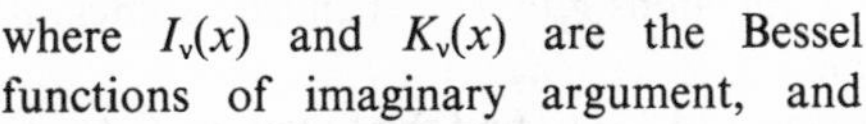

where $I_\nu(x)$ and $K_\nu(x)$ are the Bessel functions of imaginary argument, and $A, B, \ldots, F$ are arbitrary constants. Because of the asymptotic behavior of the functions $I_{i\tau}(\sigma r)$ and $K_{i\tau}(\sigma r)$ as $r \to \infty$ (see Sec. 5.11), we must set $A = 0$, which leads to the following set of particular solutions:

$$u = u_{\sigma,\tau} = [M_{\sigma,\tau} \cosh \tau\varphi + N_{\sigma,\tau} \sinh \tau\varphi]K_{i\tau}(\sigma r) \begin{matrix} \cos \sigma z \\ \sin \sigma z \end{matrix}, \tag{6.5.1}$$
$$0 \leqslant \sigma < \infty, \quad 0 \leqslant \tau < \infty.$$

We now show how to use (6.5.1) to solve the Dirichlet problem for the domain between the two planes $\varphi = \varphi_1$ and $\varphi = \varphi_2$.[8] For simplicity, suppose the functions $f_p = f_p(r, z)$ appearing in the boundary conditions

$$u|_{\varphi=\varphi_p} = f_p, \qquad p = 1, 2 \tag{6.5.2}$$

are even functions of z, which implies that the same is true of the solution $u = u(r, \varphi, z)$.[9] Assuming that each of the functions f_p can be expanded in a Fourier integral

$$f_p = f_p(r, z) = \int_0^\infty g_p(\sigma, r) \cos \sigma z \, d\sigma, \tag{6.5.3}$$

where[10]

$$g_p(\sigma, r) = \frac{2}{\pi} \int_0^\infty f_p(r, z) \cos \sigma z \, dz, \tag{6.5.4}$$

---

[8] It will be assumed that indices are assigned to φ_1, φ_2 in such a way that the domain under consideration corresponds to the interval $\varphi_1 < \varphi < \varphi_2$.

[9] The case where the f_p are odd functions of z is handled in the same way. Then the solution in the general case is represented as the sum of the solutions of the two simpler problems with the following even and odd boundary conditions:

$$u|_{\varphi=\varphi_p} = \tfrac{1}{2}[f_p(r, z) \pm f_p(r, -z)].$$

[10] G. P. Tolstov, *op. cit.*, p. 190.

we try to represent the solution of our problem as a double integral

$$u = \int_0^\infty \cos \sigma z \, d\sigma \int_0^\infty \left[G_1(\sigma, \tau) \frac{\sinh (\varphi_2 - \varphi)\tau}{\sinh (\varphi_2 - \varphi_1)\tau} + G_2(\sigma, \tau) \frac{\sinh (\varphi - \varphi_1)\tau}{\sinh (\varphi_2 - \varphi_1)\tau} \right] K_{i\tau}(\sigma r) \, d\tau, \tag{6.5.5}$$

formed by integrating solutions of the type (6.5.1) with respect to the parameters σ and τ. Clearly, the functions $G_p(\sigma, \tau)$ must satisfy the relation

$$g_p(\sigma, r) = \int_0^\infty G_p(\sigma, \tau) K_{i\tau}(\sigma r) \, d\tau, \qquad 0 < r < \infty, \tag{6.5.6}$$

and hence are the coefficients of the functions $g_p(\sigma, r)$, expanded as integrals with respect to the function $K_{i\tau}(\sigma r)$.

In some cases, we can use formula (5.14.14) to find the functions $G_p(\sigma, \tau)$. In fact, if we write

$$x = \sigma r, \qquad \xi = \sigma \rho, \qquad \sqrt{x}\, f(x) = g(\sigma, r),$$

(5.14.14) becomes

$$g(\sigma, r) = \frac{2}{\pi^2} \int_0^\infty \tau \, K_{i\tau}(\sigma r) \sinh \pi\tau \, d\tau \int_0^\infty g(\sigma, \rho) \frac{K_{i\tau}(\sigma\rho)}{\rho} \, d\rho. \tag{6.5.7}$$

The expansion theorem (6.5.7) is valid if $g(\sigma, r)$, regarded as a function of r, is piecewise continuous and of bounded variation in every finite subinterval $[r_1, r_2]$, where $0 < r_1 < r_2 < \infty$, and if the integrals

$$\int_0^{1/2} |g(\sigma, r)| r^{-1} \log \frac{1}{r} \, dr, \qquad \int_{1/2}^\infty |g(\sigma, r)| r^{-1/2} \, dr \tag{6.5.8}$$

are finite [cf. (5.14.15)]. Provided that the functions $g_p(\sigma, r)$ has these properties, a comparison of (6.5.6) and (6.5.7) shows that

$$G_p(\sigma, \tau) = \frac{2}{\pi^2} \tau \sinh \pi\tau \int_0^\infty g_p(\sigma, r) \frac{K_{i\tau}(\sigma r)}{r} \, dr, \tag{6.5.9}$$

and then (6.5.5) gives a formal solution of the problem. However, it often happens that the first of the integrals (6.5.8) is not finite, since $g_p(\sigma, r)$ generally approaches a nonzero limit $g_p(\sigma, 0)$ as $r \to 0$. To avoid this difficulty, we introduce the modified functions

$$g_p^*(\sigma, r) = g_p(\sigma, r) - g_p(\sigma, 0) e^{-\sigma r}, \qquad p = 1, 2, \tag{6.5.10}$$

and assume, as is usually the case in physical problems, that the conditions for applying formula (6.5.7) are satisfied by $g_p^*(\sigma, r)$. We then have

$$g_p^*(\sigma, r) = \int_0^\infty G_p^*(\sigma, \tau) K_{i\tau}(\sigma r) \, d\tau, \tag{6.5.11}$$

where

$$G_p^*(\sigma, \tau) = \frac{2}{\pi^2}\tau \sinh \pi\tau \int_0^\infty g_p^*(\sigma, r)\frac{K_{i\tau}(\sigma r)}{r}\,dr. \tag{6.5.12}$$

On the other hand, it is easy to prove the formula[11]

$$\frac{2}{\pi}\int_0^\infty K_{i\tau}(x)\,d\tau = e^{-x}, \qquad x > 0, \tag{6.5.13}$$

which implies

$$g_p(\sigma, 0)e^{-\sigma r} = \frac{2}{\pi} g_p(\sigma, 0) \int_0^\infty K_{i\tau}(\sigma r)\,d\tau. \tag{6.5.14}$$

Adding (6.5.11) and (6.5.14), we find the desired representation of $g_p(\sigma, r)$ as an integral with respect to $K_{i\tau}(\sigma r)$. Comparing the result with (6.5.6), we finally obtain

$$G_p(\sigma, \tau) = G_p^*(\sigma, \tau) + \frac{2}{\pi} g_p(\sigma, 0). \tag{6.5.15}$$

and then the solution is given by (6.5.5), as before.

6.6. The Field of a Point Charge near the Edge of a Conducting Sheet

We now illustrate the method developed in the preceding section, by finding the electrostatic field due to a point charge q located near the straight line edge of a thin conducting sheet held at zero potential. To avoid complicating the calculations, we assume that the charge q is at a point A in the same plane as the conducting sheet. Choosing a coordinate system whose z-axis coincides with the edge of the sheet and whose x-axis passes through the point A (see Figure 21), we represent the potential ψ of the electrostatic field as the sum of the potential ψ_0 due to the source and the potential u due to the induced charges:

$$\psi = \psi_0 + u, \qquad \psi_0 = \frac{q}{\sqrt{r^2 + a^2 + 2ar\cos\varphi + z^2}}. \tag{6.6.1}$$

Then the problem reduces to the special case of the general problem of Sec. 6.5 which corresponds to the following choice of angles and boundary conditions:

$$\varphi_1 = 0, \qquad \varphi_2 = 2\pi, \qquad f_1(r, z) = f_2(r, z) = -\frac{q}{\sqrt{(r + a)^2 + z^2}}. \tag{6.6.2}$$

---

[11] Use (5.10.23) to expand the function $e^{-x\cosh\alpha}$ in a Fourier integral with respect to $\cos \tau\alpha$, obtaining

$$e^{-x\cosh\alpha} = \frac{2}{\pi}\int_0^\infty K_{i\tau}(x)\cos\tau\alpha\,d\tau, \qquad x > 0,$$

and then set $\alpha = 0$.

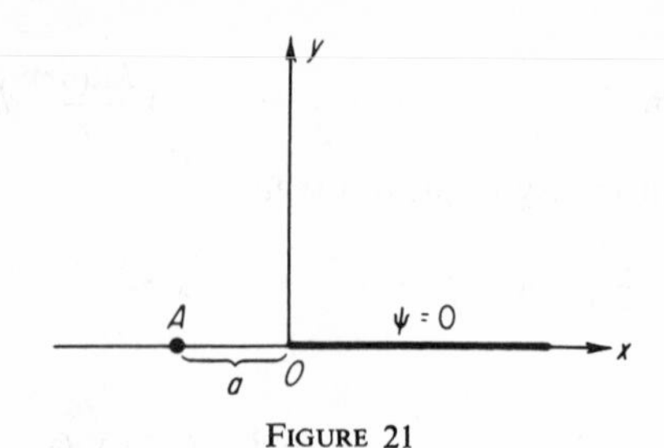

FIGURE 21

Using the integral representation given in Problem 6, formula (i), p. 140, we find that

$$g_p(\sigma, r) = -\frac{2q}{\pi}\int_0^\infty \frac{\cos \sigma z}{\sqrt{(r+a)^2+z^2}}\,dz = -\frac{2q}{\pi}K_0[\sigma(r+a)], \tag{6.6.3}$$

where $K_0(x)$ is Macdonald's function. In the present case,

$$g_p(\sigma, 0) = -\frac{2q}{\pi}K_0(\sigma a),$$

and hence, according to the method of Sec. 6.5, we must first determine the quantity

$$G_p^*(\sigma, \tau) = -\frac{4q}{\pi^3}\tau \sinh \pi\tau \int_0^\infty \frac{K_0[\sigma(r+a)] - K_0(\sigma a)e^{-\sigma r}}{r} K_{i\tau}(\sigma r)\,dr. \tag{6.6.4}$$

Since the evaluation of the integral in (6.6.4) is quite complicated, we omit the details and merely give the final result:

$$G_p^*(\sigma, \tau) = \frac{4q}{\pi^2}[K_0(\sigma a) - K_{i\tau}(\sigma a)]. \tag{6.6.5}$$

Substituting (6.6.5) into (6.5.15), we obtain

$$G_p(\sigma, \tau) = -\frac{4q}{\pi^2}K_{i\tau}(\sigma a), \tag{6.6.6}$$

and then formula (6.5.5) gives

$$u = -\frac{4q}{\pi^2}\int_0^\infty \cos \sigma z\,d\sigma \int_0^\infty \frac{\cosh(\pi-\varphi)\tau}{\cosh \pi\tau} K_{i\tau}(\sigma a)K_{i\tau}(\sigma r)\,d\tau. \tag{6.6.7}$$

The integral in (6.6.7) can be expressed in closed form in terms of elementary functions, and the final result of the calculations turns out to be

$$u = -\frac{q}{\sqrt{r^2+a^2+2ar\cos\varphi+z^2}} \times\left(1 - \frac{2}{\pi}\arctan\frac{2\sqrt{ar}\sin\frac{1}{2}\varphi}{\sqrt{r^2+a^2+2ar\cos\varphi+z^2}}\right) \tag{6.6.8}$$

(we omit the details).[12] It follows from (6.6.8) that

$$\psi = \frac{2q}{\pi\sqrt{r^2 + a^2 + 2ar\cos\varphi + z^2}} \arctan \frac{2\sqrt{ar}\sin\frac{1}{2}\varphi}{\sqrt{r^2 + a^2 + 2ar\cos\varphi + z^2}}. \tag{6.6.9}$$

Finally, we observe that the surface charge density on the sheet is given by the quantity[13]

$$-\frac{1}{4\pi r}\frac{\partial\psi}{\partial\varphi}\bigg|_{\varphi=0} = -\frac{q}{2\pi^2}\sqrt{\frac{a}{r}}\frac{1}{(r+a)^2 + z^2}. \tag{6.6.10}$$

6.7. Cooling of a Heated Cylinder

As an example of the application of cylinder functions to the nonstationary problems of mathematical physics, we now consider the problem of the cooling of an infinitely long cylinder of radius a, heated to the temperature $u_0 = f(r)$ [r is the distance from the axis] and radiating heat into the surrounding medium at zero temperature. From a mathematical point of view, the problem reduces to solving the equation of heat conduction

$$c\rho\frac{\partial u}{\partial t} = k\nabla^2 u, \tag{6.7.1}$$

subject to the boundary condition

$$\left(\frac{\partial u}{\partial r} + hu\right)\bigg|_{r=a} = 0, \tag{6.7.2}$$

and the initial condition

$$u|_{t=0} = u_0 = f(r) \tag{6.7.3}$$

where k, c, ρ, λ and $h = \lambda/k$ have the same meaning as in Sec. 2.6. Separating variables in (6.7.1) by writing $u = R(r)T(t)$, we find the equations

$$b\frac{dT}{dt} + \varkappa^2 T = 0, \qquad \frac{1}{r}\frac{d}{dr}\left(r\frac{dR}{dr}\right) + \varkappa^2 R = 0,$$

where $-\varkappa^2$ is the separation constant and $b = c\rho/k$, with solutions

$$R = AJ_0(\varkappa r) + BY_0(\varkappa r), \qquad T = Ce^{-\varkappa^2 t/b}.$$

[12] It should be noted that in the present case, the formula

$$K_0[\sigma(r+a)] = \frac{2}{\pi}\int_0^\infty K_{i\tau}(\sigma a)K_{i\tau}(\sigma r)\,d\tau$$

allows us to derive the solution (6.6.7) without recourse to the general method of expansion as an integral with respect to the functions $K_{i\tau}(\sigma r)$. To obtain this formula, set $\phi = \pi$ in formula (42), p. 55 of the Bateman Manuscript Project, *Higher Transcendental Functions, Vol. 2.*

[13] G. Joos, *op. cit.*, p. 267.

Since $J_0(\varkappa r) \to 1$, $Y_0(\varkappa r) \to \infty$ as $r \to 0$, and since R must satisfy the physical requirement of being bounded on the axis of the cylinder, the constant B must equal zero.

It follows from (6.7.2) that the parameter $\varkappa$ must satisfy the equation

$$hJ_0(\varkappa a) - \varkappa J_1(\varkappa a) = 0. \tag{6.7.4}$$

If we write $x = \varkappa a$, then (6.7.4) becomes

$$haJ_0(x) - xJ_1(x) = 0, \tag{6.7.5}$$

which has only real roots, symmetrically located with respect to the origin (see Sec. 5.13). Let $0 < x_1 < \cdots < x_n < \cdots$ be the positive roots of equation (6.7.5). Then the admissible values of the parameter $\varkappa$ are $\varkappa_n = x_n/a$, and hence the appropriate set of particular solutions of (6.7.1) is

$$u = u_n = M_n J_0\left(x_n \frac{r}{a}\right) e^{-x_n^2 t/a^2 b}, \qquad n = 1, 2, \ldots$$

Superposition of these solutions gives

$$u = \sum_{n=1}^{\infty} M_n J_0\left(x_n \frac{r}{a}\right) e^{-x_n^2 t/a^2 b}, \tag{6.7.6}$$

where, because of the initial condition (6.7.3), the coefficients M_n must be chosen to satisfy the relation

$$f(r) = \sum_{n=1}^{\infty} M_n J_0\left(x_n \frac{r}{a}\right), \qquad 0 \leqslant r < a. \tag{6.7.7}$$

This is just the problem of expanding $f(r)$ in a Dini series, which can be solved by using formulas (5.14.9–10). Thus we have

$$M_n = \frac{2}{a^2[J_0^2(x_n) + J_1^2(x_n)]} \int_0^a rf(r) J_0\left(x_n \frac{r}{a}\right) dr, \tag{6.7.8}$$

and the solution of our heat conduction problem is given by the series (6.7.6), with these values of the coefficients.

6.8 Diffraction by a Cylinder

Finally, we give an example illustrating the application of Bessel functions of the third kind. Consider the diffraction of a plane electromagnetic wave by an infinite conducting cylinder of radius a. Let (r, φ, z) be a system of cylindrical coordinates such that the z-axis coincides with the axis of the cylinder and the angle φ is measured from the direction of propagation of the incident wave. We assume that the time dependence is described by the factor $e^{i\omega t}$, where ω is the angular frequency of the incident radiation, and that the electric vector of the incident wave is parallel to the axis of the cylinder. Then the problem reduces to finding the complex amplitude of the secondary field E satisfying Helmholtz's equation

$$\frac{1}{r}\frac{\partial}{\partial r}\left(r\frac{\partial E}{\partial r}\right) + \frac{1}{r^2}\frac{\partial^2 E}{\partial \varphi^2} + k^2 E = 0, \tag{6.8.1}$$

the boundary condition

$$E|_{r=a} + E_0 e^{-ika\cos\varphi} = 0 \tag{6.8.2}$$

and the *radiation conditions*

$$E = O\left(\frac{1}{\sqrt{r}}\right), \qquad \lim_{r\to\infty} \sqrt{r}\left(\frac{\partial E}{\partial r} + ikE\right) = 0, \tag{6.8.3}$$

where $k = \omega/c$ is the wave number, and E_0 is the amplitude of the incident plane wave.[14]

Applying the method of separation of variables, we find that the particular solutions of (6.8.1), which must also be periodic in φ, are of the form

$$E = E_n = [M_n H_n^{(1)}(kr) + N_n H_n^{(2)}(kr)] \begin{matrix}\cos n\varphi,\\ \sin n\varphi\end{matrix} \qquad n = 0, 1, 2, \ldots, \tag{6.8.4}$$

where $H_n^{(1)}(kr)$, $H_n^{(2)}(kr)$ are the Hankel functions introduced in Sec. 5.6. It follows from the symmetry condition that E is an even function of φ, and hence we need only consider solutions containing $\cos n\varphi$. Moreover, examining the asymptotic behavior of the Hankel functions at infinity, we see that the radiation conditions will be satisfied only if $M_n = 0$ (no incoming waves). Therefore the solution of our problem must have the form

$$E = \sum_{n=0}^{\infty} N_n H_n^{(2)}(kr) \cos n\varphi. \tag{6.8.5}$$

It follows from the boundary condition (6.8.2) that

$$\sum_{n=0}^{\infty} N_n H_n^{(2)}(ka) \cos n\varphi + E_0 e^{-ika\cos\varphi} = 0. \tag{6.8.6}$$

Setting $z = ka$ and $t = -ie^{i\varphi}$ in formula (6.8.4), we obtain

$$e^{-ika\cos\varphi} = J_0(ka) + 2\sum_{n=1}^{\infty} (-i)^n J_n(ka) \cos n\varphi, \tag{6.8.7}$$

which, together with (6.8.5), implies

$$N_0 H_0^{(2)}(ka) = -E_0 J_0(ka), \qquad N_n H_n^{(2)}(ka) = -2E_0(-i)^n J_n(ka).$$

Therefore the required solution is given by

$$E = -E_0\left[\frac{J_0(ka)}{H_0^{(2)}(ka)} H_0^{(2)}(kr) + 2\sum_{n=1}^{\infty} (-i)^n \frac{J_n(ka)}{H_n^{(2)}(ka)} H_n^{(2)}(kr) \cos n\varphi\right]\cdot \tag{6.8.8}$$

[14] See A. N. Tikhonov and A. A. Samarski, *Differentialgleichungen der Mathematischen Physik*, VEB Deutscher Verlag der Wissenschaften, Berlin (1959), p. 497.

PROBLEMS

1. In polar coordinates r, φ, the free transverse vibrations of a stretched membrane (with equilibrium position in the $r\varphi$-plane) are described by the equation[15]

$$\nabla^2 u(r, \varphi, t) = \frac{1}{b^2}\frac{\partial^2 u(r, \varphi, t)}{\partial t^2}, \tag{i}$$

where

$$\nabla^2 = \frac{1}{r}\frac{\partial}{\partial r}\left(r\frac{\partial}{\partial r}\right) + \frac{1}{r^2}\frac{\partial^2}{\partial \varphi^2}.$$

Solve the equation of motion (i) for the case of a circular membrane of radius a, subject to the boundary condition

$$u|_{r=a} = 0$$

(fastened edge) and the initial conditions

$$u|_{t=0} = f(r), \qquad \left.\frac{\partial u}{\partial t}\right|_{t=0} = g(r).$$

2. Solve Problem 1 with the same boundary condition, but with the more general initial conditions[16]

$$u|_{t=0} = f(r, \varphi), \qquad \left.\frac{\partial u}{\partial t}\right|_{t=0} = g(r, \varphi).$$

3. In polar coordinates r, φ, the free transverse vibrations of an elastic plate (with equilibrium position in the $r\varphi$-plane) are described by the equation

$$\nabla^4 u(r, \varphi, t) = -\frac{1}{b^4}\frac{\partial^2 u(r, \varphi, t)}{\partial t^2}, \tag{ii}$$

where ∇^2 has the same meaning as in Problem 1, and $\nabla^4 = \nabla^2(\nabla^2)$. Solve the equation of motion (ii) for the case of a circular plate of radius a, subject to the boundary conditions

$$u|_{r=a} = 0, \qquad \left.\frac{\partial u}{\partial t}\right|_{r=a} = 0$$

(clamped edge), and the initial conditions

$$u|_{t=0} = f(r), \qquad \left.\frac{\partial u}{\partial t}\right|_{t=0} = g(r).$$

---

[15] For the derivation of equation (i), and equation (ii) below, see e.g., I. M. Gelfand and S. V. Fomin, *Calculus of Variations* (translated by R. A. Silverman), Prentice-Hall, Inc., Englewood Cliffs, N.J. (1963), p. 162 ff. Here we do not specify the physical meaning of the constant b. By *free* vibrations, we mean vibrations in the absence of external forces.

[16] For detailed solutions of Problems 1–2, see G. P. Tolstov, *op. cit.*, p. 288 ff.

Hint. Separate variables in (ii) by writing $u = R(r)T(t)$. The radial equation then becomes

$$\frac{1}{r}\frac{d}{dr}\left\{r\frac{d}{dr}\left[\frac{1}{r}\frac{d}{dr}\left(r\frac{dR}{dr}\right)\right]\right\} - \varkappa^4 R = 0, \tag{iii}$$

where $\varkappa^4$ is the separation constant. The general solution of (iii) which remains finite at the center of the plate is

$$R(r) = AJ_0(\varkappa r) + BI_0(\varkappa r).$$

Ans.

$$u(r, t) = \sum_{n=1}^{\infty} \frac{R_n(r)}{\int_0^a R_n^2(\rho)\rho\, d\rho}\left[\cos\frac{x_n^2 b^2 t}{a^2}\int_0^a \rho f(\rho)R_n(\rho)\, d\rho + \frac{a^2}{b^2 x_n^2}\sin\frac{x_n^2 b^2 t}{a^2}\int_0^a \rho g(\rho)R_n(\rho)\, d\rho\right],$$

where

$$R_x(r) = I_0(x)J_0\left(x\frac{r}{a}\right) - J_0(x)I_0\left(x\frac{r}{a}\right),$$

the numbers $0 < x_1 < \cdots < x_n < \cdots$ are the positive roots of the equation $R_x'(a) = 0$, and $R_n \equiv R_{x_n}$.

4. Find the stationary distribution of temperature u in a cylinder of length l and radius a whose ends are held at temperature zero, while the rest of the surface is held at temperature u_0.

5. Find the stationary distribution of temperature u in the inhomogeneous cylinder shown in Figure 22, made up of two adjacent cylindrical sections with different thermal conductivities k_1 and k_2, if the lateral surface is held at temperature u_0, while the ends are held at temperature zero.

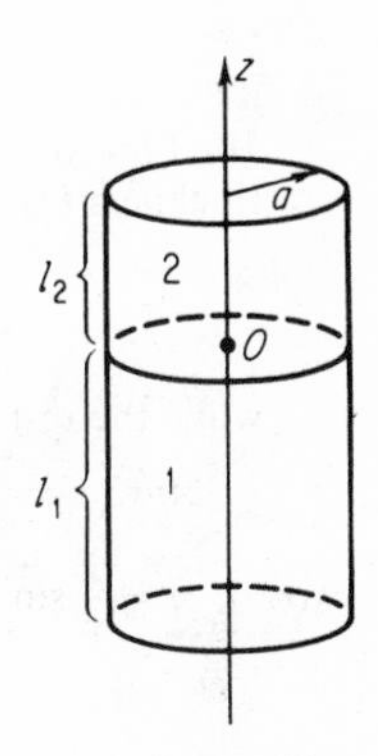

FIGURE 22

Hint. If u_1 and u_2 denote the temperatures in the sections labelled 1 and 2, respectively, then the boundary conditions are

$$u_1|_{r=a} = u_2|_{r=a} = u_0, \qquad u_1|_{z=-l_1} = u_2|_{z=l_2} = 0,$$

$$u_1|_{z=0} = u_2|_{z=0}, \qquad k_1\frac{\partial u}{\partial z}\bigg|_{z=0} = k_2\frac{\partial u}{\partial z}\bigg|_{z=0}.$$

6. Suppose an axially symmetric temperature distribution

$$u|_{t=0} = f(r)$$

is established at time $t = 0$ in an infinitely long cylinder of radius a, which transfers no heat through its surface. Find the subsequent evolution in time of the temperature distribution.

7. Find the potential ψ of the electrostatic field inside a closed cylindrical surface of length l and radius a, whose base and lateral surface are held at the potential V, while the top surface is held at zero potential.[17]

8. Find the stationary distribution of temperature u in the half-space $z > 0$, subject to the boundary condition

$$u|_{z=0} = f(r) = \begin{cases} u_0, & r < a, \\ 0, & r > a. \end{cases}$$

Ans.

$$u(r, z) = u_0 a \int_0^\infty e^{-\lambda z} J_0(\lambda r) J_1(\lambda a)\, d\lambda.$$

9. Find the potential ψ of the electrostatic field in the space between two grounded plane electrodes $z = \pm a$ due to a charge q at the point $r = 0$, $z = 0$.

Hint. Use formula (5.2.4).

Ans.

$$\psi(r, z) = \frac{q}{\sqrt{r^2 + z^2}} - q \int_0^\infty e^{-\lambda a} \frac{\cosh \lambda z}{\cosh \lambda a} J_0(\lambda r)\, d\lambda.$$

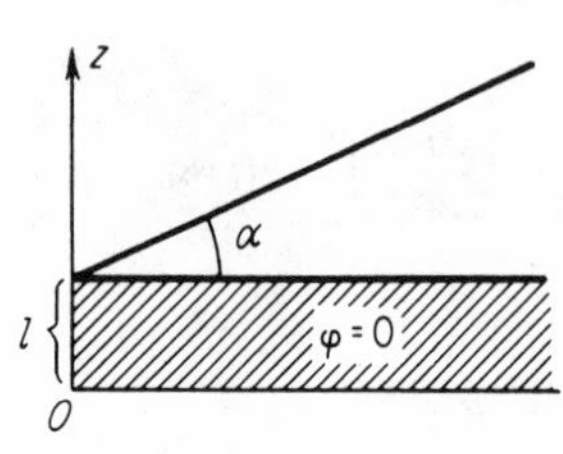

FIGURE 23

10. Find the stationary distribution of temperature u in the infinite wedge of thickness l shown in Figure 23, if the face $\varphi = \alpha$ is held at the temperature

$$u|_{\varphi=\alpha} = f(r) \sin \frac{n\pi z}{l}, \qquad n = 1, 2, \ldots,$$

while the rest of the surface is held at temperature zero.

Ans.

$$u(r, \varphi, z) = \frac{2}{\pi} \sin \frac{n\pi z}{l} \int_0^\infty \left\{ f(0) + \frac{\tau}{\pi} \sinh \pi\tau \right.$$
$$\left. \times \int_0^\infty [f(\rho) - e^{-n\pi\rho/l} f(0)] K_{i\tau}\left(\frac{n\pi\rho}{l}\right) \frac{d\rho}{\rho} \right\} \frac{\sinh \varphi\tau}{\sinh \alpha\tau} K_{i\tau}\left(\frac{n\pi r}{l}\right) d\tau.$$

11. Solve the preceding problem for an arbitrary temperature distribution

$$u|_{\varphi=\alpha} = f(r, z).$$

---

[17] One can think of the two parts of the surface as insulated from each other by an infinitely thin gasket.

7

SPHERICAL HARMONICS: THEORY

7.1. Introductory Remarks

By *spherical harmonics* we mean solutions of the linear differential equation

$$(1 - z^2)u'' - 2zu' + \left[\nu(\nu + 1) - \frac{\mu^2}{1 - z^2}\right]u = 0, \qquad (7.1.1)$$

where z is a complex variable, and μ, ν are parameters which can take arbitrary real or complex values. Equation (7.1.1) is encountered in mathematical physics when using systems of orthogonal curvilinear coordinates to solve the boundary value problems of potential theory for certain special kinds of domains (e.g., the sphere, spheroid, torus), and it is the simplest of these domains (i.e., the sphere) which gives rise to the term "spherical harmonics." In the spherical case, the variable z takes real values in the interval $(-1, 1)$, and the parameters μ and ν are nonnegative integers, but boundary value problems with more complicated geometries lead to the consideration of more general values of z, μ and ν.[1] For most applications, it is sufficient to assume (as we will do in this book) that z is either a real variable in the interval $(-1, 1)$ or a complex variable in the plane cut along the segment $[-\infty, 1]$, while ν is an arbitrary real or complex number and $\mu = m$ is a nonnegative integer ($m = 0, 1, 2, \ldots$). The reader will find a more general treatment in the references on spherical harmonics cited in the Bibliography on p. 300, especially the books by Hobson, Robin and Lense.

[1] See Chap. 8, where we consider problems in which the variable z and the parameters μ, ν take various real or complex values.

7.2. The Hypergeometric Equation and Its Series Solution

Before presenting the theory of spherical harmonics, it is appropriate to consider the problem of solving the linear differential equation

$$z(1 - z)u'' + [\gamma - (\alpha + \beta + 1)z]u' - \alpha\beta u = 0, \tag{7.2.1}$$

where z is a complex variable, and α, β, γ are parameters which can take various real or complex values. Equation (7.2.1) is called the *hypergeometric equation*, and contains as special cases many differential equations encountered in the applications. Reducing (7.2.1) to standard form by dividing it by the coefficient of u'', we obtain an equation whose coefficients are analytic functions of z in the domain $0 < |z| < 1$ and have the point $z = 0$ as a simple pole or a regular point, depending on the values of the parameters α, β and γ. It follows from the general theory of linear differential equations that (7.2.1) has a particular solution of the form

$$u = z^s \sum_{k=0}^{\infty} c_k z^k, \tag{7.2.2}$$

where $c_0 \neq 0$, s is a suitably chosen number, and the power series converges for $|z| < 1$.[2]

Substituting (7.2.2) into (7.2.1), we find that

$$\sum_{k=0}^{\infty} c_k z^{s+k-1}(s + k)(s + k - 1 + \gamma) - \sum_{k=0}^{\infty} c_k z^{s+k}(s + k + \alpha)(s + k + \beta) = 0,$$

which gives the following system of equations for determining the exponent s and the coefficients c_k:

$$c_0 s(s - 1 + \gamma) = 0,$$

$$c_k(s + k)(s + k - 1 + \gamma) - c_{k-1}(s + k - 1 + \alpha)(s + k - 1 + \beta) = 0, \qquad k = 1, 2, \ldots$$

Solving the first equation, we obtain $s = 0$ or $s = 1 - \gamma$. Suppose $\gamma \neq 0, -1, -2, \ldots$ and choose $s = 0$. Then the coefficients c_k can be calculated from the recurrence relation

$$c_k = \frac{(k - 1 + \alpha)(k - 1 + \beta)}{k(k - 1 + \gamma)} c_{k-1}, \qquad k = 1, 2, \ldots,$$

If we set $c_0 = 1$, this implies

$$c_k = \frac{(\alpha)_k(\beta)_k}{k!(\gamma)_k}, \qquad k = 0, 1, 2, \ldots,$$

where we have introduced the abbreviation

$$(\lambda)_0 = 1, \qquad (\lambda)_k = \lambda(\lambda + 1)\cdots(\lambda + k - 1), \qquad k = 1, 2, \ldots \tag{7.2.3}$$

---

[2] E. A. Coddington, *op. cit.*, Chap. 4.

as in footnote 17, p. 121. Thus, if $\gamma \neq 0, -1, -2, \ldots$, a particular solution of equation (7.2.1) is

$$u = u_1 = F(\alpha, \beta; \gamma; z) = \sum_{k=0}^{\infty} \frac{(\alpha)_k(\beta)_k}{k!(\gamma)_k} z^k, \qquad |z| < 1, \tag{7.2.4}$$

where the series on the right is known as the *hypergeometric series*.[3] The convergence of this series for $|z| < 1$ follows from the general theory of linear differential equations.[4] However, by using the ratio test, it can easily be proved without recourse to this theory that the radius of convergence of the series (7.2.4) is unity, except when one of the parameters α, β equals zero or a negative integer, in which case the series reduces to a polynomial.

Similarly, choosing $s = 1 - \gamma$ and assuming that $\gamma \neq 2, 3, 4, \ldots$, we obtain

$$c_k = \frac{(k - \gamma + \alpha)(k - \gamma + \beta)}{k(k + 1 - \gamma)} c_{k-1}, \qquad k = 1, 2, \ldots,$$

or

$$c_k = \frac{(1 - \gamma + \alpha)_k(1 - \gamma + \beta)_k}{k!(2 - \gamma)_k}, \qquad k = 0, 1, 2, \ldots,$$

if we set $c_0 = 1$. Thus, if $\gamma \neq 2, 3, 4, \ldots$, a particular solution of (7.2.1) is

$$\begin{aligned} u = u_2 &= z^{1-\gamma} \sum_{k=0}^{\infty} \frac{(1 - \gamma + \alpha)_k(1 - \gamma + \beta)_k}{k!(2 - \gamma)_k} z^k \\ &= z^{1-\gamma} F(1 - \gamma + \alpha, 1 - \gamma + \beta; 2 - \gamma; z), \\ &\qquad |z| < 1, \quad |\arg z| < \pi. \end{aligned} \tag{7.2.5}$$

Therefore, if $\gamma \neq 0, 1, 2, \ldots$, the two solutions (7.2.4–5) exist simultaneously and are linearly independent.[5] Then the general solution of (7.2.1) can be written in the form

$$u = AF(\alpha, \beta; \gamma; z) + Bz^{1-\gamma}F(1 - \gamma + \alpha, 1 - \gamma + \beta; 2 - \gamma; z), \tag{7.2.6}$$

where $|z| < 1$, $|\arg z| < \pi$, and A, B are arbitrary constants. However, if γ is an integer, this method leads to only one particular solution, and to find a second solution we must modify the method, thereby obtaining a solution which in general contains logarithmic terms.[6]

By changing variables in (7.2.1), we can obtain a number of other differential equations whose solutions can be expressed in terms of

---

[3] If γ equals zero or a negative integer, then the coefficients c_k become infinite, starting from a certain value of k, and a solution of the form (7.2.2) cannot be constructed if $s = 0$. However, it is easy to see that this situation does not arise if $s = 1 - \gamma$.

[4] E. A. Coddington, *op. cit.*, Theorem 3, p. 158.

[5] To prove the linear independence, consider the asymptotic behavior of the solutions as $z \to 0$. The two solutions coincide if $\gamma = 1$.

[6] E. A. Coddington, *op. cit.*, Theorem 4, p. 165.

hypergeometric series. Thus, for example, setting $z = t^2$, we arrive at the differential equation

$$t(1 - t^2)\frac{d^2u}{dt^2} + 2[\gamma - \tfrac{1}{2} - (\alpha + \beta + \tfrac{1}{2})t^2]\frac{du}{dt} - 4\alpha\beta t u = 0, \quad (7.2.7)$$

with particular solutions

$$u = u_1 = F(\alpha, \beta; \gamma; t^2), \qquad \gamma \neq 0, -1, -2, \ldots, \quad (7.2.8)$$

$$u = u_2 = t^{2-2\gamma}F(1 - \gamma + \alpha, 1 - \gamma + \beta; 2 - \gamma; t^2)$$
$$|t| < 1, \quad |\arg t| < \pi, \quad \gamma \neq 2, 3, 4, \ldots, \quad (7.2.9)$$

which for nonintegral γ constitute a pair of linearly independent solutions of (7.2.7) in the domain $0 < |t| < 1$.

7.3. Legendre Functions

The simplest class of spherical harmonics consists of the Legendre polynomials considered in Chapter 4, which are solutions of equation (7.1.1) for $\mu = 0$ and nonnegative integral $\nu = n$ $(n = 0, 1, 2, \ldots)$. The next class of spherical harmonics, in order of increasing complexity, consists of the *Legendre functions*, which are solutions of (7.1.1) for $\mu = 0$ and arbitrary real or complex ν, i.e., solutions of the equation

$$(1 - z^2)u'' - 2zu' + \nu(\nu + 1)u = 0, \quad (7.3.1)$$

known as *Legendre's equation*. To determine these functions, we first note that (7.3.1) can be reduced to the hypergeometric equation by making suitable changes of variables. In particular, the substitution $t = \frac{1}{2}(1 - z)$ converts (7.3.1) into the equation

$$t(1 - t)\frac{d^2u}{dt^2} + (1 - 2t)\frac{du}{dt} + \nu(\nu + 1)u = 0, \quad (7.3.2)$$

which is the special case of (7.2.1) corresponding to

$$\alpha = -\nu, \quad \beta = \nu + 1, \quad \gamma = 1,$$

while the substitution $t = z^{-2}$, $u = z^{-\nu-1}v$ converts (7.3.1) into the equation

$$t(1 - t)\frac{d^2v}{dt^2} + \left[\left(\nu + \frac{3}{2}\right) - \left(\nu + \frac{5}{2}\right)t\right]\frac{dv}{dt} - \left(\frac{\nu}{2} + 1\right)\left(\frac{\nu}{2} + \frac{1}{2}\right)v = 0, \quad (7.3.3)$$

which is the special case of (7.2.1) corresponding to

$$\alpha = \frac{\nu}{2} + 1, \quad \beta = \frac{\nu}{2} + \frac{1}{2}, \quad \gamma = \nu + \frac{3}{2}.$$

Therefore it follows from the results of the preceding section that two particular solutions of (7.3.1) are

$$u = u_1 = F\left(-\nu, \nu + 1; 1; \frac{1-z}{2}\right), \qquad |z - 1| < 2, \tag{7.3.4}$$

$$u = u_2 = \frac{\sqrt{\pi}\,\Gamma(\nu + 1)}{\Gamma(\nu + \frac{3}{2})(2z)^{\nu+1}} F\left(\frac{\nu}{2} + 1, \frac{\nu}{2} + \frac{1}{2}; \nu + \frac{3}{2}; \frac{1}{z^2}\right),$$
$$|z| > 1, \quad |\arg z| < \pi, \quad \nu \neq -1, -2, \ldots, \tag{7.3.5}$$

where $F(\alpha, \beta; \gamma; z)$ is the hypergeometric series. These solutions are called the *Legendre functions of degree* ν *of the first and second kinds*,[7] denoted by $P_\nu(z)$ and $Q_\nu(z)$, respectively. Thus we have

$$P_\nu(z) = F\left(-\nu, \nu + 1; 1; \frac{1-z}{2}\right), \qquad |z - 1| < 2, \tag{7.3.6}$$

$$Q_\nu(z) = \frac{\sqrt{\pi}\,\Gamma(\nu + 1)}{\Gamma(\nu + \frac{3}{2})(2z)^{\nu+1}} F\left(\frac{\nu}{2} + 1, \frac{\nu}{2} + \frac{1}{2}; \nu + \frac{3}{2}; \frac{1}{z^2}\right),$$
$$|z| > 1, \quad |\arg z| < \pi, \quad \nu \neq -1, -2, \ldots \tag{7.3.7}$$

The functions $P_\nu(z)$ and $Q_\nu(z)$ are defined in certain restricted regions of the complex z-plane, but, as we now show, they can be continued analytically into larger regions.[8] To make the analytic continuation of $P_\nu(z)$, the Legendre function of the first kind, we use the formula[9]

$$\frac{2}{\pi}\int_0^{\pi/2} \sin^{2k}\varphi\, d\varphi = \frac{(\frac{1}{2})_k}{k!}, \qquad k = 0, 1, 2, \ldots \tag{7.3.8}$$

to write (7.3.6) as

$$\begin{aligned} P_\nu(z) &= \sum_{k=0}^{\infty} \frac{(-\nu)_k(\nu + 1)_k}{(k!)^2}\left(\frac{1-z}{2}\right)^k \\ &= \frac{2}{\pi}\sum_{k=0}^{\infty} \frac{(-\nu)_k(\nu + 1)_k}{(\frac{1}{2})_k\, k!}\left(\frac{1-z}{2}\right)^k \int_0^{\pi/2} \sin^{2k}\varphi\, d\varphi \\ &= \frac{2}{\pi}\int_0^{\pi/2} d\varphi \sum_{k=0}^{\infty} \frac{(-\nu)_k(\nu + 1)_k}{(\frac{1}{2})_k\, k!}\left(\frac{1-z}{2}\sin^2\varphi\right)^k \\ &= \frac{2}{\pi}\int_0^{\pi/2} F\left(-\nu, \nu + 1; \frac{1}{2}; \frac{1-z}{2}\sin^2\varphi\right) d\varphi, \end{aligned} \tag{7.3.9}$$

---

[7] The term *degree* is appropriate here, since for nonnegative integral $\nu = n$, $P_n(z)$ is actually a polynomial of degree n, in fact, the nth Legendre polynomial (see Sec. 7.9).

[8] We point out that in this chapter, unlike Chapter 9, the symbol $F(\alpha, \beta; \gamma; z)$ always denotes the sum of the hypergeometric series, and hence the variable in the fourth position always has absolute value <1. This restriction disappears if we interpret $F(\alpha, \beta; \gamma; z)$ as the hypergeometric *function*. In fact, prior knowledge of the theory of the hypergeometric function leads to considerable simplification of the theory of spherical harmonics.

[9] Formula (7.3.8) is an immediate consequence of Problem 3, p. 14.

where reversing the order of summation and integration is justified because the series is uniformly convergent in the variable φ. The hypergeometric series in the right-hand side of (7.3.9) can be summed in finite form. In fact, we have the identity

$$F(-\nu, \nu + 1; \tfrac{1}{2}; -w) = \frac{(\sqrt{1+w} + \sqrt{w})^{2\nu+1} + (\sqrt{1+w} + \sqrt{w})^{-2\nu-1}}{2\sqrt{1+w}}$$

$$= f_\nu(w), \quad |w| < 1 \tag{7.3.10}$$

which is proved by noting that the function $f_\nu(w)$ is analytic in the disk $|w| < 1$ and satisfies the differential equation[10]

$$w(1 + w)f_\nu'' + (\tfrac{1}{2} + 2w)f_\nu' - \nu(\nu + 1)f_\nu = 0. \tag{7.3.11}$$

But replacing w by $-w$ converts (7.3.11) into the hypergeometric equation with parameters $\alpha = -\nu$, $\beta = \nu + 1$, $\gamma = \frac{1}{2}$. Then, since equation (7.2.1) has a unique solution which is analytic in the disk $|w| < 1$ and approaches unity as $w \to 0$, it follows that

$$f_\nu(w) \equiv F(-\nu, \nu + 1; \tfrac{1}{2}; -w),$$

as asserted.

We now substitute (7.3.10) into (7.3.9), obtaining the integral representation

$$P_\nu(z) = \frac{2}{\pi}\int_0^{\pi/2} f_\nu\left(\frac{z-1}{2}\sin^2\varphi\right) d\varphi$$

for the Legendre function of the first kind. In deriving this formula, it was assumed that $|z - 1| < 2$, but the integral in the right-hand side defines an analytic function for every z in the complex plane cut along the segment $[-\infty, -1]$. In fact, for any such z, the variable

$$w = \frac{z-1}{2}\sin^2\varphi, \qquad 0 \leqslant \varphi \leqslant \frac{\pi}{2}$$

belongs to the w-plane cut along $[-\infty, -1]$. Since $f_\nu(w)$ is analytic in this plane, our assertion follows by the usual theorem from complex variable theory.[11]

Thus the analytic continuation of $P_\nu(z)$ is given by the formula

$$P_\nu(z) = \frac{2}{\pi}\int_0^{\pi/2} f_\nu\left(\frac{z-1}{2}\sin^2\varphi\right) d\varphi, \qquad |\arg(z + 1)| < \pi. \tag{7.3.12}$$

---

[10] The point $w = 0$ is a regular point of the function $f_\nu(w)$, since $f_\nu(w)$ takes its original value after making a circuit around this point. To verify (7.3.11), it is convenient to first show that

$$\sqrt{w}\{\sqrt{w}\sqrt{1+w}[\sqrt{1+w}\,f_\nu]'\}' - (\nu + \tfrac{1}{2})^2 f_\nu = 0,$$

and then carry out the differentiation.

[11] E. C. Titchmarsh, *op. cit.*, p. 99.

The function defined by (7.3.12) is analytic in the z-plane cut along $[-\infty,-1]$ (see Figure 24), where it is a solution of the differential equation (7.3.1), by an obvious application of the principle of analytic continuation.[12] In particular, (7.3.12) implies

$$P_\nu(1) = 1. \qquad (7.3.13)$$

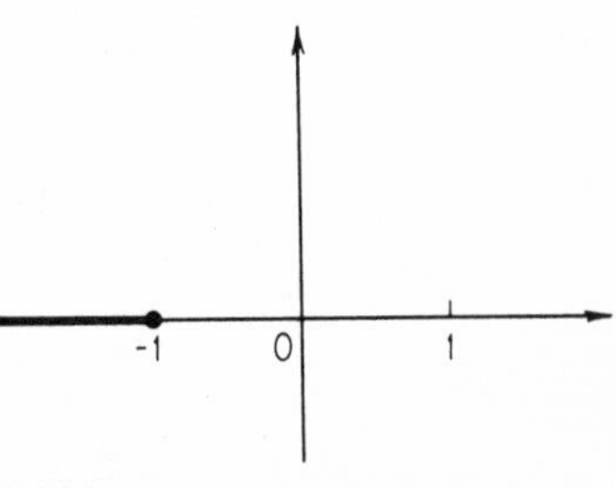

FIGURE 24

As will be shown below, every solution of (7.3.1) which is linearly independent of the solution $u = P_\nu(z)$, approaches infinity as $z \to 1$, and therefore the Legendre function of the first kind can also be defined as the solution of (7.3.1) which approaches unity as $z \to 1$. Since $f_\nu(w)$ is an entire function of the parameter ν, it follows from (7.3.12) that the same is true of $P_\nu(z)$. Moreover, it is easily verified that

$$f_{-\nu-1}(w) = f_\nu(w),$$

and hence

$$P_{-\nu-1}(z) = P_\nu(z) \qquad (7.3.14)$$

for arbitrary real or complex ν.

To make the analytic continuation of $Q_\nu(z)$, the Legendre function of the second kind, we start with the formula

$$\int_1^\infty \frac{dt}{t^{2k+\nu+3/2}\sqrt{t-1}} = \frac{\sqrt{\pi}\,\Gamma(\nu+1)\left(\frac{\nu}{2}+\frac{1}{2}\right)_k\left(\frac{\nu}{2}+1\right)_k}{\Gamma\left(\nu+\frac{3}{2}\right)\left(\frac{\nu}{2}+\frac{3}{4}\right)_k\left(\frac{\nu}{2}+\frac{5}{4}\right)_k}, \qquad (7.3.15)$$

$$\operatorname{Re}\nu > -1, \quad k = 0, 1, 2, \ldots,$$

which is easily proved by making the substitution $t = s^{-1}$ and using formulas (1.5.2), (1.5.6) and (1.2.3) from the theory of the gamma function. Then, using (7.3.15) and the definition of $Q_\nu(z)$, and assuming that

$$|z| > 1, \quad |\arg z| < \pi, \quad \operatorname{Re}\nu > -1,$$

---

[12] Let $f(z)$ be analytic in a domain D, and suppose $Lf(z) = 0$ for all z in a smaller domain D^* contained in D, where L is a linear differential operator whose coefficients are analytic in D. [In the present case,

$$L = (1 - z^2)\frac{d^2}{dz^2} - 2z\frac{d}{dz} + \nu(\nu + 1).]$$

Then $Lf(z) = 0$ for all z in D. Cf. footnote 6, p. 3.

we have

$$
\begin{aligned}
Q_\nu(z) &= \frac{\sqrt{\pi}\,\Gamma(\nu+1)}{\Gamma\left(\nu+\frac{3}{2}\right)(2z)^{\nu+1}} \sum_{k=0}^{\infty} \frac{\left(\frac{\nu}{2}+1\right)_k \left(\frac{\nu}{2}+\frac{1}{2}\right)_k}{\left(\nu+\frac{3}{2}\right)_k k!} \frac{1}{z^{2k}} \\
&= \frac{1}{(2z)^{\nu+1}} \sum_{k=0}^{\infty} \frac{\left(\frac{\nu}{2}+\frac{3}{4}\right)_k \left(\frac{\nu}{2}+\frac{5}{4}\right)_k}{\left(\nu+\frac{3}{2}\right)_k k!} \frac{1}{z^{2k}} \int_1^\infty \frac{dt}{t^{2k+\nu+3/2}\sqrt{t-1}} \\
&= \frac{1}{(2z)^{\nu+1}} \int_1^\infty \frac{dt}{t^{\nu+3/2}\sqrt{t-1}} \sum_{k=0}^{\infty} \frac{\left(\frac{\nu}{2}+\frac{3}{4}\right)_k \left(\frac{\nu}{2}+\frac{5}{4}\right)_k}{\left(\nu+\frac{3}{2}\right)_k k!} \frac{1}{(zt)^{2k}} \\
&= \frac{1}{(2z)^{\nu+1}} \int_1^\infty F\left(\frac{\nu}{2}+\frac{3}{4}, \frac{\nu}{2}+\frac{5}{4}; \nu+\frac{3}{2}; \frac{1}{z^2t^2}\right) \frac{dt}{t^{\nu+3/2}\sqrt{t-1}},
\end{aligned}
\tag{7.3.16}
$$

where reversing the order of summation and integration can be justified by an absolute convergence argument. The rest of the derivation is based on the formula

$$
F\left(\frac{\nu}{2}+\frac{3}{4}, \frac{\nu}{2}+\frac{5}{4}; \nu+\frac{3}{2}; w\right) = \frac{1}{\sqrt{1-w}} \left(\frac{1+\sqrt{1-w}}{2}\right)^{-\nu-1/2} = g_\nu(w).
\tag{7.3.17}
$$

To prove (7.3.17), it is sufficient to show that the right-hand side satisfies equation (7.2.1) for the values[13]

$$
\alpha = \frac{\nu}{2}+\frac{3}{4}, \quad \beta = \frac{\nu}{2}+\frac{5}{4}, \quad \gamma = \nu+\frac{3}{2}, \quad z = w.
$$

Together, (7.3.16) and (7.3.17) imply

$$
Q_\nu(z) = \frac{1}{(2z)^{\nu+1}} \int_1^\infty g_\nu\left(\frac{1}{z^2t^2}\right) \frac{dt}{t^{\nu+3/2}\sqrt{t-1}},
$$

$$
|z| > 1, \quad |\arg z| < \pi, \quad \operatorname{Re}\nu > -1.
\tag{7.3.18}
$$

---

[13] To simplify the calculation, which is a bit tedious, it is convenient to first show that

$$
(\sqrt{1-w}\,g_\nu)' = \frac{\left(\frac{\nu}{2}+\frac{1}{4}\right)g_\nu}{1+\sqrt{1-w}}, \qquad [\sqrt{1-w}(\sqrt{1-w}\,g_\nu)']' = \frac{\left(\frac{\nu}{2}+\frac{1}{4}\right)\left(\frac{\nu}{2}+\frac{3}{4}\right)g_\nu}{(1+\sqrt{1-w})^2},
$$

Then multiply the first equation by $(\nu+\frac{3}{2})\sqrt{1-w}$ and the second by w, carry out the differentiation, and add the resulting equations. Formula (7.3.17) can also be derived from the second of the formulas (9.8.3) by setting

$$
\alpha = \frac{\nu}{2}+\frac{3}{4}, \qquad z = w.
$$

We now assume temporarily that z is a real number greater than 1, and introduce a new variable of integration by setting

$$zt = 1 + (z - 1)\cosh^2 \psi.$$

Then (7.3.18) takes the form[14]

$$Q_\nu(z) = \int_0^\infty h_\nu\left(\frac{z-1}{2}\cosh^2\psi\right) d\psi,$$

where

$$h_\nu(w) = \frac{(\sqrt{1+w} + \sqrt{w})^{-2\nu-1}}{\sqrt{1+w}}, \qquad |\arg w| < \pi, \quad |\arg(1+w)| < \pi. \tag{7.3.19}$$

Although this formula for $Q_\nu(z)$ has been derived under the assumption that $z > 1$, it is not hard to see that the integral on the right has meaning in a larger region. In fact, for z in the plane cut along $[-\infty, 1]$ and ψ in the interval $[0, \infty]$,[15] the integrand is continuous in ψ for every z and analytic in z for every ψ. Moreover, if $\operatorname{Re}\nu > -1$, the integral converges uniformly in every region

$$0 < \rho \leqslant |z - 1| \leqslant R < \infty, \qquad |\arg(z-1)| \leqslant \pi - \delta,$$

and hence, by the usual argument,[16] represents an analytic function in the plane cut along $[-\infty, 1]$. Thus the analytic continuation of $Q_\nu(z)$ is given by the formula

$$Q_\nu(z) = \int_0^\infty h_\nu\left(\frac{z-1}{2}\cosh^2\psi\right) d\psi, \qquad |\arg(z-1)| < \pi, \quad \operatorname{Re}\nu > -1. \tag{7.3.20}$$

To obtain the analytic continuation of $Q_\nu(z)$ for the case $\operatorname{Re}\nu \leqslant -1$, we first observe that $Q_\nu(z)$ satisfies the recurrence relation

$$Q_\nu(z) = \frac{2\nu+3}{\nu+1} zQ_{\nu+1}(z) - \frac{\nu+2}{\nu+1} Q_{\nu+2}(z), \tag{7.3.21}$$

which can be verified by direct substitution of the series (7.3.7). If p is any

---

[14] In the course of the calculations, we use the familiar identity

$$\sqrt{A + \sqrt{B}} = \sqrt{\frac{A + \sqrt{A^2 - B}}{2}} + \sqrt{\frac{A - \sqrt{A^2 - B}}{2}}.$$

[15] For these values of z and ψ, the variable

$$w = \frac{z-1}{2}\cosh^2\psi$$

belongs to the plane cut along $[-\infty, 0]$, where $h_\nu(w)$ is analytic.

[16] E. C. Titchmarsh, *op. cit.*, pp. 99–100.

positive integer, we can use (7.3.21) to write the function $Q_\nu(z)$ with arbitrary index $\nu \neq -1, -2, \ldots$ in the form

$$Q_\nu(z) = a_p(z, \nu)Q_{\nu+p}(z) + b_p(z, \nu)Q_{\nu+p+1}(z), \tag{7.3.22}$$

where $a_p(z, \nu)$ and $b_p(z, \nu)$ are polynomials in z. Then, choosing p so large that $\text{Re}\,\nu > -(p + 1)$, we can use (7.3.20) to make the analytic continuation of each of the Legendre functions in the right-hand side of (7.3.22), and substituting the corresponding expressions into (7.3.22), we obtain a function which is analytic in the z-plane cut along $[-\infty, 1]$ (see Figure 25). It follows that $Q_\nu(z)$ is analytic in this cut plane, for arbitrary complex $\nu \neq -1, -2, \ldots$ Like $P_\nu(z)$, the function $Q_\nu(z)$ satisfies the differential equation (7.3.1) [cf. footnote 12, p. 167]. Moreover, (7.3.20) implies

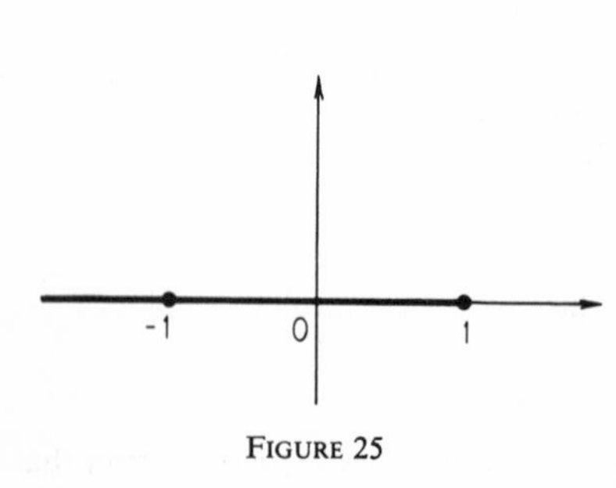

FIGURE 25

$$\lim_{z\to 1+} Q_\nu(z) = \infty. \tag{7.3.23}$$

Comparing (7.3.23) and (7.3.13), we see that $P_\nu(z)$ and $Q_\nu(z)$ are linearly independent solutions of (7.3.1).

We now study $Q_\nu(z)$ as a function of the degree ν, and show that for every fixed z, the ratio

$$q_\nu(z) = \frac{Q_\nu(z)}{\Gamma(\nu + 1)} \tag{7.3.24}$$

is an entire function of ν. For $|z| > 1$, this fact is an immediate consequence of (7.3.7). To give a proof which is valid for every z in the plane cut along $[-\infty, 1]$, we use the integral representation (7.3.20) and the recurrence relation

$$q_\nu(z) = (2\nu + 3)zq_{\nu+1}(z) - (\nu + 2)^2 q_{\nu+2}(z), \tag{7.3.25}$$

implied by (7.3.21). It follows from (7.3.20) that $q_\nu(z)$ is an analytic function of ν in the half-plane $\text{Re}\,\nu > -1$.[17] Repeated application of (7.3.25) leads to the expression

$$q_\nu(z) = \alpha_p(\nu, z)q_{\nu+p}(z) + \beta_p(\nu, z)q_{\nu+p+1}(z), \tag{7.3.26}$$

where p is a positive integer, and $\alpha_p(\nu, z)$, $\beta_p(\nu, z)$ are polynomials in ν. It follows that $q_\nu(z)$ is analytic in the half-plane $\text{Re}\,\nu > -(p + 1)$. Since p can be

---

[17] Note that $h_\nu(w)$ is an entire function of ν, while the integral (7.3.20) is uniformly convergent in ν in the region $\text{Re}\,\nu \geqslant -1 + \delta$, where $\delta > 0$ is arbitrarily small. Therefore the usual theorem concerning analytic functions defined by integrals is applicable.

chosen arbitrarily large, we conclude that $q_\nu(z)$ is an entire function of ν. Therefore, according to (7.3.24), $Q_\nu(z)$ is a meromorphic function of ν, with simple poles at the points $\nu = -1, -2, \ldots$

The general solution u of the differential equation (7.3.1) can be written as a linear combination of Legendre functions of the first and second kinds, i.e.,

$$u = AP_\nu(z) + BQ_\nu(z), \tag{7.3.27}$$

where $|\arg(z - 1)| < \pi$, $\nu \neq -1, -2, \ldots$ In the applications, it is often necessary to find a general solution of (7.3.1) for the case where x is a real number in the interval $(-1, 1)$. Since $P_\nu(z)$ is defined for such x, we need only construct a second linearly independent solution. It is not hard to see that such a solution is given by the function

$$Q_\nu(x) = \tfrac{1}{2}[Q_\nu(x + i0) + Q_\nu(x - i0)], \tag{7.3.28}$$

equal to half the sum of the values of $Q_\nu(z)$ on the upper and lower edges of the cut (cf. Sec. 7.7).[18] Thus, if $z = x$ $(-1 < x < 1)$, the general solution of (7.3.1) is

$$u = AP_\nu(x) + BQ_\nu(x), \qquad \nu \neq -1, -2, \ldots \tag{7.3.29}$$

7.4. Integral Representations of the Legendre Functions

The Legendre functions have various integral representations in terms of definite integrals and contour integrals containing the variables z and ν as parameters. As a rule, the most general representations of this type involve contour integrals, but for practical purposes, representations involving integrals along segments of the real axis are of greatest importance. For this reason, we will only consider representations of this type, referring the reader elsewhere for integral representations of other kinds.[19]

We begin by deriving an integral representation of the function $P_\nu(z)$. Assuming that $z = \cosh\alpha$ $(\alpha > 0)$ and introducing a new variable of integration in (7.3.12) by setting

$$\sinh\frac{\theta}{2} = \sinh\frac{\alpha}{2}\sin\varphi,$$

we find that

$$P_\nu(\cosh\alpha) = \frac{2}{\pi}\int_0^\alpha \frac{\cosh(\nu + \frac{1}{2})\theta}{\sqrt{2\cosh\alpha - 2\cosh\theta}}\,d\theta \tag{7.4.1}$$

---

[18] In the German literature, the symbols $P_\nu(z)$ and $Q_\nu(z)$ are used to denote the solutions of (7.3.1) for $-1 < z < 1$, and the corresponding Gothic letters are used for all other cases.

[19] E. W. Hobson, *op. cit.*, and E. W. Barnes, *On generalized Legendre functions*, Quart. J. Math., **39**, 97 (1908).

for any real or complex value of the degree ν. Writing (7.4.1) in the form

$$P_\nu(\cosh \alpha) = \frac{1}{\pi}\int_{-\alpha}^{\alpha} \frac{e^{-(\nu+\frac{1}{2})\theta}}{\sqrt{2\cosh\alpha - 2\cosh\theta}}\, d\theta$$

and then setting

$$e^\theta = \cosh\alpha + \sinh\alpha\cos\psi,$$

we arrive at another integral representation of the Legendre function of the first kind, i.e.,

$$P_\nu(\cosh\alpha) = \frac{1}{\pi}\int_0^\pi \frac{d\psi}{(\cosh\alpha + \sinh\alpha\cos\psi)^{\nu+1}}, \tag{7.4.2}$$

where ν is arbitrary. Replacing ν by $-\nu - 1$ in (7.4.2) and using (7.3.14), we obtain

$$P_\nu(\cosh\alpha) = \frac{1}{\pi}\int_0^\pi (\cosh\alpha + \sinh\alpha\cos\psi)^\nu\, d\psi. \tag{7.4.3}$$

Two other useful integral representations of the function $P_\nu(\cosh\alpha)$ can

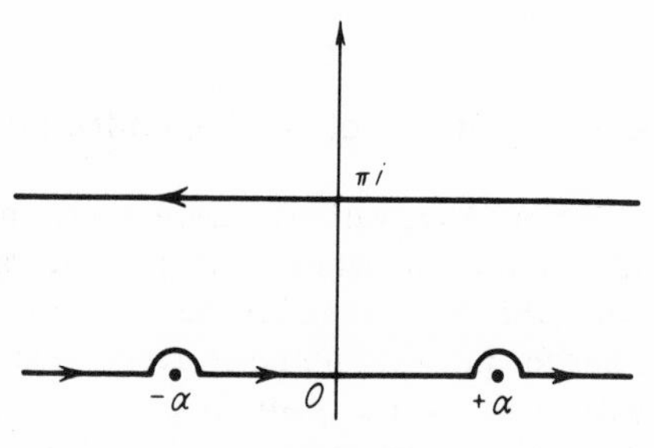

FIGURE 26

be derived from (7.4.1) by using contour integration, provided that $-1 < \operatorname{Re}\nu < 0$. We begin by considering the integral

$$\frac{1}{\pi}\int_C \frac{e^{(\nu+\frac{1}{2})t}}{\sqrt{2\cosh\alpha - 2\cosh t}}\, dt,$$

evaluated along the contour C consisting of the segments $(-\infty, -\alpha - \rho)$, $(-\alpha + \rho, \alpha - \rho)$ and $(\alpha + \rho, \infty)$ of the real axis, two semicircles of small radius ρ bypassing the two branch points $t = \pm\alpha$, and the line $\operatorname{Im} t = \pi$ (see Figure 26). Let $f(t)$ be the single-valued branch of $\sqrt{2\cosh\alpha - 2\cosh t}$ such that the values of $\arg f$ along the segment $(-\alpha + \rho, -\alpha - \rho)$, the segment $(\alpha + \rho, \infty)$, the line $\operatorname{Im} t = \pi$ and the segment $(-\infty, \alpha - \rho)$ are 0, $-\pi/2$, 0 and $\pi/2$, respectively. Then $f(t)$ is analytic inside C, and if $-1 < \operatorname{Re}\nu < 0$, the integrals along the segments $\operatorname{Re} t = \pm N$, needed to close the contour, approach zero as $N \to \infty$. Therefore, passing to the limit as $\rho \to 0$, and taking account of the change of arg f along the path of integration, we obtain

$$\frac{1}{\pi}\int_{-\alpha}^{\alpha}\frac{e^{(\nu+\frac{1}{2})\theta}}{\sqrt{2\cosh\alpha-2\cosh\theta}}\,d\theta-\frac{1}{\pi i}\int_{\alpha}^{\infty}\frac{e^{(\nu+\frac{1}{2})\theta}}{\sqrt{2\cosh\theta-2\cosh\alpha}}\,d\theta$$

$$+\frac{1}{\pi}\int_{\infty}^{-\infty}\frac{e^{(\nu+\frac{1}{2})(\theta+\pi i)}}{\sqrt{2\cosh\theta+2\cosh\alpha}}\,d\theta+\frac{1}{\pi i}\int_{-\infty}^{-\alpha}\frac{e^{(\nu+\frac{1}{2})\theta}}{\sqrt{2\cosh\theta-2\cosh\alpha}}\,d\theta=0,$$

which after some simple transformations becomes

$$P_\nu(\cosh\alpha)=\frac{2}{\pi}e^{(\nu+\frac{1}{2})\pi i}\int_0^{\infty}\frac{\cosh(\nu+\frac{1}{2})\theta}{\sqrt{2\cosh\theta+2\cosh\alpha}}\,d\theta$$
$$+\frac{2}{\pi i}\int_{\alpha}^{\infty}\frac{\sinh(\nu+\frac{1}{2})\theta}{\sqrt{2\cosh\theta-2\cosh\alpha}}\,d\theta,\qquad -1<\operatorname{Re}\nu<0. \tag{7.4.4}$$

Replacing ν by $-\nu-1$ in (7.4.4) and recalling (7.3.14), we find that

$$P_\nu(\cosh\alpha)=\frac{2}{\pi}e^{-(\nu+\frac{1}{2})\pi i}\int_0^{\infty}\frac{\cosh(\nu+\frac{1}{2})\theta}{\sqrt{2\cosh\theta+2\cosh\alpha}}\,d\theta$$
$$-\frac{2}{\pi i}\int_{\alpha}^{\infty}\frac{\sinh(\nu+\frac{1}{2})\theta}{\sqrt{2\cosh\theta-2\cosh\alpha}}\,d\theta, \tag{7.4.5}$$

where again $-1<\operatorname{Re}\nu<0$. Adding (7.4.4) and (7.4.5), and then subtracting (7.4.5) from (7.4.4), we obtain

$$2P_\nu(\cosh\alpha)=\frac{4}{\pi}\cos(\nu+\tfrac{1}{2})\pi\int_0^{\infty}\frac{\cosh(\nu+\frac{1}{2})\theta}{\sqrt{2\cosh\theta+2\cosh\alpha}}\,d\theta,$$

$$0=\frac{4i}{\pi}\sin(\nu+\tfrac{1}{2})\pi\int_0^{\infty}\frac{\cosh(\nu+\frac{1}{2})\theta}{\sqrt{2\cosh\theta+2\cosh\alpha}}\,d\theta$$
$$+\frac{4}{\pi i}\int_{\alpha}^{\infty}\frac{\sinh(\nu+\frac{1}{2})\theta}{\sqrt{2\cosh\theta-2\cosh\alpha}}\,d\theta,$$

which imply the desired integral representations

$$P_\nu(\cosh\alpha)=\frac{2}{\pi}\cos(\nu+\tfrac{1}{2})\pi\int_0^{\infty}\frac{\cosh(\nu+\frac{1}{2})\theta}{\sqrt{2\cosh\theta+2\cosh\alpha}}\,d\theta,$$
$$\alpha>0,\quad -1<\operatorname{Re}\nu<0, \tag{7.4.6}$$

$$P_\nu(\cosh\alpha)=\frac{2}{\pi}\cot(\nu+\tfrac{1}{2})\pi\int_{\alpha}^{\infty}\frac{\sinh(\nu+\frac{1}{2})\theta}{\sqrt{2\cosh\theta-2\cosh\alpha}}\,d\theta,$$
$$\alpha>0,\quad -1<\operatorname{Re}\nu<0. \tag{7.4.7}$$

Next we derive integral representations of $Q_\nu(z)$, the Legendre function of the second kind. Assuming that $z = \cosh \alpha\ (\alpha > 0)$ and introducing a new variable of integration in (7.3.20) by setting

$$\sinh \frac{\theta}{2} = \sinh \frac{\alpha}{2} \cosh \psi,$$

we find that

$$Q_\nu(\cosh \alpha) = \int_\alpha^\infty \frac{e^{-(\nu+\frac{1}{2})\theta}}{\sqrt{2 \cosh \theta - 2 \cosh \alpha}}\, d\theta \tag{7.4.8}$$

for $\operatorname{Re} \nu > -1$. Then writing

$$e^\theta = \cosh \alpha + \sinh \alpha \cosh \varphi,$$

we reduce (7.4.8) to the form

$$Q_\nu (\cosh \alpha) = \int_0^\infty \frac{d\varphi}{(\cosh \alpha + \sinh \alpha \cosh \varphi)^{\nu+1}}, \qquad \alpha > 0, \quad \operatorname{Re} \nu > -1. \tag{7.4.9}$$

Formulas (7.4.1–9) were derived under the assumption that $\alpha > 0$, i.e., that $z = \cosh \alpha > 1$, but, according to the principle of analytic continuation, they remain valid in any region of the complex α-plane where both sides of a given formula represent an analytic function. For example, (7.4.2) holds in the region $\operatorname{Re} \cosh \alpha > 0$, while (7.4.6) holds in the whole z-plane cut along $[-\infty, -1]$.

Finally, we derive an integral representation of the function $P_\nu(z)$ which is valid in the interval $-1 < z < 1$. In this case we set

$$z = \cos \beta\ (0 < \beta < \pi), \qquad \sin \frac{\theta}{2} = \sin \frac{\beta}{2} \sin \varphi$$

in formula (7.3.12), obtaining

$$P_\nu (\cos \beta) = \frac{2}{\pi} \int_0^\beta \frac{\cos (\nu + \frac{1}{2})\theta}{\sqrt{2 \cos \theta - 2 \cos \beta}}\, d\theta \tag{7.4.10}$$

for arbitrary values of the degree ν.

7.5. Some Relations Satisfied by the Legendre Functions

The differential equation (7.3.1) does not change if we replace ν by $-\nu - 1$ or z by $-z$, and hence it has solutions $P_{-\nu-1}(z)$, $Q_{-\nu-1}(z)$, $P_\nu(-z)$ and $Q_\nu(-z)$, as well as $P_\nu(z)$ and $Q_\nu(z)$. Since every three solutions of a second-order linear differential equation are linearly dependent, there must be certain functional relations between the solutions just enumerated. The simplest such relation is the formula

$$P_{-\nu-1}(z) = P_\nu(z), \tag{7.5.1}$$

proved in Sec. 7.3. To obtain a relation connecting $P_\nu(z)$, $Q_\nu(z)$ and $Q_{-\nu-1}(z)$, we assume temporarily that $z > 1$ and $-1 < \operatorname{Re} \nu < 0$. In this case, $-1 < \operatorname{Re}(-\nu - 1) < 0$, and using formulas (7.4.7–8), we have

$$Q_\nu(\cosh \alpha) - Q_{-\nu-1}(\cosh \alpha) = \pi \cot \nu\pi \, P_\nu(\cosh \alpha),$$

or

$$\sin \nu\pi [Q_\nu(z) - Q_{-\nu-1}(z)] = \pi \cos \nu\pi \, P_\nu(z). \tag{7.5.2}$$

Formula (7.5.2) remains valid for all z in the plane cut along $[-\infty, 1]$, since in this region both sides are analytic functions of z. Moreover, for all z in the cut plane, both sides of (7.5.2) are analytic functions of ν, except when ν is an integer, and therefore (7.5.2) holds for all $\nu \neq 0, \pm 1, \pm 2, \ldots$ Setting $\nu = n - \frac{1}{2}$ $(n = 0, \pm 1, \pm 2, \ldots)$ in (7.5.2), we find that

$$Q_{n-\frac{1}{2}}(z) = Q_{-n-\frac{1}{2}}(z). \tag{7.5.3}$$

We now derive another relation between the solutions of (7.3.1), assuming temporarily that $|z| > 1$ and $|\arg z| < \pi$. Then formula (7.3.7) gives

$$Q_\nu(-z) = -e^{\pm \nu\pi i} Q_\nu(z), \qquad \nu \neq -1, -2, \ldots, \tag{7.5.4}$$

where the upper sign corresponds to $\operatorname{Im} z > 0$ and the lower sign to $\operatorname{Im} z < 0$. Using the principle of analytic continuation, we can drop the condition $|z| > 1$, thereby establishing the validity of (7.5.4) for arbitrary z in the plane cut along $[-\infty, 1]$ and arbitrary $\nu \neq -1, -2, \ldots$ Finally, combining (7.5.2) and (7.5.4), we obtain

$$-\sin \nu\pi [e^{\pm \nu\pi i} Q_\nu(z) + e^{\mp \nu\pi i} Q_{-\nu-1}(z)] = \pi \cos \nu\pi \, P_\nu(-z),$$

and then using (7.5.2) to eliminate $Q_{-\nu-1}(z)$, we find that

$$\frac{2 \sin \nu\pi}{\pi} Q_\nu(z) = P_\nu(z) e^{\mp \nu\pi i} - P_\nu(-z), \tag{7.5.5}$$

where $\nu \neq -1, -2, \ldots$, and the upper sign is chosen if $\operatorname{Im} z > 0$ and the lower sign if $\operatorname{Im} z < 0$.

The relations (7.5.1–5) play an important role in the theory of spherical harmonics. In particular, it follows from (7.5.5) that

$$\begin{aligned} \frac{2 \sin \nu\pi}{\pi} Q_\nu(x + i0) &= P_\nu(x) e^{-\nu\pi i} - P_\nu(-x), \\ \frac{2 \sin \nu\pi}{\pi} Q_\nu(x - i0) &= P_\nu(x) e^{\nu\pi i} - P_\nu(-x), \end{aligned} \tag{7.5.6}$$

if $-1 < x < 1$. This implies

$$Q_\nu(x + i0) - Q_\nu(x - i0) = -i\pi P_\nu(x), \qquad -1 < x < 1, \tag{7.5.7}$$

and shows why the cut must be extended to the point $z = 1$ in the case of a Legendre function of the second kind.

7.6. Series Representations of the Legendre Functions

The Legendre functions defined in Sec. 7.3 are analytic functions of the complex variable z in the plane cut along $[-\infty, -1]$ in the case of $P_\nu(z)$, and along $[-\infty, 1]$ in the case of $Q_\nu(z)$. In restricted regions of these cut planes, the Legendre functions can be represented by hypergeometric series with various choices of α, β, γ and z, examples of which are given by the series (7.3.6–7). A simple method for constructing all expansions of this type is due to Barnes,[20] and is based on transformations of the contour integrals used to define the Legendre functions, but most of these results can be obtained by more elementary means. We begin by deriving formulas suitable for representing the Legendre functions in the domain $|z| > 1$, $|\arg z| < \pi$. According to (7.3.7), we have

$$Q_\nu(z) = \frac{\sqrt{\pi}\,\Gamma(\nu + 1)}{\Gamma(\nu + \frac{3}{2})(2z)^{\nu+1}} F\left(\frac{\nu}{2} + 1, \frac{\nu}{2} + \frac{1}{2}; \nu + \frac{3}{2}; \frac{1}{z^2}\right) \tag{7.6.1}$$

for z in this domain and arbitrary $\nu \neq -1, -2, \ldots$ To obtain the corresponding series expansion of the Legendre function of the first kind, we assume temporarily that 2ν is not an integer and use the relation (7.5.2), which can then be written in the form

$$P_\nu(z) = \frac{\tan \nu\pi}{\pi}\,[Q_\nu(z) - Q_{-\nu-1}(z)]. \tag{7.6.2}$$

Substituting the series (7.6.1) into (7.6.2), and using formula (1.2.2) to transform the ratios of gamma functions we obtain

$$\begin{aligned} P_\nu(z) = {} & \frac{\Gamma(\nu + \frac{1}{2})}{\sqrt{\pi}\,\Gamma(\nu + 1)} (2z)^\nu F\left(\frac{1}{2} - \frac{\nu}{2}, -\frac{\nu}{2}; \frac{1}{2} - \nu; \frac{1}{z^2}\right) \\ & + \frac{\Gamma(-\nu - \frac{1}{2})}{\sqrt{\pi}\,\Gamma(-\nu)} (2z)^{-\nu-1} F\left(\frac{\nu}{2} + 1, \frac{\nu}{2} + \frac{1}{2}; \nu + \frac{3}{2}; \frac{1}{z^2}\right), \\ & |z| > 1, \quad |\arg z| < \pi. \end{aligned} \tag{7.6.3}$$

The condition imposed on the parameter ν can be replaced by the weaker condition $2\nu \neq 2p + 1$ ($p = 0, \pm 1, \pm 2, \ldots$), since both sides of (7.6.3) remain analytic at points $\nu = p$. Therefore formula (7.6.3) holds for any $\nu \neq \pm\frac{1}{2}, \pm\frac{3}{2}, \ldots$

To derive expansions of the Legendre functions which hold in the part of

---

[20] E. W. Barnes, *op. cit.* The reader familiar with the theory of the hypergeometric function can derive the formulas of this section as special cases of the general relations of Secs. 9.5–6. A compilation of representations of the Legendre functions in terms of hypergeometric series is given in the Bateman Manuscript Project, *Higher Transcendental Functions, Vol. 1*, pp. 124–139.

the cut plane where $|z| < 1$, we first note that the substitution $t = z^2$ transforms the differential equation (7.3.1) into

$$t(1-t)\frac{d^2u}{dt^2} + \left(\frac{1}{2} - \frac{3}{2}t\right)\frac{du}{dt} + \frac{\nu}{2}\left(\frac{\nu}{2} + \frac{1}{2}\right)u = 0, \tag{7.6.4}$$

which is the special case of the hypergeometric equation (7.2.1) corresponding to the values

$$\alpha = -\frac{\nu}{2}, \quad \beta = \frac{\nu}{2} + \frac{1}{2}, \quad \gamma = \frac{1}{2}.$$

According to Sec. 7.2, the general solution of (7.6.4) for $|z| < 1$ can be written in the form

$$u = AF\left(\frac{\nu}{2} + \frac{1}{2}, -\frac{\nu}{2}; \frac{1}{2}; z^2\right) + BzF\left(\frac{1}{2} - \frac{\nu}{2}, \frac{\nu}{2} + 1; \frac{3}{2}; z^2\right), \tag{7.6.5}$$

where A and B are arbitrary constants. In particular, if the values of these constants are chosen to be $A = P_\nu(0)$, $B = P'_\nu(0)$, then $u \equiv P_\nu(z)$, and to obtain the desired expansion, we need only calculate the values of the Legendre function $P_\nu(z)$ and its derivative at the point $z = 0$.

With this aim, we set $z = 0$ in the series (7.3.6), obtaining

$$\begin{aligned} P_\nu(0) &= F(-\nu, \nu+1; 1; \tfrac{1}{2}) = \sum_{k=0}^{\infty} \frac{(-\nu)_k(\nu+1)_k}{(1)_k k!}\frac{1}{2^k} \\ &= \frac{1}{\Gamma(-\nu)\Gamma(\nu+1)}\sum_{k=0}^{\infty}\frac{\Gamma(k-\nu)\Gamma(k+\nu+1)}{2^k k!^2} \\ &= -\frac{\sin \nu\pi}{\pi}\sum_{k=0}^{\infty}\frac{\Gamma(k-\nu)\Gamma(k+\nu+1)}{2^k k!^2}, \end{aligned}$$

where we have used formula (1.2.2) from the theory of the gamma function. If we temporarily assume that $-1 < \operatorname{Re}\nu < 0$, then (see Sec. 1.5)

$$\frac{\Gamma(k-\nu)\Gamma(\nu+1)}{\Gamma(k+1)} = B(k-\nu, \nu+1) = \int_0^1 t^{k-\nu-1}(1-t)^\nu\,dt, \qquad k = 0, 1, 2, \ldots,$$

and hence

$$\begin{aligned} P_\nu(0) &= \frac{\sin \nu\pi}{\pi}\sum_{k=0}^{\infty}\frac{\Gamma(k+\nu+1)}{2^k k!\Gamma(\nu+1)}\int_0^1 t^{k-\nu-1}(1-t)^\nu\,dt \\ &= -\frac{\sin \nu\pi}{\pi}\int_0^1 t^{-\nu-1}(1-t)^\nu\,dt\sum_{k=0}^{\infty}\frac{\Gamma(k+\nu+1)}{k!\Gamma(\nu+1)}\left(\frac{t}{2}\right)^k \\ &= -\frac{\sin \nu\pi}{\pi}\int_0^1 t^{-\nu-1}(1-t)^\nu\left(1-\frac{t}{2}\right)^{-\nu-1}dt, \end{aligned}$$

where the reversal of the order of summation and integration is justified by an absolute convergence argument. Setting $1 - t = \sqrt{s}$, we find that

$$P_\nu(0) = -\frac{2^\nu \sin \nu\pi}{\pi} \int_0^1 s^{\frac{1}{2}(\nu-1)}(1-s)^{-\nu-1}\, ds = -\frac{2^\nu \sin \nu\pi}{\pi} \frac{\Gamma(-\nu)\Gamma\left(\frac{\nu}{2}+\frac{1}{2}\right)}{\Gamma\left(\frac{1}{2}-\frac{\nu}{2}\right)},$$

or

$$P_\nu(0) = \frac{\sqrt{\pi}}{\Gamma\left(\frac{1}{2}-\frac{\nu}{2}\right)\Gamma\left(\frac{\nu}{2}+1\right)}, \tag{7.6.6}$$

where we have used formulas (1.2.2–3). Since both sides of (7.6.6) are entire functions of ν, our result holds for arbitrary values of ν. Using (1.2.2), we can also write (7.6.6) in the form

$$P_\nu(0) = \frac{\Gamma\left(\frac{\nu}{2}+\frac{1}{2}\right)}{\sqrt{\pi}\,\Gamma\left(\frac{\nu}{2}+1\right)} \cos\frac{\nu\pi}{2}. \tag{7.6.7}$$

Once we have found $P_\nu(0)$, we can easily deduce $P'_\nu(0)$ by using the recurrence relation (7.8.5). This gives

$$P'_\nu(0) = \nu P_{\nu-1}(0) = \nu \frac{\Gamma\left(\frac{\nu}{2}\right)}{\sqrt{\pi}\,\Gamma\left(\frac{\nu}{2}+\frac{1}{2}\right)} \sin\frac{\nu\pi}{2},$$

or

$$P'_\nu(0) = \frac{2\Gamma\left(\frac{\nu}{2}+1\right)}{\sqrt{\pi}\,\Gamma\left(\frac{\nu}{2}+\frac{1}{2}\right)} \sin\frac{\nu\pi}{2}, \tag{7.6.8}$$

where we take account of formula (1.2.1). Combining (7.6.5, 7–8), we obtain the following series expansion of the Legendre function of the first kind, valid for $|z| < 1$ and arbitrary ν:

$$\begin{aligned} P_\nu(z) = {} & \frac{\Gamma\left(\frac{\nu}{2}+\frac{1}{2}\right)}{\sqrt{\pi}\,\Gamma\left(\frac{\nu}{2}+1\right)} \cos\frac{\nu\pi}{2}\, F\left(\frac{\nu}{2}+\frac{1}{2}, -\frac{\nu}{2}; \frac{1}{2}; z^2\right) \\ & + \frac{2\Gamma\left(\frac{\nu}{2}+1\right)}{\sqrt{\pi}\,\Gamma\left(\frac{\nu}{2}+\frac{1}{2}\right)} \sin\frac{\nu\pi}{2}\, zF\left(\frac{1}{2}-\frac{\nu}{2}, \frac{\nu}{2}+1; \frac{3}{2}; z^2\right). \end{aligned} \tag{7.6.9}$$

The corresponding expansion for the Legendre function of the second kind is obtained from (7.6.9) and (7.5.5). After some simple transformations, we find that

$$Q_\nu(z) = e^{\mp \nu\pi i/2}\left[\frac{\Gamma\left(\frac{\nu}{2}+1\right)\sqrt{\pi}}{\Gamma\left(\frac{\nu}{2}+\frac{1}{2}\right)} zF\left(\frac{1}{2}-\frac{\nu}{2}, \frac{\nu}{2}+1; \frac{3}{2}; z^2\right)\right.$$
$$\left. \mp i\frac{\Gamma\left(\frac{\nu}{2}+\frac{1}{2}\right)\sqrt{\pi}}{2\Gamma\left(\frac{\nu}{2}+1\right)} F\left(\frac{\nu}{2}+\frac{1}{2}, -\frac{\nu}{2}; \frac{1}{2}; z^2\right)\right], \tag{7.6.10}$$

where $|z| < 1, \nu \neq -1, -2, \ldots$, and the upper sign is chosen if $\operatorname{Im} z > 0$ and the lower sign if $\operatorname{Im} z < 0$. A formula of even greater practical interest is the series expansion of $Q_\nu(x)$, obtained from (7.6.10) and (7.3.28):

$$Q_\nu(x) = \frac{\Gamma\left(\frac{\nu}{2}+1\right)\sqrt{\pi}\cos\frac{\nu\pi}{2}}{\Gamma\left(\frac{\nu}{2}+\frac{1}{2}\right)} xF\left(\frac{1}{2}-\frac{\nu}{2}, \frac{\nu}{2}+1; \frac{3}{2}; x^2\right)$$
$$-\frac{\Gamma\left(\frac{\nu}{2}+\frac{1}{2}\right)\sqrt{\pi}\sin\frac{\nu\pi}{2}}{2\Gamma\left(\frac{\nu}{2}+1\right)} F\left(\frac{\nu}{2}+\frac{1}{2}, -\frac{\nu}{2}; \frac{1}{2}; x^2\right) \tag{7.6.11}$$
$$-1 < x < 1, \quad \nu \neq -1, -2, \ldots$$

To obtain another important class of expansions of Legendre functions, we temporarily assume that z is a real number greater than 1 and that $\operatorname{Re} \nu > -1$. Writing $z = \cosh\alpha$ $(\alpha > 0)$ and using the integral representation (7.4.9), we obtain

$$Q_\nu(\cosh\alpha) = \int_0^\infty \frac{d\varphi}{(\cosh\alpha + \sinh\alpha\cosh\varphi)^{\nu+1}} = \int_0^\infty \frac{d\varphi}{\left(e^\alpha\cosh^2\frac{\varphi}{2} - e^{-\alpha}\sinh^2\frac{\varphi}{2}\right)^{\nu+1}}$$

$$= e^{-(\nu+1)\alpha}\int_0^\infty \frac{d\varphi}{\left(1 - e^{-2\alpha}\tanh^2\frac{\varphi}{2}\right)^{\nu+1}\cosh^{2\nu+2}\frac{\varphi}{2}}$$

$$= e^{-(\nu+1)\alpha}\int_0^\infty \frac{d\varphi}{\cosh^{2\nu+2}\frac{\varphi}{2}}\sum_{k=0}^\infty \frac{\Gamma(\nu+k+1)}{\Gamma(\nu+1)k!} e^{-2k\alpha}\tanh^{2k}\frac{\varphi}{2}$$

$$= e^{-(\nu+1)\alpha}\sum_{k=0}^\infty \frac{\Gamma(\nu+k+1)}{\Gamma(\nu+1)k!} e^{-2k\alpha}\int_0^\infty \frac{\tanh^{2k}\frac{\varphi}{2}}{\cosh^{2\nu+2}\frac{\varphi}{2}}\, d\varphi,$$

where the reversal of the order of summation and integration is easily justified. Then setting $t = \tanh^2(\varphi/2)$, we find that

$$\int_0^\infty \frac{\tanh^{2k}\frac{\varphi}{2}}{\cosh^{2\nu+2}\frac{\varphi}{2}}\,d\varphi = \int_0^1 t^{k-1/2}(1-t)^\nu\,dt = \frac{\Gamma(k+\frac{1}{2})\Gamma(\nu+1)}{\Gamma(k+\nu+\frac{3}{2})},$$

which implies

$$Q_\nu(\cosh\alpha) = e^{-(\nu+1)\alpha}\sum_{k=0}^\infty \frac{\Gamma(\nu+k+1)\Gamma(k+\frac{1}{2})}{k!\Gamma(k+\nu+\frac{3}{2})}e^{-2k\alpha}$$

$$= e^{-(\nu+1)\alpha}\frac{\Gamma(\nu+1)\Gamma(\frac{1}{2})}{\Gamma(\nu+\frac{3}{2})}\sum_{k=0}^\infty \frac{(\nu+1)_k(\frac{1}{2})_k}{k!(\nu+\frac{3}{2})_k}e^{-2k\alpha}.$$

Therefore we have

$$Q_\nu(\cosh\alpha) = \frac{\sqrt{\pi}\,\Gamma(\nu+1)}{\Gamma(\nu+\frac{3}{2})}e^{-(\nu+1)\alpha}F(\nu+1,\tfrac{1}{2};\nu+\tfrac{3}{2};e^{-2\alpha}), \qquad (7.6.12)$$

or, if we return to the variable z,

$$Q_\nu(z) = \frac{\sqrt{\pi}\,\Gamma(\nu+1)}{\Gamma(\nu+\frac{3}{2})}(z-\sqrt{z^2-1})^{\nu+1}F\{\nu+1,\tfrac{1}{2};\nu+\tfrac{3}{2};(z-\sqrt{z^2-1})^2\}. \qquad (7.6.13)$$

Let z be a complex number belonging to the domain $|\arg(z-1)| < \pi$. Then

$$w = z - \sqrt{z^2-1} = z - \sqrt{z-1}\sqrt{z+1}$$

belongs to the domain $|w| < 1$, $|\arg w| < \pi$, and is an analytic function of z (we choose the branch of $\sqrt{z^2-1}$ which is positive when z is real and greater than 1). Since both sides of (7.6.13) are analytic functions, this formula, just proved for real $z > 1$, remains valid in the whole domain $|\arg(z-1)| < \pi$. Using the principle of analytic continuation, we can also easily get rid of the condition $\operatorname{Re}\nu > -1$, replacing it by the single requirement that $\nu \neq -1, -2, \ldots$ Therefore (7.6.13) holds throughout the domain of definition of $Q_\nu(z)$, which explains the particular importance of this formula.

To derive a series expansion of the function $P_\nu(z)$ from (7.6.13), we use the relation (7.6.2). Assuming temporarily that 2ν is not an integer, we find after a simple calculation based on (1.2.2) that

$$P_\nu(z) = \frac{\Gamma(\nu+1)}{\sqrt{\pi}\,\Gamma(\nu+\frac{3}{2})}\tan\nu\pi\,(z-\sqrt{z^2-1})^{\nu+1}F\{\nu+1,\tfrac{1}{2};\nu+\tfrac{3}{2};(z-\sqrt{z^2-1})^2\}$$

$$+\frac{\Gamma(\nu+\frac{1}{2})}{\sqrt{\pi}\,\Gamma(\nu+1)}(z-\sqrt{z^2-1})^{-\nu}F\{-\nu,\tfrac{1}{2};\tfrac{1}{2}-\nu;(z-\sqrt{z^2-1})^2\},$$

$$|\arg(z-1)| < \pi. \qquad (7.6.14)$$

The condition imposed on the parameter ν can be replaced by the weaker condition $2\nu \neq 2p + 1$ ($p = 0, \pm 1, \pm 2, \ldots$), since both sides of (7.6.14) remain analytic at the points $\nu = p$. Therefore formula (7.6.14) holds for all $\nu \neq \pm\frac{1}{2}, \pm\frac{3}{2}, \ldots$ and for all z in the plane cut along $[-\infty, 1]$. For $\nu = \pm\frac{1}{2}, \pm\frac{3}{2}, \ldots$, the formula becomes indeterminate, and a passage to the limit is required to obtain the corresponding analytic expression for $P_\nu(z)$.

7.7. Wronskians of Pairs of Solutions of Legendre's Equation

Let $u_1(z)$ and $u_2(z)$ be a pair of solutions of Legendre's equation, with Wronskian $W\{u_1(z), u_2(z)\}$ [see Sec. 5.9]. Then

$$\frac{d}{dz}[(1 - z^2)u_1'] + \nu(\nu + 1)u_1 = 0,$$

$$\frac{d}{dz}[(1 - z^2)u_2'] + \nu(\nu + 1)u_2 = 0,$$

and subtracting the first equation multiplied by u_2 from the second equation multiplied by u_1, we obtain

$$\frac{d}{dz}[(1 - z^2)W\{u_1(z), u_2(z)\}] = 0,$$

which implies

$$W\{u_1(z), u_2(z)\} = \frac{C}{1 - z^2}.$$

In particular, choosing $u_1(z) = Q_\nu(z)$, $u_2(z) = Q_{-\nu-1}(z)$, assuming temporarily that 2ν is not an integer, and letting $|z| \to \infty$ in formula (7.6.1), we find that

$$u_1(z) = \frac{\sqrt{\pi}\,\Gamma(\nu + 1)}{\Gamma(\nu + \frac{3}{2})(2z)^{\nu+1}}[1 + O(|z|^{-2})],$$

$$u_2(z) = \frac{\sqrt{\pi}\,\Gamma(-\nu)}{\Gamma(\frac{1}{2} - \nu)}(2z)^\nu[1 + O(|z|^{-2})],$$

$$u_1'(z) = -\frac{\sqrt{\pi}(\nu + 1)\Gamma(\nu + 1)}{\Gamma(\nu + \frac{3}{2})(2z)^{\nu+1}z}[1 + O(|z|^{-2})],$$

$$u_2'(z) = \frac{\sqrt{\pi}\,\nu\Gamma(-\nu)}{\Gamma(\frac{1}{2} - \nu)}\frac{(2z)^\nu}{z}[1 + O(|z|^{-2})].$$

Therefore

$$\begin{aligned} W\{u_1(z), u_2(z)\} &= \frac{\pi\Gamma(\nu + 1)\Gamma(-\nu)}{2\Gamma(\frac{1}{2} - \nu)\Gamma(\nu + \frac{3}{2})}\frac{2\nu + 1}{z^2}[1 + O(|z|^{-2})] \\ &= -\pi\cot\nu\pi\frac{1}{z^2}[1 + O(|z|^{-2})], \end{aligned}$$

where we have used formulas (1.2.1–2) from the theory of the gamma function. A comparison of these results shows that for our choice of u_1 and u_2, the constant C equals $\pi \cot \nu\pi$, and hence

$$W\{Q_\nu(z), Q_{-\nu-1}(z)\} = \frac{\pi \cot \nu\pi}{1 - z^2}, \qquad |\arg (z - 1)| < \pi. \tag{7.7.1}$$

Formula (7.7.1) is valid for arbitrary $\nu \neq 0, \pm 1, \pm 2, \ldots$, since both sides are still analytic at the points $\nu = n - \frac{1}{2}$ $(n = 0, \pm 1, \pm 2, \ldots)$. It follows from (7.7.1) that for all nonintegral ν, $Q_\nu(z)$ and $Q_{-\nu-1}(z)$ are a pair of linearly independent solutions of equation (7.3.1), except for the case of half-integral ν, where the Wronskian vanishes and $Q_\nu(z)$, $Q_{-\nu-1}(z)$ are connected by the linear relation (7.5.3).

Next let $u_1 = P_\nu(z)$, $u_2 = Q_\nu(z)$. To calculate the Wronskian of this pair of solutions, we use (7.6.2), assuming once again that 2ν is not an integer. This gives

$$W\{P_\nu(z), Q_\nu(z)\} = \frac{\tan \nu\pi}{\pi} W\{Q_\nu(z), Q_{-\nu-1}(z)\} = \frac{1}{1 - z^2},$$
$$|\arg (z - 1)| < \pi. \tag{7.7.2}$$

According to the principle of analytic continuation, (7.7.2) is valid for arbitrary $\nu \neq -1, -2, \ldots$, and therefore the functions $P_\nu(z)$, $Q_\nu(z)$ are a pair of linearly independent solutions of equation (7.3.1) for any ν such that both functions are meaningful.

Similarly, using the relation (7.5.5), we find that

$$W\{P_\nu(z), P_\nu(-z)\} = -\frac{2 \sin \nu\pi}{\pi} W\{P_\nu(z), Q_\nu(z)\} = -\frac{2 \sin \nu\pi}{\pi} \frac{1}{1 - z^2},$$
$$|\arg (1 \pm z)| < \pi, \tag{7.7.3}$$

for arbitrary values of ν. Thus the solutions $P_\nu(z)$ and $P_\nu(-z)$ are linearly independent if ν is not an integer. Finally we point out that in the interval $-1 < x < 1$ we have the formula

$$W\{P_\nu(x), Q_\nu(x)\} = \frac{1}{1 - x^2}, \tag{7.7.4}$$

where $Q_\nu(x)$ is the function defined by (7.3.28), and $\nu \neq -1, -2, \ldots$

The results obtained in this section show that the general solution of Legendre's equation (7.3.1) can be written in any of the three equivalent forms

$$u = AP_\nu(z) + BQ_\nu(z), \qquad |\arg (z - 1)| < \pi, \quad \nu \neq -1, -2, \ldots, \tag{7.7.5}$$

$$u = CP_\nu(z) + DP_\nu(-z), \qquad |\arg (1 \pm z)| < \pi, \quad \nu \neq 0, \pm 1, \pm 2, \ldots, \tag{7.7.6}$$

$$u = EQ_\nu(z) + FQ_{-\nu-1}(z), \qquad |\arg (z - 1)| < \pi, \quad 2\nu \neq 0, \pm 1, \pm 2, \ldots, \tag{7.7.7}$$

where $A, B, \ldots, F$ are arbitrary constants. The same formulas can be written for real $z = x$ in the interval $(-1, 1)$, if $Q_\nu(x)$ is taken to be the function defined by (7.3.28).

7.8. Recurrence Relations for the Legendre Functions

The Legendre functions satisfy simple recurrence relations connecting functions with consecutive indices. To derive these relations, we set $z = \cosh \alpha$ $(\alpha > 0)$, assuming for the time being that z is a real number greater than 1. Then, using the integral representation (7.4.1), we have

$$
\begin{aligned}
&P_{\nu+1}(\cosh \alpha) + P_{\nu-1}(\cosh \alpha) \\
&= \frac{4}{\pi}\int_0^\alpha \frac{\cosh(\nu + \frac{1}{2})\theta \cosh\theta}{\sqrt{2\cosh\alpha - 2\cosh\theta}}\, d\theta \\
&= \frac{4}{\pi}\int_0^\alpha \frac{\cosh\alpha\cosh(\nu+\frac{1}{2})\theta}{\sqrt{2\cosh\alpha - 2\cosh\theta}}\, d\theta - \frac{2}{\pi}\int_0^\alpha \sqrt{2\cosh\alpha - 2\cosh\theta}\cosh(\nu+\tfrac{1}{2})\theta\, d\theta \\
&= 2\cosh\alpha\, P_\nu(\cosh\alpha) - \frac{4}{(2\nu+1)\pi}\int_0^\alpha \sqrt{2\cosh\alpha - 2\cosh\theta}\, d\sinh(\nu+\tfrac{1}{2})\theta \\
&= 2\cosh\alpha\, P_\nu(\cosh\alpha) - \frac{4}{(2\nu+1)\pi}\int_0^\alpha \frac{\sinh(\nu+\frac{1}{2})\theta\sinh\theta}{\sqrt{2\cosh\alpha - 2\cosh\theta}}\, d\theta \\
&= 2\cosh\alpha\, P_\nu(\cosh\alpha) - \frac{2}{(2\nu+1)\pi}\int_0^\alpha \frac{\cosh(\nu+\frac{3}{2})\theta - \cosh(\nu-\frac{1}{2})\theta}{\sqrt{2\cosh\alpha - 2\cosh\theta}}\, d\theta \\
&= 2\cosh\alpha\, P_\nu(\cosh\alpha) - \frac{1}{2\nu+1}[P_{\nu+1}(\cosh\alpha) - P_{\nu-1}(\cosh\alpha)],
\end{aligned}
$$

which implies

$$
(\nu + 1)P_{\nu+1}(z) - (2\nu + 1)zP_\nu(z) + \nu P_{\nu-1}(z) = 0. \tag{7.8.1}
$$

According to the principle of analytic continuation, formula (7.8.1) holds for arbitrary z in the plane with a cut along the segment $[-\infty, -1]$. In the same way, we find that

$$
\begin{aligned}
&P_{\nu+1}(\cosh\alpha) - P_{\nu-1}(\cosh\alpha) \\
&\qquad = \frac{4}{\pi}\int_0^\alpha \frac{\sinh(\nu+\frac{1}{2})\theta\sinh\theta}{\sqrt{2\cosh\alpha - 2\cosh\theta}}\, d\theta \\
&\qquad = -\frac{4}{\pi}\int_0^\alpha \sinh(\nu+\tfrac{1}{2})\theta\, d\sqrt{2\cosh\alpha - 2\cosh\theta} \\
&\qquad = (2\nu+1)\frac{2}{\pi}\int_0^\alpha \sqrt{2\cosh\alpha - 2\cosh\theta}\cosh(\nu+\tfrac{1}{2})\,\theta\, d\theta.
\end{aligned}
$$

After differentiation with respect to α, this becomes

$$P'_{\nu+1}(\cosh\alpha) - P'_{\nu-1}(\cosh\alpha) = (2\nu+1)\frac{2}{\pi}\int_0^\alpha \frac{\cosh(\nu+\frac{1}{2})\theta}{\sqrt{2\cosh\alpha - 2\cosh\theta}}\,d\theta$$
$$= (2\nu+1)P_\nu(\cosh\alpha),$$

or

$$P'_{\nu+1}(z) - P'_{\nu-1}(z) = (2\nu+1)P_\nu(z), \tag{7.8.2}$$

where the result holds in the whole plane cut along $[-\infty, -1]$.

The rest of the recurrence relations satisfied by the function $P_\nu(z)$ can be deduced from formulas (7.8.1–2). For example, differentiating (7.8.1) with respect to z and using (7.8.2) to eliminate first $P'_{\nu-1}(z)$ and then $P'_{\nu+1}(z)$ from the resulting equation, we arrive at the relations

$$P'_{\nu+1}(z) - zP'_\nu(z) = (\nu+1)P_\nu(z), \tag{7.8.3}$$

$$zP'_\nu(z) - P'_{\nu-1}(z) = \nu P_\nu(z). \tag{7.8.4}$$

Moreover, replacing ν by $\nu - 1$ in (7.8.3) and eliminating $P'_{\nu-1}(z)$, we have

$$(1-z^2)P'_\nu(z) = \nu P_{\nu-1}(z) - \nu z P_\nu(z). \tag{7.8.5}$$

Recurrence relations for $Q_\nu(z)$, the Legendre function of the second kind, can be obtained in just the same way, starting from the integral representation (7.4.8). It turns out that these recurrence relations are exactly the same as for the function $P_\nu(z)$:

$$(\nu+1)Q_{\nu+1}(z) - (2\nu+1)zQ_\nu(z) + \nu Q_{\nu-1}(z) = 0, \tag{7.8.6}$$

$$Q'_{\nu+1}(z) - Q'_{\nu-1}(z) = (2\nu+1)Q_\nu(z), \tag{7.8.7}$$

$$Q'_{\nu+1}(z) - zQ'_\nu(z) = (\nu+1)Q_\nu(z), \tag{7.8.8}$$

$$zQ'_\nu(z) - Q'_{\nu-1}(z) = \nu Q_\nu(z), \tag{7.8.9}$$

$$(1-z^2)Q'_\nu(z) = \nu Q_{\nu-1}(z) - \nu z Q_\nu(z). \tag{7.8.10}$$

Formulas (7.8.6–10) hold for any complex z in the plane cut along $[-\infty, 1]$ and for arbitrary $\nu \neq -1, -2, \ldots$[21] It is easily verified that these formulas remain valid for the functions $Q_\nu(x)$ defined by (7.3.28).

7.9. Legendre Functions of Nonnegative Integral Degree and Their Relation to Legendre Polynomials

An important class of spherical harmonics, frequently encountered in the applications, consists of the Legendre functions of nonnegative integral

---

[21] Note that $\nu Q_{\nu-1}(z) \to 1$, $Q'_{\nu-1}(z) \to \dfrac{z}{1-z^2}$ as $\nu \to 0$.

degree $\nu = n$ $(n = 0, 1, 2, \ldots)$. Since for $\nu = n$, equation (7.3.1) coincides with equation (4.3.8), which has the Legendre polynomial of degree n as a particular solution, it is natural to expect that there is a simple connection between this class of functions and the Legendre polynomials. To establish the connection, we first observe that substitution of $\nu = 0, 1$ into (7.3.10) gives $f_0(w) = 1$, $f_1(w) = 1 + 4w$, and then (7.3.12) implies $P_0(z) = 1$, $P_1(z) = z$. Since the recurrence relation (7.8.1) for the Legendre functions coincides with the recurrence relation (4.3.1) for the Legendre polynomials, it follows that the functions $P_\nu(z)$ of nonnegative integral degree $\nu = n$ $(n = 0, 1, 2, \ldots)$ are identical with the Legendre polynomials considered in Chap. 4.

The Legendre functions of the second kind of nonnegative integral degree $\nu = n$ can also be expressed in closed form in terms of elementary functions. To prove this, we set $\nu = 0, 1$ in (7.3.7), assuming temporarily that z is a positive number greater than 1. After some simple calculations, this leads to

$$\begin{aligned} Q_0(z) &= \sum_{k=0}^{\infty} \frac{1}{2k+1} \frac{1}{z^{2k+1}} = \frac{1}{2} \log \frac{z+1}{z-1}, \\ Q_1(z) &= \sum_{k=0}^{\infty} \frac{1}{2k+3} \frac{1}{z^{2k+2}} = \frac{z}{2} \log \frac{z+1}{z-1} - 1, \end{aligned} \tag{7.9.1}$$

where, according to the principle of analytic continuation, the formulas (7.9.1) are valid in the whole z-plane cut along $[-\infty, 1]$. The corresponding expressions for the remaining functions $Q_n(z)$ can be derived from (7.9.1) and the recurrence relation (7.8.6). By using mathematical induction, it is easily verified that the result can be written in the form

$$Q_n(z) = \frac{1}{2} P_n(z) \log \frac{z+1}{z-1} - f_{n-1}(z), \qquad n = 0, 1, 2, \ldots, \tag{7.9.2}$$

where $P_n(z)$ is the Legendre polynomial of degree n, and $f_{n-1}(z)$ is a polynomial of degree $n - 1$ $[f_{-1}(z) \equiv 0]$. Formula (7.9.2) shows that the Legendre functions of the second kind of nonnegative integral degree have logarithmic singularities at the points $z = \pm 1$. Bearing in mind that

$$\log \frac{z+1}{z-1} = \log \frac{1+x}{1-x} \mp \pi i,$$

for $z = x \pm i0$ $(-1 < x < 1)$, and using the definition (7.3.28) of $Q_\nu(x)$, we find that

$$\begin{aligned} Q_0(x) &= \frac{1}{2} \log \frac{1+x}{1-x}, \qquad Q_1(x) = \frac{x}{2} \log \frac{1+x}{1-x} - 1, \\ Q_n(x) &= \frac{P_n(x)}{2} \log \frac{1+x}{1-x} - f_{n-1}(x), \end{aligned} \tag{7.9.3}$$

which, in particular, shows that $Q_n(x) \to \pm \infty$ as $x \to \pm 1$.

7.10. Legendre Functions of Half-Integral Degree

Another special class of functions encountered in practice consists of the Legendre functions of half-integral degree $\nu = n - \frac{1}{2}$ $(n = 0, 1, 2, \ldots)$.[22] This class of functions is also of theoretical interest, since the case $\nu = n - \frac{1}{2}$ occupies a special position in the theory of spherical harmonics, and many formulas need modification when $\nu = n - \frac{1}{2}$. In the present section, we assume that the variable z is greater than 1, setting $z = \cosh \alpha$ $(\alpha > 0)$. This is the case of greatest practical interest (cf. Sec. 8.11).

To obtain a general formula for the function $Q_{n-\frac{1}{2}}(\cosh \alpha)$, we use (7.6.12), which for $\nu = n - \frac{1}{2}$ becomes

$$Q_{n-\frac{1}{2}}(\cosh \alpha) = \frac{\sqrt{\pi}\,\Gamma(n + \frac{1}{2})}{\Gamma(n + 1)} e^{-(n+\frac{1}{2})\alpha} F(n + \tfrac{1}{2}, \tfrac{1}{2}; n + 1; e^{-2\alpha}), \quad (7.10.1)$$

where $\alpha > 0$, $n = 0, 1, 2, \ldots$ A similar representation of $P_{n-\frac{1}{2}}(\cosh \alpha)$ cannot be written down directly from (7.6.14), since this formula becomes indeterminate for $\nu = n - \frac{1}{2}$. However, the required expansion can be deduced from the relation (7.6.2) by using L'Hospital's rule to pass to the limit $\nu \to n - \frac{1}{2}$. This gives

$$P_{n-\frac{1}{2}}(\cosh \alpha) = \frac{1}{\pi^2}\left\{\left[\frac{\partial Q_{-\nu-1}(\cosh \alpha)}{\partial \nu}\right]_{\nu=n-\frac{1}{2}} - \left[\frac{\partial Q_{\nu}(\cosh \alpha)}{\partial \nu}\right]_{\nu=n-\frac{1}{2}}\right\}. \quad (7.10.2)$$

Writing formula (7.6.12) in the form

$$Q_\nu(\cosh \alpha) = \sum_{k=0}^{\infty} \frac{\Gamma(k + \nu + 1)\Gamma(k + \frac{1}{2})}{\Gamma(k + \nu + \frac{3}{2})\Gamma(k + 1)} e^{-\alpha(2k+\nu+1)}, \quad (7.10.3)$$

we find that

$$\frac{\partial Q_\nu(\cosh \alpha)}{\partial \nu} = \sum_{k=0}^{\infty} \frac{\Gamma(k + \nu + 1)\Gamma(k + \frac{1}{2})}{\Gamma(k + \nu + \frac{3}{2})\Gamma(k + 1)} \times [\psi(k + \nu + 1) - \psi(k + \nu + \tfrac{3}{2}) - \alpha]e^{-\alpha(2k+\nu+1)}, \quad (7.10.4)$$

$$\frac{\partial Q_{-\nu-1}(\cosh \alpha)}{\partial \nu} = -\sum_{k=0}^{\infty} \frac{\Gamma(k - \nu)\Gamma(k + \frac{1}{2})}{\Gamma(k - \nu + \frac{1}{2})\Gamma(k + 1)} \times [\psi(k - \nu) - \psi(k - \nu + \tfrac{1}{2}) - \alpha]e^{-\alpha(2k-\nu)}, \quad (7.10.5)$$

where $\psi(z)$ is the logarithmic derivative of the gamma function (see Sec. 1.3).

If we set $\nu = n - \frac{1}{2}$ $(n = 1, 2, \ldots)$, the first n terms of the series (7.10.5) become indeterminate, since

$$\Gamma(k - n + 1) = \infty, \qquad \psi(k - n + 1) = \infty, \qquad k = 0, 1, \ldots, n - 1.$$

---

[22] Because of (7.5.1, 3) there is no need to consider the case $n = -1, -2, \ldots$ separately.

However, using formulas (1.2.2) and (1.3.4), we obtain

$$\lim_{\nu\to n-\frac{1}{2}} \frac{\psi(k-\nu+\frac{1}{2})}{\Gamma(k-\nu+\frac{1}{2})} = (-1)^{n-k}\Gamma(n-k), \qquad k = 0, 1, \ldots, n-1,$$

which implies

$$\left[\frac{\partial Q_{-\nu-1}(\cosh\alpha)}{\partial\nu}\right]_{\nu=n-\frac{1}{2}}$$
$$= \sum_{k=0}^{n-1} \frac{(-1)^{n-k}\Gamma(n-k)\Gamma(k-n+\frac{1}{2})}{\Gamma(k+1)} \Gamma(k+\tfrac{1}{2})e^{-\alpha(2k-n+\frac{1}{2})}$$
$$- \sum_{k=0}^{\infty} \frac{\Gamma(k+n+\frac{1}{2})\Gamma(k+\frac{1}{2})}{\Gamma(k+n+1)\Gamma(k+1)} [\psi(k+\tfrac{1}{2}) - \psi(k+1) - \alpha]e^{-\alpha(2k+n+\frac{1}{2})}, \tag{7.10.6}$$

if we introduce a new summation index in the series

$$\sum_{k=n}^{\infty} \ldots,$$

by replacing k by $k + n$. For $n = 0$ the first term in (7.10.6) must be set equal to zero. Moreover, it follows at once that

$$\left[\frac{\partial Q_{\nu}(\cosh\alpha)}{\partial\nu}\right]_{\nu=n-\frac{1}{2}}$$
$$= \sum_{k=0}^{\infty} \frac{\Gamma(k+n+\frac{1}{2})\Gamma(k+\frac{1}{2})}{\Gamma(k+n+1)\Gamma(k+1)} \tag{7.10.7}$$
$$\times [\psi(k+n+\tfrac{1}{2}) - \psi(k+n+1) - \alpha]e^{-\alpha(2k+n+\frac{1}{2})}.$$

Substituting (7.10.6–7) into (7.10.2), and noting that

$$(-1)^{n-k}\Gamma(k-n+\tfrac{1}{2}) = \frac{\pi}{\Gamma(n+\frac{1}{2}-k)},$$

according to (1.2.2), we find that

$$P_{n-\frac{1}{2}}(\cosh\alpha)$$
$$= \frac{e^{\alpha(n-\frac{1}{2})}}{\pi} \sum_{k=0}^{n-1} \frac{\Gamma(n-k)\Gamma(k+\frac{1}{2})}{\Gamma(k+1)\Gamma(n+\frac{1}{2}-k)} e^{-2k\alpha}$$
$$+ \frac{e^{-\alpha(n+\frac{1}{2})}}{\pi^2} \sum_{k=0}^{\infty} \frac{\Gamma(k+n+\frac{1}{2})\Gamma(k+\frac{1}{2})}{\Gamma(k+n+1)\Gamma(k+1)} \tag{7.10.8}$$
$$\times [2\alpha + \psi(k+1) - \psi(k+\tfrac{1}{2}) + \psi(k+n+1) - \psi(k+n+\tfrac{1}{2})]e^{-2k\alpha},$$

where $\alpha > 0$, $n = 0, 1, 2, \ldots$, and the first term must be omitted if $n = 0$. Formula (7.10.8) is the desired series representation of the function

$P_{n-1/2}(\cosh \alpha)$. To find the values of the logarithmic derivative of the gamma function appearing in (7.10.8), we use formulas (1.3.6–9). Thus we have

$$\begin{aligned} \psi(1) &= -\gamma, \qquad \psi(m+1) = -\gamma + 1 + \frac{1}{2} + \cdots + \frac{1}{m}, \\ \psi(\tfrac{1}{2}) &= -\gamma - 2\log 2, \\ \psi(m+\tfrac{1}{2}) &= -\gamma - 2\log 2 + 2\left(1 + \frac{1}{3} + \cdots + \frac{1}{2m-1}\right), \end{aligned} \tag{7.10.9}$$

where $\gamma = 0.57721566\ldots$, and $n = 1, 2, \ldots$

Integral representations of the Legendre functions of half-integral degree can be obtained by setting $\nu = n - \frac{1}{2}$ in the appropriate formulas of Sec. 7.4. In addition, there are some special integral representations valid only for this class of spherical harmonics. For example,

$$Q_{n-1/2}(\cosh \alpha) = \int_0^\pi \frac{\cos n\varphi}{\sqrt{2\cosh \alpha - 2\cos \varphi}}\, d\varphi, \qquad n = 0, 1, 2, \ldots, \tag{7.10.10}$$

which is easily proved by expanding the right-hand side in a series of negative powers of $\cosh \alpha$, carrying out the integration and comparing the result with (7.3.7).[23]

Finally, we point out that the Legendre functions of half-integral degree can be expressed in terms of the complete elliptic integrals of the first and second kinds

$$K(k) = \int_0^{\pi/2} \frac{d\varphi}{\sqrt{1 - k^2 \sin^2 \varphi}}, \qquad E(k) = \int_0^{\pi/2} \sqrt{1 - k^2 \sin^2 \varphi}\, d\varphi \tag{7.10.11}$$

with modulus $0 \leqslant k < 1$, a fact of some interest, since there exist detailed tables of $K(k)$ and $E(k)$.[24] To derive these expressions, we use the integral representations (7.4.1) and (7.10.10) and reduce the resulting elliptic integrals to the standard form (7.10.11). For example, we have[25]

$$P_{-1/2}(\cosh \alpha) = \frac{2}{\pi \cosh \frac{\alpha}{2}} K\left(\tanh \frac{\alpha}{2}\right), \qquad Q_{-1/2}(\cosh \alpha) = 2e^{-\alpha/2} K(e^{-\alpha}), \tag{7.10.12}$$

and so on.

---

[23] See footnote 17, p. 121, and use the easily verified formula

$$\int_0^\pi \cos n\varphi \cos^{n+2k} \varphi \, d\varphi = \frac{\pi(n+2k)!}{2^{n+2k} k!(n+k)!}, \qquad k = 0, 1, 2, \ldots .$$

[24] A. Fletcher, *A table of complete elliptic integrals*, Phil. Mag., **30**, 516 (1940).

[25] To prove the first formula, make the preliminary substitution

$$\sinh \frac{\theta}{2} = \sinh \frac{\alpha}{2} \sin \varphi$$

in (7.4.1), and then use the fourth entry in Table 4, p. 319 of the Bateman Manuscript Project, *Higher Transcendental Functions, Vol. 2.* To prove the second formula, use the sixth entry in the same table.

7.11. Asymptotic Representations of the Legendre Functions for Large $|\nu|$

The study of the asymptotic behavior of the Legendre functions as $|z| \to \infty$ for fixed ν is an elementary problem, whose solution is an immediate consequence of the various series representations of $P_\nu(z)$, $Q_\nu(z)$ given above. A less trivial problem, and one of great practical importance, is to find asymptotic representations of the Legendre functions as $|\nu| \to \infty$ for fixed z. In this section, it will be assumed that z is a real number greater than 1 and $|\arg \nu| \leqslant \frac{1}{2}\pi - \delta$ (see, however, the remark on p. 192). For asymptotic formulas valid under more general assumptions concerning z and ν, we refer the reader to the special literature on spherical harmonics.[26]

To derive an asymptotic representation of $P_\nu(z)$, we begin with the integral representation (7.4.1), which we write in the form

$$P_\nu(\cosh\alpha) = \frac{1}{\pi}\int_0^\alpha (2\cosh\alpha - 2\cosh\theta)^{-1/2} e^{(\nu+\frac{1}{2})\theta}\, d\theta$$
$$+ \frac{1}{\pi}\int_0^\alpha (2\cosh\alpha - 2\cosh\theta)^{-1/2} e^{-(\nu+\frac{1}{2})\theta}\, d\theta = \mathscr{I}_1 + \mathscr{I}_2. \tag{7.11.1}$$

Making the substitution $t = \alpha - \theta$ in the integral $\mathscr{I}_1$, we obtain

$$\mathscr{I}_1 = \frac{e^{(\nu+\frac{1}{2})\alpha}}{\pi(2\sinh\alpha)^{1/2}}\int_0^\alpha \frac{e^{-(\nu+\frac{1}{2})t}}{(\sinh t)^{1/2}}\left(1 - \tanh\frac{t}{2}\coth\alpha\right)^{-1/2} dt$$
$$= \frac{e^{(\nu+\frac{1}{2})\alpha}}{\pi(2\sinh\alpha)^{1/2}}\left\{\int_0^\infty \frac{e^{-(\nu+\frac{1}{2})t}}{(\sinh t)^{1/2}}\,dt + \int_0^\alpha \frac{e^{-(\nu+\frac{1}{2})t}}{(\sinh t)^{1/2}}\right.$$
$$\left.\times\left[\left(1 - \tanh\frac{t}{2}\coth\alpha\right)^{-1/2} - 1\right] dt - \int_\alpha^\infty \frac{e^{-(\nu+\frac{1}{2})t}}{(\sinh t)^{1/2}}\,dt\right\}$$
$$= \frac{e^{(\nu+\frac{1}{2})\alpha}}{\pi(2\sinh\alpha)^{1/2}}\,[\mathscr{I}_3 + \mathscr{I}_4 - \mathscr{I}_5]. \tag{7.11.2}$$

The integral $\mathscr{I}_3$ can be expressed in terms of the gamma function, and in fact

$$\mathscr{I}_3 = 2^{1/2}\int_0^\infty e^{-(\nu+1)t}(1 - e^{-2t})^{-1/2}\,dt = 2^{-1/2}B\left(\frac{1}{2}, \frac{\nu+1}{2}\right)$$
$$= \left(\frac{\pi}{2}\right)^{1/2}\frac{\Gamma\left(\dfrac{\nu+1}{2}\right)}{\Gamma\left(\dfrac{\nu}{2}+1\right)}$$

---

[26] E. W. Hobson, *op. cit.*, E. W. Barnes, *op. cit.*, and G. N. Watson, *Asymptotic expansions of hypergeometric functions*, Trans. Camb. Phil. Soc., **22**, 277 (1918). The last reference gives the most detailed treatment of the problem.

(see Sec. 1.5), which implies

$$\mathscr{I}_3 = \left(\frac{\pi}{\nu}\right)^{1/2} [1 + O(|\nu|^{-1}), \tag{7.11.3}$$

because of the asymptotic behavior of the gamma function for $|\nu| \to \infty$, $|\arg \nu| \leqslant \frac{1}{2}\pi - \delta$ (see Sec. 1.4).

To estimate the integral $\mathscr{I}_4$, we use the inequality

$$(1 - x)^{-1/2} - 1 \leqslant x(1 - a)^{-1/2}, \qquad 0 \leqslant x \leqslant a < 1,$$

which implies

$$\left(1 - \tanh\frac{t}{2}\coth\alpha\right)^{-1/2} - 1 \leqslant 2^{1/2}\cosh\frac{\alpha}{2}\tanh\frac{t}{2}\coth\alpha, \qquad 0 \leqslant t \leqslant \alpha.$$

From now on, we assume that

$$0 < \alpha_0 \leqslant \alpha \leqslant \alpha_1 < \infty.$$

It follows that

$$\begin{aligned} |\mathscr{I}_4| &\leqslant 2^{1/2}\cosh\frac{\alpha}{2}\coth\alpha \int_0^\alpha e^{-(|\nu|\sin\delta + \frac{1}{2})t}(\sinh t)^{-1/2}\tanh\frac{t}{2}\,dt \\ &\leqslant 2^{1/2}\cosh\frac{\alpha}{2}\coth\alpha_0 \int_0^\infty e^{-(|\nu|\sin\delta + \frac{1}{2})t}(\sinh t)^{-1/2}\tanh\frac{t}{2}\,dt \\ &= O(1)\int_0^\infty e^{-|\nu|t\sin\delta}\,t^{1/2}\,dt = O(|\nu|^{-3/2}, \end{aligned} \tag{7.11.4}$$

where we use (1.5.1). Finally we have

$$\begin{aligned} |\mathscr{I}_5| &\leqslant \int_\alpha^\infty e^{-(|\nu|\sin\delta + \frac{1}{2})t}(\sinh t)^{-1/2}\,dt \leqslant (\alpha\sinh\alpha)^{-1/2}\int_\alpha^\infty e^{-(|\nu|\sin\delta + \frac{1}{2})t}\,t^{1/2}\,dt \\ &\leqslant (\alpha_0\sinh\alpha_0)^{-1/2}\int_0^\infty e^{-|\nu|t\sin\delta}\,t^{1/2}\,dt = O(|\nu|^{-3/2}). \end{aligned} \tag{7.11.5}$$

It follows from (7.11.2–5) that

$$\mathscr{I}_1 = \frac{e^{(\nu + \frac{1}{2})\alpha}}{(2\nu\pi\sinh\alpha)^{1/2}}[1 + O(|\nu|^{-1})]. \tag{7.11.6}$$

To estimate $\mathscr{I}_2$ is an easier matter. We see at once that

$$\begin{aligned} |\mathscr{I}_2| &\leqslant \frac{1}{\pi}\int_0^\alpha (2\cosh\alpha - 2\cosh\theta)^{-1/2}\,d\theta \\ &< \frac{1}{\pi}\int_0^\alpha (2\cosh\alpha - 2\cosh\theta)^{-1/2}\cosh\frac{\theta}{2}\,d\theta = \frac{1}{2}, \end{aligned}$$

and hence

$$\mathscr{I}_2 = O(1). \tag{7.11.7}$$

Combining (7.11.6–7), we obtain the desired asymptotic representation of the Legendre function of the first kind:

$$P_\nu(\cosh\alpha) = \frac{e^{(\nu+\frac{1}{2})\alpha}}{(2\nu\pi\sinh\alpha)^{1/2}}\,[1 + O(|\nu|^{-1})], \tag{7.11.8}$$
$$|\nu| \to \infty, \quad |\arg\nu| \leqslant \frac{\pi}{2} - \delta, \quad 0 < \alpha_0 \leqslant \alpha \leqslant \alpha_1 < \infty.$$

To derive an asymptotic representation of $Q_\nu(z)$, under the same assumptions, we begin with the integral representation (7.4.8), making the substitution $\theta = \alpha + t$:

$$\begin{aligned} Q_\nu(\cosh\alpha) &= \frac{e^{-(\nu+\frac{1}{2})\alpha}}{(2\sinh\alpha)^{1/2}} \int_0^\infty \frac{e^{-(\nu+\frac{1}{2})t}}{(\sinh t)^{1/2}} \left(1 + \coth\alpha\tanh\frac{t}{2}\right)^{-1/2} dt \\ &= \frac{e^{-(\nu+\frac{1}{2})\alpha}}{(2\sinh\alpha)^{1/2}} \left\{ \int_0^\infty \frac{e^{-(\nu+\frac{1}{2})t}}{(\sinh t)^{1/2}}\, dt \right. \\ &\qquad \left. - \int_0^\infty \frac{e^{-(\nu+\frac{1}{2})t}}{(\sinh t)^{1/2}} \left[1 - \left(1 + \coth\alpha\tanh\frac{t}{2}\right)^{-1/2}\right] dt \right\} \\ &= \frac{e^{-(\nu+\frac{1}{2})\alpha}}{(2\sinh\alpha)^{1/2}}\,[\mathscr{I}_3 + \mathscr{I}_6]. \end{aligned} \tag{7.11.9}$$

The integral $\mathscr{I}_3$ has already been estimated in (7.11.3). To estimate the integral $\mathscr{I}_6$, we use the inequality

$$1 - (1 + x)^{-1/2} \leqslant \tfrac{1}{2}x, \qquad x \geqslant 0,$$

which implies

$$1 - (1 + \coth\alpha\tanh t)^{-1/2} \leqslant \tfrac{1}{2}\coth\alpha\tanh t, \qquad t \geqslant 0.$$

Therefore

$$|\mathscr{I}_6| \leqslant O(1) \int_0^\infty e^{-|\nu| t \sin\delta}\, t^{1/2}\, dt = O(|\nu|^{-3/2}), \tag{7.11.10}$$

provided that $\alpha \geqslant \alpha_0 > 0$. Combining these results, we obtain the desired asymptotic representation of the Legendre function of the second kind:

$$Q_\nu(\cosh\alpha) = \frac{\pi^{1/2}}{(2\nu\sinh\alpha)^{1/2}}\, e^{-(\nu+\frac{1}{2})\alpha}[1 + O(|\nu|^{-1})], \tag{7.11.11}$$
$$|\nu| \to \infty, \quad |\arg\nu| \leqslant \frac{\pi}{2} - \delta, \quad 0 < \alpha_0 \leqslant \alpha < \infty.$$

Remark. By similar methods, one can derive asymptotic representations of $P_\nu(z)$ and $Q_\nu(z)$ for the case where z belongs to the interval $(-1, 1)$ and $\arg \nu = 0$. It is found that[27]

$$P_\nu(\cos\theta) = \left(\frac{2}{\nu\pi \sin\theta}\right)^{1/2} \sin\left[(\nu + \tfrac{1}{2})\theta + \tfrac{1}{4}\pi\right]\cdot\left[1 + O(|\nu|^{-1})\right],$$

$$Q_\nu(\cos\theta) = \left(\frac{\pi}{2\nu \sin\theta}\right)^{1/2} \cos\left[(\nu + \tfrac{1}{2})\theta + \tfrac{1}{4}\pi\right]\cdot\left[1 + O(|\nu|^{-1})\right],$$

$$\nu \to \infty, \quad \delta \leqslant \theta \leqslant \pi - \delta. \quad (7.11.12)$$

7.12. Associated Legendre Functions

The next class of spherical harmonics, in order of increasing complexity, consists of the *associated Legendre functions*, which are solutions of the differential equation

$$(1 - z^2)u'' - 2zu' + \left[\nu(\nu + 1) - \frac{m^2}{1 - z^2}\right] u = 0, \quad (7.12.1)$$

for arbitrary ν and integral $m = 0, 1, 2, \ldots$. These functions generalize the functions $P_\nu(z)$ and $Q_\nu(z)$ considered in Secs. 7.3–11, and reduce to these functions for $m = 0$.

To define the associated Legendre functions, we assume that z is an arbitrary complex number belonging to the plane cut along $[-\infty, 1]$, and we introduce a new function v related to u by the formula

$$u = (z^2 - 1)^{m/2} v = (z - 1)^{m/2} (z + 1)^{m/2} v.$$

Then equation (7.12.1) takes the form

$$(1 - z^2)v'' - 2(m + 1)zv' + (\nu - m)(\nu + m + 1)v = 0. \quad (7.12.2)$$

Let w be a solution of Legendre's equation

$$(1 - z^2)w'' - 2zw' + \nu(\nu + 1)w = 0. \quad (7.12.3)$$

Then it is easily verified that the function $v = w^{(m)}$ satisfies equation (7.12.3).[28] It follows that the solutions of (7.12.1) are given by

$$P_\nu^m(z) = (z^2 - 1)^{m/2} \frac{d^m P_\nu(z)}{dz^m},$$

$$Q_\nu^m(z) = (z^2 - 1)^{m/2} \frac{d^m Q_\nu(z)}{dz^m}, \quad m = 0, 1, 2, \ldots, \quad (7.12.4)$$

where $P_\nu(z)$ and $Q_\nu(z)$ are the Legendre functions defined earlier. The functions

---

[27] See J. Lense, *Kugelfunktionen*, second edition, Akademische Verlagsgesellschaft, Geest & Portig K.-G., Leipzig (1954), p. 168 ff., and E. W. Hobson, *op. cit.*, p. 293 ff.

[28] Use Leibniz's rule (D. V. Widder, *op. cit.*, p. 483) to calculate the derivatives $(z^2v'')^{(m)}$ and $(zv')^{(m)}$.

$P_\nu^m(z)$ and $Q_\nu^m(z)$ are called the *associated Legendre functions of the first and second kinds, respectively*. It follows from (7.12.4) and the results of Sec. 7.3 that $P_\nu^m(z)$ and $Q_\nu^m(z)$ are entire functions of z in the plane cut along $[-\infty, 1]$. Moreover, $P_\nu^m(z)$ is an entire function of ν, while $Q_\nu^m(z)$ is a meromorphic function of ν, with poles at the points $\nu = -1, -2, \ldots$

In the applications, it is often necessary to find the solution of equation (7.12.1) for real $z = x$ belonging to the interval $(-1, 1)$. To this end, we first note that values of the associated Legendre functions on the upper and lower edges of the cut are

$$P_\nu^m(x \pm i0) = e^{\pm(m\pi i/2)}(1 - x^2)^{m/2} \frac{d^m P_\nu(x)}{dx^m},$$

$$Q_\nu^m(x \pm i0) = e^{\pm(m\pi i/2)}(1 - x^2)^{m/2} \frac{d^m Q_\nu(x \pm i0)}{dx^m}.$$

Then we introduce two new functions $\mathrm{P}_\nu^m(x)$ and $\mathrm{Q}_\nu^m(x)$ by writing

$$\begin{aligned} \mathrm{P}_\nu^m(x) &= e^{m\pi i/2} P_\nu^m(x + i0) = e^{-m\pi i/2} P_\nu^m(x - i0) \\ &= (-1)^m (1 - x^2)^{m/2} \frac{d^m P_\nu(x)}{dx^m}, \\ \mathrm{Q}_\nu^m(x) &= \frac{(-1)^m}{2} [e^{-m\pi i/2} Q_\nu^m(x + i0) + e^{m\pi i/2} Q_\nu^m(x - i0)] \\ &= (-1)^m (1 - x^2)^{m/2} \frac{d^m \mathrm{Q}_\nu(x)}{dx^m}, \end{aligned} \tag{7.12.5}$$

where $-1 < x < 1$, ν is arbitrary [except that $\nu \neq -1, -2, \ldots$ in the case of $\mathrm{Q}_\nu^m(z)$], $m = 0, 1, 2, \ldots$, and $\mathrm{Q}_\nu(x)$ is the function defined by (7.3.28). The functions $\mathrm{P}_\nu^m(x)$ and $\mathrm{Q}_\nu^m(x)$, which are easily seen to satisfy equation (7.12.1) for real $z = x$ $(-1 < x < 1)$, will simply be called the *associated Legendre functions for the interval* $(-1, 1)$.[29]

In the special case where $\nu = n$ is a nonnegative integer $(n = 0, 1, 2, \ldots)$, $P_\nu(z) = P_n(z)$, where $P_n(z)$ is the Legendre polynomial of degree n. Then, according to (4.2.1), we have

$$P_n^m(z) = (z^2 - 1)^{m/2} \frac{1}{2^n n!} \frac{d^{m+n}}{dz^{m+n}} (z^2 - 1)^n, \qquad m = 0, 1, 2, \ldots, \quad n = 0, 1, 2, \ldots, \tag{7.12.6}$$

and obviously $P_n^m(z) \equiv 0$ if $m > n$. If $m \leqslant n$, the function $P_n^m(z)$ is the product

---

[29] Some authors define $\mathrm{P}_\nu^m(x)$ and $\mathrm{Q}_\nu^m(x)$, $-1 < x < 1$ by the formulas

$$\mathrm{P}_\nu^m(x) = (1 - x^2)^{m/2} \frac{d^m P_\nu(x)}{dx^m}, \qquad \mathrm{Q}_\nu^m(x) = (1 - x^2)^{m/2} \frac{d^m \mathrm{Q}_\nu(x)}{dx^m},$$

differing from (7.12.5) by the constant factor $(-1)^m$, a fact which should be kept in mind when consulting handbooks and tables involving these functions.

of $(z^2 - 1)^{m/2}$ and a polynomial of degree $n - m$. In the interval $-1 < x < 1$, the analogue of formula (7.12.6) is

$$P_n^m(x) = (-1)^m(1 - x^2)^{m/2} \frac{1}{2^n n!} \frac{d^{m+n}}{dx^{m+n}} (x^2 - 1)^n. \qquad (7.12.7)$$

If we set $v = (d/dz)^m P_\nu(z)$ in (7.12.2) and multiply the result by $(z^2 - 1)^{m/2}$, we obtain the recurrence relation

$$P_\nu^{m+2}(z) + \frac{2(m+1)z}{(z^2-1)^{1/2}} P_\nu^{m+1}(z) - (\nu - m)(\nu + m + 1) P_\nu^m(z) = 0,$$
$$m = 0, 1, 2, \ldots, \qquad (7.12.8)$$

which can be used to calculate the function $P_\nu^m(z)$ step by step, starting from

$$P_\nu^0(z) = P_\nu(z),$$

$$P_\nu^1(z) = (z^2 - 1)^{1/2} P_\nu'(z) = \frac{-\nu}{(z^2-1)^{1/2}} P_{\nu-1}(z) + \frac{\nu z}{(z^2-1)^{1/2}} P_\nu(z).$$

In just the same way, we find that

$$Q_\nu^{m+2}(z) + \frac{2(m+1)z}{(z^2-1)^{1/2}} Q_\nu^{m+1}(z) - (\nu - m)(\nu + m + 1) Q_\nu^m(z) = 0,$$
$$m = 0, 1, 2, \ldots \qquad (7.12.9)$$

Similarly, using the definitions (7.12.5), we can easily deduce recurrence relations for the functions $P_\nu^m(x)$ and $Q_\nu^m(x)$, obtaining

$$P_\nu^{m+2}(x) + \frac{2(m+1)x}{(1-x^2)^{1/2}} P_\nu^{m+1}(x) + (\nu - m)(\nu + m + 1) P_\nu^m(x) = 0,$$
$$Q_\nu^{m+2}(x) + \frac{2(m+1)x}{(1-x^2)^{1/2}} Q_\nu^{m+1}(x) + (\nu - m)(\nu + m + 1) Q_\nu^m(x) = 0, \qquad (7.12.10)$$

where $-1 < x < 1$, ν is arbitrary [except that $\nu \neq -1, -2, \ldots$ in the case of $Q_\nu^m(x)$] and $m = 0, 1, 2, \ldots$

The associated Legendre functions also satisfy recurrence relations of another type, involving functions with the same superscript m but different subscripts ν. To derive these formulas, which generalize the corresponding formulas of Sec. 7.8, we first differentiate (7.8.2) m times with respect to z and use (7.12.4), obtaining

$$P_{\nu+1}^{m+1}(z) - P_{\nu-1}^{m+1}(z) = (z^2 - 1)^{1/2}(2\nu + 1) P_\nu^m(z). \qquad (7.12.11)$$

Then, differentiating (7.8.1) m times with respect to z and again using (7.12.4), we find that

$$(\nu+1)P_{\nu+1}^m(z) - (2\nu+1)zP_\nu^m(z) - (2\nu+1)m(z^2-1)^{1/2}P_\nu^{m-1}(z) + \nu P_{\nu-1}^m(z) = 0,$$

which together with (7.12.11) implies

$$(\nu - m + 1)P_{\nu+1}^m(z) - (2\nu + 1)zP_\nu^m(z) + (\nu + m)P_{\nu-1}^m(z) = 0,$$
$$m = 0, 1, 2, \ldots \qquad (7.12.12)$$

This recurrence relation is the first of the type mentioned, and reduces to (7.8.1) for $m = 0$. To obtain two other such recurrence relations, we differentiate (7.8.2) and (7.8.3) m times with respect to z and replace $(d/dz)^m P_\nu(z)$ by $(z^2 - 1)^{-m/2} P_\nu^m(z)$, obtaining

$$\frac{dP_{\nu+1}^m(z)}{dz} - \frac{dP_{\nu-1}^m(z)}{dz} - \frac{mz}{z^2 - 1}[P_{\nu+1}^m(z) - P_{\nu-1}^m(z)] = (2\nu + 1)P_\nu^m(z), \tag{7.12.13}$$

$$\frac{dP_{\nu+1}^m(z)}{dz} - z\frac{dP_\nu^m(z)}{dz} + \frac{mz}{z^2 - 1}[zP_\nu^m(z) - P_{\nu+1}^m(z)] = (\nu + m + 1)P_\nu^m(z), \tag{7.12.14}$$

where $m = 0, 1, 2, \ldots$ Subtraction of (7.12.14) from (7.12.13) then gives

$$z\frac{dP_\nu^m(z)}{dz} - \frac{dP_{\nu-1}^m(z)}{dz} - \frac{mz}{z^2 - 1}[zP_\nu^m(z) - P_{\nu-1}^m(z)] = (\nu - m)P_\nu^m(z). \tag{7.12.15}$$

For $m = 0$, formulas (7.12.13–15) reduce to formulas (7.8.2–4), respectively. Finally, replacing ν by $\nu - 1$ in (7.12.14) and using (7.12.15) to eliminate $(d/dz)P_{\nu-1}^m(z)$, we obtain the following generalization of formula (7.8.5):[30]

$$(z^2 - 1)\frac{dP_\nu^m(z)}{dz} = \nu z P_\nu^m(z) - (\nu + m)P_{\nu-1}^m(z), \qquad m = 0, 1, 2, \ldots \tag{7.12.16}$$

Recurrence relations for the functions $Q_\nu^m(z)$ can be derived in exactly the same way, starting from formulas (7.8.6–10), and obviously must be identical with the corresponding recurrence relations for the functions $P_\nu^m(z)$. In the case of the associated Legendre functions for the interval $(-1, 1)$, recurrence relations can be derived by using (7.12.5). For example, we have

$$(\nu - m + 1)P_{\nu+1}^m(x) - (2\nu + 1)xP_\nu^m(x) + (\nu + m)P_{\nu-1}^m(x) = 0,$$

$$(x^2 - 1)\frac{dP_\nu^m(x)}{dx} = \nu x P_\nu^m(x) - (\nu + m)P_{\nu-1}^m(x) = 0, \qquad m = 0, 1, 2, \ldots,$$

and so on.

A closely related result is the formula giving the Wronskian of the pair of solutions $P_\nu^m(z)$, $Q_\nu^m(z)$ of equation (7.12.1). To derive this formula, we first differentiate each of the equations (7.12.4) with respect to z, and then use (7.12.4) again to eliminate the derivatives. This gives

$$\begin{aligned} \frac{dP_\nu^m(z)}{dz} &= \frac{1}{z^2 - 1}[(z^2 - 1)^{1/2}P_\nu^{m+1}(z) + mzP_\nu^m(z)], \\ \frac{dQ_\nu^m(z)}{dz} &= \frac{1}{z^2 - 1}[(z^2 - 1)^{1/2}Q_\nu^{m+1}(z) + mzQ_\nu^m(z)]. \end{aligned} \tag{7.12.17}$$

[30] In Hobson's treatise (*op. cit.*, p. 290), this formula is given incorrectly.

Substituting (7.12.17) into the expression for the Wronskian, we obtain

$$W\{P_\nu^m(z), Q_\nu^m(z)\} = \frac{1}{(z^2 - 1)^{1/2}} [Q_\nu^{m+1}(z)P_\nu^m(z) - P_\nu^{m+1}(z)Q_\nu^m(z)].$$

Next we observe that (7.12.8) and (7.12.9) imply the identity

$$Q_\nu^{m+1}(z)P_\nu^m(z) - P_\nu^{m+1}(z)Q_\nu^m(z)$$
$$= (\nu + m)(m - \nu - 1)[Q_\nu^m(z)P_\nu^{m-1}(z) - P_\nu^m(z)Q_\nu^{m-1}(z)],$$

and therefore the Wronskian becomes

$$W\{P_\nu^m(z), Q_\nu^m(z)\} = (\nu + m)(m - \nu - 1)W\{P_\nu^{m-1}(z), Q_\nu^{m-1}(z)\},$$
$$m = 1, 2, \ldots$$

Repeatedly applying this formula and using (7.7.2), we find that

$$W\{P_\nu^m(z), Q_\nu^m(z)\} = \frac{\Gamma(\nu + m + 1)}{\Gamma(\nu + 1)} \frac{\Gamma(m - \nu)}{\Gamma(-\nu)} \frac{1}{1 - z^2},$$

or, after taking account of (1.2.2),

$$W\{P_\nu^m(z), Q_\nu^m(z)\} = \frac{\Gamma(\nu + m + 1)}{\Gamma(\nu - m + 1)} \frac{(-1)^m}{1 - z^2}, \qquad (7.12.18)$$

where

$$|\arg (z - 1)| < \pi, \quad \nu \neq -1, -2, \ldots, \qquad m = 0, 1, 2, \ldots$$

This result generalizes (7.7.2) and shows that $P_\nu^m(z)$, $Q_\nu^m(z)$ are a pair of linearly independent solutions of equation (7.12.1), except when $\nu = 0, 1, \ldots, m - 1$, in which case both sides of (7.12.18) vanish identically. Thus, apart from this degenerate case, the general solution of (7.12.1) can be written in the form

$$u = AP_\nu^m(z) + BQ_\nu^m(z). \qquad (7.12.19)$$

It follows from (7.12.18) and the definition (7.12.5) of the associated Legendre functions for the interval $(-1, 1)$ that

$$W\{P_\nu^m(x), Q_\nu^m(x)\} = \frac{(-1)^m}{2} W\{P_\nu^m(x + i0), Q_\nu^m(x + i0)\}$$
$$+ W\{P_\nu^m(x - i0), Q_\nu^m(x - i0)\}$$
$$= \frac{\Gamma(\nu + m + 1)}{\Gamma(\nu - m + 1)} \frac{1}{1 - x^2}. \qquad (7.12.20)$$

We also observe that the differential equation (7.12.1) does not change if we replace ν by $-\nu - 1$ or z by $-z$, and hence it has solutions $P_{-\nu-1}^m(z)$, $Q_{-\nu-1}^m(z)$, $P_\nu^m(-z)$ and $Q_\nu^m(-z)$, as well as $P_\nu^m(z)$ and $Q_\nu^m(z)$. Since every three solutions of a second-order linear differential equation are linearly dependent, there must be certain functional relations between the solutions just enumerated. These relations can be obtained directly by differentiating each of the relations

(7.5.1–2, 4–5) m times with respect to z, and then using the definitions (7.12.4). This gives

$$P^m_{-\nu-1}(z) = P^m_\nu(z), \tag{7.12.21}$$

$$\sin \nu\pi[Q^m_\nu(z) - Q^m_{-\nu-1}(z)] = \pi \cos \nu\pi \, P^m_\nu(z), \tag{7.12.22}$$

$$Q^m_\nu(-z) = -e^{\pm \nu\pi i} Q^m_\nu(z), \tag{7.12.23}$$

$$P^m_\nu(z)e^{\mp \nu\pi i} - P^m_\nu(-z) = \frac{2}{\pi} \sin \nu\pi \, Q^m_\nu(z), \tag{7.12.24}$$

where $m = 0, 1, 2, \ldots$, and the upper sign is chosen if $\operatorname{Im} z > 0$ and the lower sign if $\operatorname{Im} z < 0$.

The associated Legendre functions can be represented by hypergeometric series in suitably restricted regions of the z-plane cut along $[-\infty, 1]$. The problem of deriving all expansions of this type lies behind the scope of this book. At this point we consider only the simplest examples, referring the reader interested in a more detailed treatment to the sources cited in footnote 20, p. 176.

An expansion of $P^m_\nu(z)$ valid in the domain $|z - 1| < 2$, $|\arg (z - 1)| < \pi$ can be obtained by m-fold differentiation of the series (7.3.6). First we note that

$$\begin{aligned}\frac{d}{dx} F(\alpha, \beta; \gamma; x) &= \sum_{k=1}^{\infty} \frac{(\alpha)_k(\beta)_k}{(\gamma)_k \, k!} kx^{k-1} = \sum_{k=0}^{\infty} \frac{(\alpha)_{k+1}(\beta)_{k+1}}{(\gamma)_{k+1}k!} x^k \\ &= \frac{\alpha\beta}{\gamma} \sum_{k=0}^{\infty} \frac{(\alpha + 1)_k(\beta + 1)_k}{(\gamma + 1)_k \, k!} x^k \\ &= \frac{\alpha\beta}{\gamma} F(\alpha + 1, \beta + 1; \gamma + 1; x)\end{aligned} \tag{7.12.25}$$

for $|x| < 1$, since $(\lambda)_{k+1} = \lambda(\lambda + 1)_k$ by definition. Repeated application of this formula gives

$$\frac{d^m}{dx^m} F(\alpha, \beta; \gamma; x) = \frac{(\alpha)_m(\beta)_m}{(\gamma)_m} F(\alpha + m, \beta + m; \gamma + m; x), \qquad m = 0, 1, 2, \ldots \tag{7.12.26}$$

It follows that

$$\begin{aligned}P^m_\nu(z) &= (z^2 - 1)^{m/2} \frac{d^m}{dz^m} F\left(-\nu, \nu + 1; 1; \frac{1 - z}{2}\right) \\ &= \frac{(z^2 - 1)^{m/2}(-1)^m}{2^m} \frac{(-\nu)_m(\nu + 1)_m}{(1)_m} F\left(m - \nu, \nu + m + 1; m + 1; \frac{1 - z}{2}\right).\end{aligned}$$

Moreover, according to (1.2.2),

$$(\nu + 1)_m = \frac{\Gamma(\nu + m + 1)}{\Gamma(\nu + 1)},$$

$$(-\nu)_m = \frac{\Gamma(m - \nu)}{\Gamma(-\nu)} = (-1)^m \frac{\Gamma(\nu + 1)}{\Gamma(\nu - m + 1)},$$

and hence

$$P_\nu^m(z) = \frac{\Gamma(\nu + m + 1)}{2^m\Gamma(m + 1)\Gamma(\nu - m + 1)}(z^2 - 1)^{m/2} \times F\left(m - \nu, \nu + m + 1; m + 1; \frac{1 - z}{2}\right), \tag{7.12.27}$$

where $|z - 1| < 2$, $|\arg(z - 1)| < \pi$, ν is arbitrary, and $m = 0, 1, 2, \ldots$ This expansion generalizes formula (7.3.6), to which it reduces for $m = 0$. To obtain the corresponding formula for the interval $-1 < x < 1$, we use (7.12.5) and (7.12.27), obtaining

$$P_\nu^m(x) = \frac{(-1)^m\Gamma(\nu + m + 1)}{2^m\Gamma(m + 1)\Gamma(\nu - m + 1)}(1 - x^2)^{m/2} \times F\left(m - \nu, \nu + m + 1; m + 1; \frac{1 - x}{2}\right). \tag{7.12.28}$$

Next we derive the formula generalizing the basic expansion (7.3.7) of the function $Q_\nu(z)$. Using the duplication formula (1.2.3), we write (7.3.7) in the form

$$Q_\nu(z) = \frac{1}{2}\sum_{k=0}^{\infty} \frac{\Gamma\left(k + \frac{\nu + 2}{2}\right)\Gamma\left(k + \frac{\nu + 1}{2}\right)}{k!\Gamma(k + \nu + \frac{3}{2})} z^{-(2k+\nu+1)}, \qquad |z| < 1,$$

and then differentiate this series m times with respect to z. According to (1.2.1, 3), we have

$$\begin{aligned}\frac{d^m}{dz^m} z^{-(2k+\nu+1)} &= (-1)^m(2k + \nu + 1)(2k + \nu + 2)\cdots(2k + \nu + m)z^{-(2k+\nu+m+1)} \\ &= (-1)^m \frac{\Gamma(2k + \nu + m + 1)}{\Gamma(2k + \nu + 1)} z^{-(2k+\nu+m+1)} \\ &= (-1)^m \frac{2^m\Gamma\left(k + \frac{\nu + m + 2}{2}\right)\Gamma\left(k + \frac{\nu + m + 1}{2}\right)}{\Gamma\left(k + \frac{\nu + 2}{2}\right)\Gamma\left(k + \frac{\nu + 1}{2}\right)} z^{-(2k+\nu+m+1)},\end{aligned}$$

and therefore

$$\begin{aligned}\frac{d^m Q_\nu(z)}{dz^m} &= (-1)^m 2^{m-1} z^{-(\nu+m+1)} \times \sum_{k=0}^{\infty} \frac{\Gamma\left(k + \frac{\nu + m + 2}{2}\right)\Gamma\left(k + \frac{\nu + m + 1}{2}\right)}{k!\Gamma(k + \nu + \frac{3}{2})} \frac{1}{z^{2k}} \\ &= (-1)^m \frac{\sqrt{\pi}\Gamma(\nu + m + 1)}{2^{\nu+1}\Gamma(\nu + \frac{3}{2})} z^{-(\nu+m+1)} \times F\left(\frac{\nu + m + 2}{2}, \frac{\nu + m + 1}{2}; \nu + \frac{3}{2}; \frac{1}{z^2}\right),\end{aligned}$$

which implies

$$Q_\nu^m(z) = \frac{(-1)^m\sqrt{\pi}\,\Gamma(\nu + m + 1)}{2^{\nu+1}\Gamma(\nu + \frac{3}{2})z^{\nu+m+1}}(z^2 - 1)^{m/2} \times F\left(\frac{\nu + m + 2}{2}, \frac{\nu + m + 1}{2}; \nu + \frac{3}{2}; \frac{1}{z^2}\right), \tag{7.12.29}$$

where

$$|z| > 1, \quad |\arg(z - 1)| < \pi, \quad m = 0, 1, 2, \ldots, \quad \nu \neq -1, -2, \ldots.$$

We conclude this outline of the theory of the associated Legendre functions by citing the following integral representations which generalize the corresponding formulas of Sec. 7.4:[31]

$$P_\nu^m(z) = \frac{\Gamma(\nu + m + 1)(z^2 - 1)^{m/2}}{2^m\sqrt{\pi}\,\Gamma(m + \frac{1}{2})\Gamma(\nu - m + 1)} \times \int_0^\pi (z + \sqrt{z^2 - 1}\cos\psi)^{\nu - m}\sin^{2m}\psi\,d\psi,$$

$$\operatorname{Re} z > 0, \quad m = 0, 1, 2, \ldots, \tag{7.12.30}$$

$$P_\nu^m(z) = \frac{\Gamma(\nu + m + 1)}{\pi\Gamma(\nu + 1)}\int_0^\pi (z + \sqrt{z^2 - 1}\cos\psi)^\nu \cos m\psi\,d\psi,$$

$$\operatorname{Re} z > 0, \quad m = 0, 1, 2, \ldots, \tag{7.12.31}$$

$$P_\nu^m(\cos\beta) = \frac{(-1)^m 2\Gamma(\nu + m + 1)}{\sqrt{\pi}\,\Gamma(m + \frac{1}{2})\Gamma(\nu - m + 1)}\frac{1}{(2\sin\beta)^m} \times \int_0^\beta \frac{\cos(\nu + \frac{1}{2})\theta}{(2\cos\theta - 2\cos\beta)^{\frac{1}{2} - m}}\,d\theta,$$

$$0 < \beta < \pi, \quad m = 0, 1, 2, \ldots \tag{7.12.32}$$

PROBLEMS

1. Prove the formulas

$$P_\nu(-x + i0) - P_\nu(-x - i0) = 2i\sin\nu\pi\, P_\nu(x),$$

$$Q_\nu(-x + i0) - Q_\nu(-x - i0) = 2i\sin\nu\pi\, Q_\nu(x),$$

where $x > 1$.

---

[31] The parameter ν is arbitrary in (7.12.30–32). For these and many other integral representations, with suggestions as to proofs, see the Bateman Manuscript Project, *Higher Transcendental Functions, Vol. 1*, p. 155 ff.

2. Derive the following representations of the Legendre function $P_\nu(z)$ in terms of hypergeometric series:

$$P_\nu(z) = F\left(\frac{\nu+1}{2}, -\frac{\nu}{2}; 1; 1 - z^2\right), \qquad |1 - z^2| < 1, \quad |\arg(z+1)| < \pi,$$

$$P_\nu(z) = \left(\frac{z+1}{2}\right)^\nu F\left(-\nu, -\nu; 1; \frac{z-1}{z+1}\right), \qquad \operatorname{Re} z > 0.$$

Hint. Apply the method used to derive (7.6.9).

3. Derive the following formulas:

$$P_\nu(z) = (z + \sqrt{z^2-1})^\nu F\left(-\nu, \frac{1}{2}; 1; \frac{2\sqrt{z^2-1}}{z + \sqrt{z^2-1}}\right),$$

$$\left|\frac{2(\sqrt{z^2-1})}{z+\sqrt{z^2-1}}\right| < 1, \quad |\arg(z-1)| < \pi,$$

$$P_\nu(z) = (z - \sqrt{z^2-1})^\nu F\left(-\nu, \frac{1}{2}; 1; -\frac{2\sqrt{z^2-1}}{z - \sqrt{z^2-1}}\right),$$

$$\left|\frac{2\sqrt{z^2-1}}{z-\sqrt{z^2-1}}\right| < 1, \quad |\arg(z-1)| < \pi,$$

$$P_\nu(z) = z^\nu F\left(-\frac{\nu}{2}, \frac{1-\nu}{2}; 1; 1 - \frac{1}{z^2}\right), \qquad \operatorname{Re} z^2 > \frac{1}{2}, \quad |\arg z| < \pi.$$

Hint. Expand the integrand of (7.4.3) in series of powers of $\sin^2(\psi/2)$, $\cos^2(\psi/2)$ and $\cos\psi$, and then integrate term by term.

4. Derive the following formulas

$$Q_\nu(z) = \frac{\sqrt{\pi}\,\Gamma(\nu+1)}{2^{\nu+1}\Gamma(\nu+\frac{3}{2})}(z-1)^{-\nu-1} F\left(1+\nu, 1+\nu; 2+2\nu; \frac{2}{1-z}\right),$$

$$|z-1| > 2, \quad |\arg(z-1)| < \pi, \quad \nu \neq -1, -2, \ldots,$$

$$Q_\nu(z) = \frac{\sqrt{\pi}\,\Gamma(\nu+1)}{2^{\nu+1}\Gamma(\nu+\frac{3}{2})}(z+1)^{-\nu-1} F\left(1+\nu, 1+\nu; 2+2\nu; \frac{2}{1+z}\right),$$

$$|z+1| > 2, \quad |\arg(z+1)| < \pi, \quad \nu \neq -1, -2, \ldots,$$

$$Q_\nu(z) = \frac{\sqrt{\pi}\,\Gamma(\nu+1)}{2^{\nu+1}\Gamma(\nu+\frac{3}{2})}(z^2-1)^{-\frac{1}{2}(\nu+1)} F\left(\frac{\nu+1}{2}, \frac{\nu+1}{2}; \nu+\frac{3}{2}; \frac{1}{1-z^2}\right),$$

$$|z^2-1| > 1, \quad |\arg(z-1)| < \pi, \quad \nu \neq -1, -2, \ldots$$

Hint. Apply the method used to derive (7.6.9).

5. Prove the formula

$$Q_\nu(z) = \sqrt{\frac{\pi}{2}}\,\frac{\Gamma(\nu+1)}{\Gamma(\nu+\frac{3}{2})}(z^2-1)^{-1/4}(z - \sqrt{z^2-1})^{\nu+\frac{1}{2}}$$

$$\times F\left(\frac{1}{2}, \frac{1}{2}; \nu+\frac{3}{2}; -\frac{z-\sqrt{z^2-1}}{2\sqrt{z^2-1}}\right),$$

$$\left|\frac{z-\sqrt{z^2-1}}{2\sqrt{z^2-1}}\right| < 1, \quad |\arg(z-1)| < \pi, \quad \nu \neq -1, -2, \ldots$$

Hint. Introduce the new variable of integration $t = \theta - \alpha$ in (7.4.8), and then expand in powers of $1 - e^{-t}$.

6. Prove that if ν is not an integer, then the asymptotic behavior as $z \to -1$ of the Legendre function of the first kind and its derivative is described by the formulas[32]

$$P_\nu(z) \approx \frac{\sin \nu\pi}{\pi} \log \frac{z+1}{2}, \qquad P'_\nu(z) \approx \frac{\sin \nu\pi}{\pi} \frac{1}{1+z}, \qquad z \to -1.$$

7. Using the result of the preceding problem and the functional relations connecting the Legendre functions of the first and second kinds, show that for any ν, the function $Q_\nu(z)$, $|\arg (z-1)| < \pi$ has a logarithmic singularity at $z = 1$, while the function $Q_\nu(x)$, $-1 < x < 1$ has logarithmic singularities at both end points of the interval $(-1, 1)$.

8. Derive the integral representations

$$P_\nu (\cosh \alpha) = \frac{1}{\Gamma(\nu+1)} \int_0^\infty e^{-t \cosh \alpha} I_0(t \sinh \alpha) t^\nu \, dt,$$

$$Q_\nu (\cosh \alpha) = \frac{1}{\Gamma(\nu+1)} \int_0^\infty e^{-t \cosh \alpha} K_0(t \sinh \alpha) t^\nu \, dt,$$

$$P_\nu(\cos \theta) = \frac{1}{\Gamma(\nu+1)} \int_0^\infty e^{-t \cos \theta} J_0(t \sin \theta) t^\nu \, dt,$$

where

$$|\mathrm{Im}\, \alpha| \leqslant \frac{\pi}{2}, \quad \mathrm{Re}\, \nu > -1, \qquad 0 \leqslant \theta < \pi,$$

and $J_0(x)$, $I_0(x)$ and $K_0(x)$ are Bessel functions.

9. Derive the integral representations

$$P_{\nu - \frac{1}{2}} (\cosh \alpha) = \frac{\cos \nu\pi}{\pi} \sqrt{\frac{2}{\pi}} \int_0^\infty e^{-t \cosh \alpha} \frac{K_\nu(t)}{\sqrt{t}} \, dt, \qquad |\mathrm{Re}\, \nu| < \tfrac{1}{2}, \quad \alpha \geqslant 0,$$

$$Q_{\nu - \frac{1}{2}} (\cosh \alpha) = \sqrt{\frac{\pi}{2}} \int_0^\infty e^{-t \cosh \alpha} \frac{I_\nu(t)}{\sqrt{t}} \, dt, \qquad \mathrm{Re}\, \nu > -\tfrac{1}{2}, \quad \alpha > 0,$$

where $I_\nu(t)$ and $K_\nu(t)$ are Bessel functions of imaginary argument (see Sec. 5.7).[33]

10. Prove the formulas

$$\int_{-1}^1 P_l^m(x) P_n^m(x) \, dx = 0, \qquad \int_{-1}^1 [P_n^m(x)]^2 \, dx = \frac{2}{2n+1} \frac{(n+m)!}{(n-m)!},$$

$$m = 0, 1, 2, \ldots, \quad l = m, m+1, \ldots, \quad n = m, m+1, \ldots,$$

generalizing the results of Sec. 4.5.

---

[32] A possible approach is to use the expansion of $P_\nu(z)$ given by E. W. Hobson, *op. cit.*, p. 225.

[33] Proof of the formulas given in Problems 8–9 can be found in Watson's treatise (*op. cit.*, p. 387).

Comment. These formulas play an important role in the theory of series expansions with respect to the functions $P_n^m(x)$.

11. Prove the following addition theorem for the Legendre polynomials:

$$P_n(zz' - \sqrt{z^2-1}\sqrt{z'^2-1}\cos\varphi)$$
$$= P_n(z)P_n(z') + 2\sum_{m=1}^{n}(-1)^m\frac{(n-m)!}{(n+m)!}P_n^m(z)P_n^m(z')\cos m\varphi.$$

Prove the analogous theorem for the Legendre functions:[34]

$$P_\nu(zz' - \sqrt{z^2-1}\sqrt{z'^2-1}\cos\varphi)$$
$$= P_\nu(z)P_\nu(z') + 2\sum_{m=1}^{\infty}(-1)^m\frac{\Gamma(\nu-m+1)}{\Gamma(\nu+m+1)}P_\nu^m(z)P_\nu^m(z')\cos m\varphi,$$
$$|\arg(z-1)| < \pi, \quad |\arg(z'-1)| < \pi, \quad \operatorname{Re} z > 0, \quad \operatorname{Re} z' > 0.$$

12. Prove that the Legendre functions of complex degree $\nu = -\frac{1}{2} + i\tau$ satisfy the integral equation

$$P_{-\frac{1}{2}+i\tau}(x) = \frac{\cosh\pi\tau}{\pi}\int_1^\infty \frac{P_{-\frac{1}{2}+i\tau}(y)}{x+y}\,dy, \qquad 1 \leqslant x < \infty.$$

13. Derive the following integral representation of the square of the function $P_{-\frac{1}{2}+i\tau}(x)$:

$$[P_{-\frac{1}{2}+i\tau}(x)]^2 = \frac{\cosh\pi\tau}{\pi}\int_1^\infty \frac{P_{-\frac{1}{2}+i\tau}(y)}{\sqrt{1+y}\sqrt{2x^2-1+y}}\,dy, \qquad 1 \leqslant x < \infty.$$

14. Derive the following asymptotic formulas for the Legendre functions of complex degree $\nu = -\frac{1}{2} + i\tau$:[35]

$$P_{-\frac{1}{2}+i\tau}(\cos\theta) \approx \frac{e^{\tau\theta}}{\sqrt{2\pi\tau\sin\theta}}, \qquad \tau\to\infty, \quad \delta \leqslant \theta \leqslant \pi - \delta,$$

$$P_{-\frac{1}{2}+i\tau}(\cosh\alpha) \approx \frac{\sqrt{2}}{\sqrt{\pi\tau\sinh\alpha}}\sin(\alpha\tau + \tfrac{1}{4}\pi), \qquad \tau\to\infty, \quad \delta \leqslant \alpha \leqslant a < \infty.$$

15. Prove that

$$\int_{-1}^{1} x^{2m}P_{2n}(x)\,dx = 2^{2n+1}\frac{\Gamma(2m+1)\Gamma(m+n+1)}{\Gamma(m-n+1)\Gamma(2m+2n+2)}.$$

16. Prove the formulas

$$\int_{-1}^{1}\frac{P_{2n}(x)}{\sqrt{\cosh^2\alpha - x^2}}\,dx = 2iP_{2n}(0)Q_{2n}(i\sinh\alpha), \tag{i}$$

$$\int_{-1}^{1}\frac{P_{2n}(x)}{\sqrt{\sinh^2\alpha + x^2}}\,dx = 2P_{2n}(0)Q_{2n}(\cosh\alpha). \tag{ii}$$

---

[34] For the proof of these and similar formulas, see E. W. Hobson, *op. cit.*, Chap. 8.

[35] These formulas are important in connection with the problems of mathematical physics considered in Secs. 8.5, 8.9, 12–13. They are special cases of general asymptotic formulas given in Barnes' paper (*op. cit.*).

Hint. The substitution $\alpha \to \alpha - \frac{1}{2}\pi i$ converts (i) into (ii). To prove (i), expand $\sqrt{\cosh^2 \alpha - x^2}$ in a power series and integrate term by term, using the result of the preceding problem. Also anticipate formula (9.5.2), and use (7.3.7) and (7.6.7).

8

SPHERICAL HARMONICS: APPLICATIONS

8.1. Introductory Remarks

The present chapter is devoted to the study of some boundary value problems of mathematical physics which can be solved by the use of spherical harmonics. Except for Sec. 8.14 (dealing with Helmholtz's equation), we will be concerned exclusively with potential theory, i.e., with solutions of Laplace's equation. In fact, we will confine our attention to the *Dirichlet problem*, which, according to Sec. 6.3, can be stated as follows: *Given a domain τ with boundary σ, and a function f defined on σ, find the function u such that* 1) *u is harmonic in τ and continuous in the closed domain $\tau + \sigma$, and* 2) *u coincides with f on σ.* In the case of an unbounded domain, this statement of the problem must be supplemented by a condition characterizing the behavior of the function u at infinity.

An effective general method for solving boundary value problems is to find a system S of orthogonal curvilinear coordinates α, β, γ such that

1. The surface σ corresponds to a constant value of one of the coordinates α, β, γ;
2. Variables can be separated in Laplace's equation, after it has been transformed to the system S by using the formulas

$$x = x(\alpha, \beta, \gamma), \quad y = y(\alpha, \beta, \gamma), \quad z = z(\alpha, \beta, \gamma). \tag{8.1.1}$$

If such a coordinate system S can be found, then a solution of the problem can usually be obtained by superposition of particular solutions of Laplace's equation written in the system S (cf. Sec. 6.3). In this regard, we remind the reader of the following fact from advanced calculus:[1] If the square of the element of arc length in the system S is given by

$$ds^2 = h_\alpha^2\, d\alpha^2 + h_\beta^2\, d\beta^2 + h_\gamma^2\, d\gamma^2, \tag{8.1.2}$$

in terms of the *metric coefficients* $h_\alpha, h_\beta, h_\gamma$, then in the system S, the Laplacian operator takes the form

$$\nabla^2 u = \frac{1}{h_\alpha h_\beta h_\gamma}\left[\frac{\partial}{\partial\alpha}\left(\frac{h_\beta h_\gamma}{h_\alpha}\frac{\partial u}{\partial\alpha}\right) + \frac{\partial}{\partial\beta}\left(\frac{h_\gamma h_\alpha}{h_\beta}\frac{\partial u}{\partial\beta}\right) + \frac{\partial}{\partial\gamma}\left(\frac{h_\alpha h_\beta}{h_\gamma}\frac{\partial u}{\partial\gamma}\right)\right]. \tag{8.1.3}$$

8.2. Solution of Laplace's Equation in Spherical Coordinates

One of the most important systems of orthogonal curvilinear coordinates permitting separation of variables in Laplace's equation is the system of spherical coordinates r, θ, φ, related to the rectangular coordinates x, y, z by the formulas

$$x = r\sin\theta\cos\varphi, \quad y = r\sin\theta\sin\varphi, \quad z = r\cos\theta, \tag{8.2.1}$$

where

$$0 \leqslant r < \infty, \quad 0 \leqslant \theta \leqslant \pi, \quad -\pi < \varphi \leqslant \pi.$$

The corresponding triply orthogonal system of surfaces consists of the spheres $r = \text{const}$, the circular cones $\theta = \text{const}$ and the planes $\varphi = \text{const}$ passing through the z-axis. Moreover, the square of the element of arc length is

$$ds^2 = dr^2 + r^2\, d\theta^2 + r^2 \sin^2\theta\, d\varphi^2, \tag{8.2.2}$$

and hence, according to (8.1.2), the metric coefficients are

$$h_r = 1, \quad h_\theta = r, \quad h_\varphi = r\sin\theta,$$

and Laplace's equation takes the form [cf. (8.1.3)]

$$\nabla^2 u = \frac{1}{r^2}\frac{\partial}{\partial r}\left(r^2\frac{\partial u}{\partial r}\right) + \frac{1}{r^2\sin\theta}\frac{\partial}{\partial\theta}\left(\sin\theta\frac{\partial u}{\partial\theta}\right) + \frac{1}{r^2\sin^2\theta}\frac{\partial^2 u}{\partial\varphi^2} = 0. \tag{8.2.3}$$

It is easy to see that if we look for particular solutions of (8.2.3) of the form

$$u = R(r)\Theta(\theta)\Phi(\varphi), \tag{8.2.4}$$

then variables can be separated, so that the problem of determining each factor in (8.2.4) reduces to the solution of an ordinary differential equation.

[1] F. B. Hildebrand, *op. cit.*, p. 302.

In fact, substituting (8.2.4) into (8.2.3), multiplying by $r^2 \sin^2 \theta$ and dividing by $R\Theta\Phi$, we find that

$$\left[\frac{1}{R}\frac{d}{dr}\left(r^2\frac{dR}{dr}\right) + \frac{1}{\Theta \sin\theta}\frac{d}{d\theta}\left(\sin\theta\frac{d\Theta}{d\theta}\right)\right]\sin^2\theta = -\frac{1}{\Phi}\frac{d^2\Phi}{d\varphi^2},$$

which is possible only if both sides equal a constant, which we denote by μ^2. This leads to two equations

$$\frac{d^2\Phi}{d\varphi^2} + \mu^2\Phi = 0,$$
$$\frac{1}{R}\frac{d}{dr}\left(r^2\frac{dR}{dr}\right) = \frac{\mu^2}{\sin^2\theta} - \frac{1}{\Theta\sin\theta}\left(\frac{d}{d\theta}\sin\theta\frac{d\Theta}{d\theta}\right). \tag{8.2.5}$$

The same reasoning shows that both sides of the last equation must equal a constant, which this time it is convenient to denote by $\nu(\nu + 1)$. As a result, we obtain the equations

$$\frac{1}{\sin\theta}\frac{d}{d\theta}\left(\sin\theta\frac{d\Theta}{d\theta}\right) + \left[\nu(\nu+1) - \frac{\mu^2}{\sin^2\theta}\right]\Theta = 0, \tag{8.2.6}$$

$$\frac{d}{dr}\left(r^2\frac{dR}{dr}\right) - \nu(\nu+1)R = 0. \tag{8.2.7}$$

Thus, determining the factors in the product (8.2.4) reduces to the relatively simple problem of solving the ordinary differential equations (8.2.5–7). The corresponding particular solutions (8.2.4) of Laplace's equation depend on two parameters μ and ν (in general, complex),[2] which can be used to construct solutions of boundary values problems of mathematical physics involving various special domains (spheres, cones, etc.). The parameters μ, ν and the corresponding solutions of equations (8.2.5–7) must be chosen in such a way that each particular solution (8.2.4) is harmonic in the given domain, and an appropriate superposition of particular solutions solves the given boundary value problem.

8.3. The Dirichlet Problem for a Sphere

As a simple example of the application of the superposition method, we consider the interior Dirichlet problem for a spherical domain. To keep things as simple as possible, we assume that the boundary function f and the solution u are independent of the angle φ. Choosing the origin at the center of the sphere (of radius a) and the z-axis along the axis of symmetry, we can formulate our problem as follows: *Find the function* $u = u(r, \theta)$ *such that* 1) u

[2] Without loss of generality, we can assume that $\operatorname{Re}\mu \geqslant 0$ and $\operatorname{Re}\nu \geqslant -\frac{1}{2}$, since replacing μ by $-\mu$ or ν by $-\nu - 1$ does not affect the separation constants μ^2 and $\nu(\nu + 1)$.

is harmonic in the domain $r < a$ *and continuous in the closed domain* $r \leqslant a$, *and* 2) u *satisfies the boundary condition* $u|_{r=a} = f(\theta)$, *where* $f(\theta)$ *is continuous in the interval* $0 \leqslant \theta \leqslant \pi$.[3]

The rotational symmetry of the problem corresponds to setting $\Phi = 1$ in (8.2.4) and $\mu = 0$ in (8.2.6). Then (8.2.6) reduces to the differential equation (7.3.1) for the Legendre functions of argument $x = \cos\theta$, which for $-1 < x < 1$ has the general solution [cf. (7.3.29)]

$$\Theta = AP_\nu(\cos\theta) + BQ_\nu(\cos\theta), \tag{8.3.1}$$

where $P_\nu(x)$ and $Q_\nu(x)$ are Legendre functions of the first and second kinds, and ν is an arbitrary complex number such that $\operatorname{Re}\nu \geqslant -\frac{1}{2}$.[4] Since the variable $x = \cos\theta$ actually ranges over the closed interval $[-1, 1]$, and since as $x \to 1$, $Q_\nu(x) \to \infty$ while $P_\nu(x)$ remains bounded [cf. (7.3.13, 23) and Problem 7, p. 201] we must set $B = 0$ if the solution is to remain bounded inside the sphere. Moreover, since $P_\nu(x) \to \infty$ as $x \to -1$ unless ν is a nonnegative integer [cf. (4.2.6) and Problem 6, p. 201], the same reason compels us to choose $\nu = n$ $(n = 0, 1, 2, \ldots)$. Therefore, the only solutions of (8.2.6) for $\mu = 0$ which remain bounded in the closed interval $0 \leqslant \theta \leqslant \pi$ correspond to nonnegative integral ν and are of the form

$$\Theta = AP_n(\cos\theta), \qquad n = 0, 1, 2, \ldots, \tag{8.3.2}$$

where $P_n(x)$ is the Legendre polynomial of degree n. As for the radial equation (8.2.6), it is an Euler equation, with general solution (for $\nu \neq -\frac{1}{2}$)[5]

$$R = Cr^\nu + Dr^{-\nu-1}. \tag{8.3.3}$$

In the present case $\nu = n$, and the requirement that the solution be bounded at the center of the sphere compels us to choose $D = 0$. It follows that

$$R = Cr^n, \qquad n = 0, 1, 2, \ldots, \tag{8.3.4}$$

and hence the appropriate set of particular solutions of Laplace's equation inside the sphere is

$$u = u_n = M_n r^n P_n(\cos\theta), \qquad n = 0, 1, 2, \ldots \tag{8.3.5}$$

We can now solve our boundary value problem by superposition of the solutions (8.3.5). In fact, suppose the boundary function $f(\theta)$ can be expanded in a series of Legendre polynomials (see Sec. 4.7), i.e.,

$$f(\theta) = \sum_{n=0}^{\infty} f_n P_n(\cos\theta), \qquad 0 \leqslant \theta \leqslant \pi, \tag{8.3.6}$$

---

[3] The statement of the problem must be suitably modified if f has discontinuities.

[4] As already noted (see footnote 2), this is the only case that need be considered.

[5] E. A. Coddington, *op. cit.*, Theorem 1, p. 147.

where

$$f_n = (n + \tfrac{1}{2}) \int_0^{\pi} f(\theta) P_n (\cos \theta) \sin \theta \, d\theta, \tag{8.3.7}$$

and suppose the series (8.3.6) converges uniformly in the interval $[0, \pi]$. Then, choosing $M_n = f_n a^{-n}$ and summing the solutions (8.3.5), we obtain the series

$$u = \sum_{n=0}^{\infty} f_n \left(\frac{r}{a}\right)^n P_n (\cos \theta), \tag{8.3.8}$$

which, according to *Harnack's theorem* on sequences of harmonic functions,[6] converges uniformly for $0 \leqslant r \leqslant a$ to a harmonic function with boundary values

$$u|_{r=a} = f(\theta),$$

i.e., (8.3.8) solves the Dirichlet problem for a sphere.[7]

Remark 1. The solutions of the Neumann problem and the mixed problem, involving the boundary conditions (6.3.1b) and (6.3.1c), can be obtained by similar methods.

Remark 2. In the case of the more general problem where $f = f(\theta, \varphi)$ is a function of both angular coordinates, it turns out that the appropriate set of particular solutions of Laplace's equation in the domain $r < a$ has the form[8]

$$\begin{gathered} u = u_{mn} = [M_{mn} \cos m\varphi + N_{mn} \sin m\varphi] r^n P_n^m (\cos \theta), \\ m = 0, 1, 2, \ldots, \quad n = m, m+1, m+2, \ldots, \end{gathered} \tag{8.3.9}$$

in terms of the associated Legendre functions $P_n^m (\cos \theta)$. Moreover, by replacing the factor r^n in (8.3.9) or (8.3.5) by the linear combination $Cr^n + Dr^{-n-1}$, we obtain particular solutions which can be used to solve boundary value problems for a spherical shell, or for the domain lying outside a sphere (in the latter case, we must set $C = 0$ to prevent the solution from becoming infinite as $r \to \infty$).

8.4. The Field of a Point Charge inside a Hollow Conducting Sphere

As an application of the results of the preceding section, consider the problem of determining the electrostatic field due to a point charge q inside a

---

[6] See W. J. Sternberg and T. L. Smith, *The Theory of Potential and Spherical Harmonics*, University of Toronto Press, Toronto (1952), pp. 216, 247, and R. Courant and D. Hilbert, *Methods of Mathematical Physics, Vol. 2*, Interscience Publishers, New York (1962), p. 273, where the result is called *Weierstrass' convergence theorem.*

[7] One can also solve the interior Dirichlet problem for a sphere in the case where $f(\theta)$ is only piecewise continuous. See the analogous treatment of the interior Dirichlet problem for a circle, given in A. N. Tikhonov and A. A. Samarski, *op. cit.*, pp.. 284, 301.

[8] See E. T. Whittaker and G. N. Watson, *A Course of Modern Analysis*, fourth edition, Cambridge University Press, London (1963), p. 392.

hollow conducting sphere of radius a, held at zero potential. Choose the origin O at the center of the sphere, and let the z-axis pass through the position A of the charge, which is at distance b from O (see Figure 27). To eliminate the singularity at A, we write the potential ψ of the electrostatic field as a sum of the potential of the source and the potential u of the secondary field due to the charges induced on the inner surface of the sphere, i.e.,

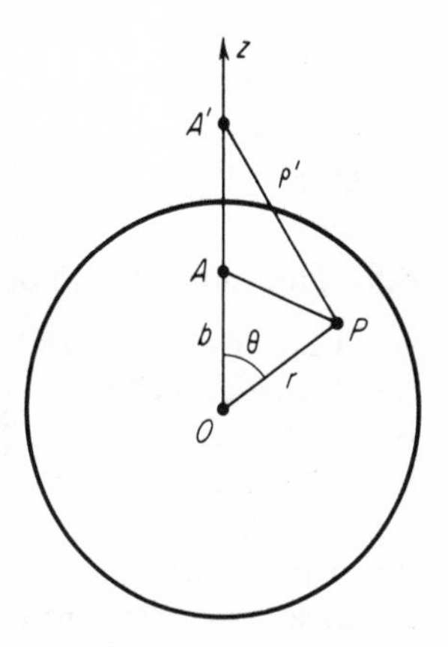

FIGURE 27

$$\psi = \frac{q}{\rho} + u, \tag{8.4.1}$$

where

$$\rho = AP = \sqrt{r^2 + b^2 - 2br\cos\theta}$$

is the distance from A to a variable point P, with coordinates r, θ.[9] Since ψ must vanish on the surface of the sphere, determination of the function $u = u(r, \theta)$ reduces to solving the Dirichlet problem with the boundary condition

$$u|_{r=a} = -\frac{q}{\sqrt{a^2 + b^2 - 2ab\cos\theta}} = f(\theta). \tag{8.4.2}$$

The right-hand side of (8.4.2) can easily be expanded in a series of Legendre polynomials, and in fact there is no need to evaluate the integral (8.3.7). Instead, we use formula (4.2.3) which immediately implies

$$u|_{r=a} = -\frac{q}{a}\sum_{n=0}^{\infty}\left(\frac{b}{a}\right)^n P_n(\cos\theta). \tag{8.4.3}$$

Moreover, since $b < a$ it follows from the estimate (4.4.2) that the series (8.4.3) is uniformly convergent in the interval $[0, \pi]$. Therefore, according to Sec. 8.3, the function u is given by the formula

$$u = -\frac{q}{a}\sum_{n=0}^{\infty}\left(\frac{br}{a^2}\right)^n P_n(\cos\theta). \tag{8.4.4}$$

Using (4.2.3) again, we find that the sum of the series (8.4.3) is

$$u = -\frac{q}{a}\frac{1}{\sqrt{1 - 2\left(\frac{br}{a^2}\right)\cos\theta + \left(\frac{br}{a^2}\right)^2}} = \frac{q'}{\rho'}, \tag{8.4.5}$$

where

$$q' = -q\frac{a}{b}, \quad b' = \frac{a^2}{b}, \qquad \rho' = \sqrt{r^2 + b'^2 - 2b'r\cos\theta}.$$

[9] Since the problem is rotationally symmetric, u is independent of the angle φ.

Thus the potential ψ can be written as a sum

$$\psi = \frac{q}{\rho} + \frac{q'}{\rho'}, \tag{8.4.6}$$

where the first term is the potential of the charge q in the absence of the conducting sphere, and the second term is the potential of the *image charge* q' at the *image point* A', which takes account of the influence of the sphere.[10]

8.5. The Dirichlet Problem for a Cone

The ability to separate variables in Laplace's equation written in spherical coordinates also allows us to solve boundary value problems for the domain bounded by the surface of an infinite circular cone. Choose the origin at the vertex of the cone, and let the z-axis lie along the axis of symmetry of the cone (see Figure 28). Then the equation of the cone is $\theta = \theta_0$ ($\theta_0 < \pi$), and the Dirichlet problem for the case of axially symmetric boundary conditions can be stated as follows: *Find the functions* $u = u(r, \theta)$ *such that* 1) u *is harmonic in the domain* $0 < r < \infty, 0 \leqslant \theta < \theta_0$ *and continuous in the closed domain* $0 \leqslant r < \infty, 0 \leqslant \theta \leqslant \theta_0$, *and* 2) u *satisfies the boundary condition* $u|_{\theta=\theta_0} = f(r)$ *and the condition at infinity* $u|_{r\to\infty} \to 0$ *uniformly in* θ,[11] *where* $f(r)$ *is continuous in the interval* $0 \leqslant r < \infty$ *and* $f(r)|_{r\to\infty} = 0$.

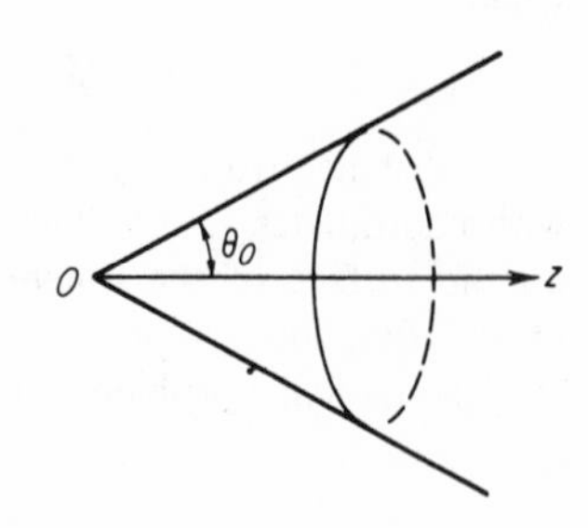

FIGURE 28

In applying the method of separation of variables to this problem, we must set $B = 0$ in (8.3.1), if the solution is to remain bounded on the axis of the cone. However, in the present case, there is no reason to choose ν to be a nonnegative integer, since $P_\nu(\cos\theta)$ is bounded for arbitrary ν if $0 \leqslant \theta \leqslant \theta_0$. In fact, with some extra restrictions on the function $f(r)$, the problem can be solved by choosing

$$\nu = -\tfrac{1}{2} + i\tau, \qquad \tau \geqslant 0,$$

which corresponds to the following set of particular solutions of Laplace's equation:

$$u = u_\tau = [M_\tau \cos(\tau \log r) + N_\tau \sin(\tau \log r)]r^{-1/2}P_{-\frac{1}{2}+i\tau}(\cos\theta). \tag{8.5.1}$$

Here M_τ and N_τ are arbitrary continuous functions ($\tau \geqslant 0$), and the solutions

---

[10] Note that $\rho' = \overline{A'P}$, i.e., ρ' is the distance between the image point A' and the variable point P (see Figure 27).

[11] The second condition is necessary for the uniqueness of the function u. See A. N. Tikhonov and A. A. Samarski, *op. cit.*, p. 288.

depend continuously on the parameter τ. Using (7.3.6), we find that the Legendre functions of complex degree appearing in (8.5.1) have the series expansion

$$\begin{aligned} P_{-\frac{1}{2}+i\tau}(\cos\theta) &= F\left(\tfrac{1}{2}+i\tau, \tfrac{1}{2}-i\tau; 1; \sin^2\frac{\theta}{2}\right) \\ &= 1 + \frac{\frac{1}{4}+\tau^2}{(1!)^2}\sin^2\frac{\theta}{2} + \frac{(\frac{1}{4}+\tau^2)(\frac{9}{4}+\tau^2)}{(2!)^2}\sin^4\frac{\theta}{2} + \cdots \end{aligned} \tag{8.5.2}$$

It follows from (8.5.2) that $P_{-\frac{1}{2}+i\tau}(\cos\theta)$ is real and satisfies the inequalities

$$\begin{aligned} 1 \leqslant P_{-\frac{1}{2}+i\tau}(\cos\theta), &\qquad 0 \leqslant \theta \leqslant \pi, \\ P_{-\frac{1}{2}+i\tau}(\cos\theta) \leqslant P_{-\frac{1}{2}+i\tau}(\cos\theta_0), &\qquad 0 \leqslant \theta \leqslant \theta_0. \end{aligned} \tag{8.5.3}$$

Now suppose that $f(r)$ is such that $\varphi(r) = r^{1/2}f(r)$ has a Fourier expansion of the form[12]

$$g(r) = r^{1/2}f(r) = \int_0^\infty [G_c(\tau)\cos(\tau\log r) + G_s(\tau)\sin(\tau\log r)]\,d\tau, \qquad 0 < r < \infty,$$

$$G_c(\tau) = \frac{1}{\pi}\int_0^\infty f(r)r^{-1/2}\cos(\tau\log r)\,dr, \quad G_s(\tau) = \frac{1}{\pi}\int_0^\infty f(r)r^{-1/2}\sin(\tau\log r)\,dr, \tag{8.5.4}$$

where the integral is uniformly convergent in every finite subinterval $[r_1, r_2]$ such that $0 < r_1 < r_2 < \infty$. Then, choosing

$$M_\tau = \frac{G_c(\tau)}{P_{-\frac{1}{2}+i\tau}(\cos\theta_0)}, \qquad N_\tau = \frac{G_s(\tau)}{P_{-\frac{1}{2}+i\tau}(\cos\theta_0)}$$

in (8.5.1), and integrating with respect to the parameter τ from 0 to ∞, we obtain the function

$$u = r^{-1/2}\int_0^\infty [G_c(\tau)\cos(\tau\log r) + G_s(\tau)\sin(\tau\log r)]\frac{P_{-\frac{1}{2}+i\tau}(\cos\theta)}{P_{-\frac{1}{2}+i\tau}(\cos\theta_0)}\,d\tau, \tag{8.5.5}$$

which gives the solution of our problem, at least formally.

---

[12] The expansion (8.5.4), which reduces to the standard form of the Fourier integral if we make the substitution $\log r = \xi\ (-\infty < \xi < \infty)$, is valid if $f(r)$ is continuous and of bounded variation in every finite subinterval $[r_1, r_2]$, where $0 < r_1 < r_2 < \infty$, and if the integral

$$\int_0^\infty |f(r)|r^{-1/2}\,dr$$

is finite. See E. C. Titchmarsh, *Introduction to the Theory of Fourier Integrals*, second edition, Oxford University Press, London (1950), Theorem 3, p. 13.

Example. *Find the electrostatic field due to a point charge q on the axis of a hollow conducting cone, held at zero potential, if the charge is at distance a from the vertex of the cone.*

As in Sec. 8.4, we write the potential ψ as a sum

$$\psi = \frac{q}{\rho} + u, \tag{8.5.6}$$

where $\rho = \sqrt{r^2 + a^2 - 2ar\cos\theta}$. Then u satisfies the boundary condition

$$u|_{\theta=\theta_0} = f(r) = -\frac{q}{\sqrt{r^2 + a^2 - 2ar\cos\theta_0}}. \tag{8.5.7}$$

Using the integral representation (7.4.6), we find that

$$\begin{aligned} G_c(\tau) &= -\frac{q}{\pi}\int_0^\infty \frac{\cos(\tau\log r)}{\sqrt{r}\sqrt{r^2+a^2-2ar\cos\theta_0}}\,dr \\ &= -\frac{q}{\pi\sqrt{a}}\int_0^\infty \frac{\cos(\tau\log r)}{\sqrt{\frac{r}{a}+\frac{a}{r}-2\cos\theta_0}}\frac{dr}{r} \\ &= -\frac{q}{\pi\sqrt{a}}\int_{-\infty}^\infty \frac{\cos[\tau(s+\log a)]}{\sqrt{2\cosh s - 2\cos\theta_0}}\,ds \\ &= -\frac{2q\cos(\tau\log a)}{\pi\sqrt{a}}\int_0^\infty \frac{\cos\tau s}{\sqrt{2\cosh s - 2\cos\theta_0}}\,ds \\ &= -\frac{q}{\sqrt{a}}\frac{\cos(\tau\log a)}{\cosh\pi\tau}P_{-\frac{1}{2}+i\tau}(-\cos\theta_0), \end{aligned} \tag{8.5.8}$$

and similarly,

$$G_s(\tau) = -\frac{q}{\sqrt{a}}\frac{\sin(\tau\log a)}{\cosh\pi\tau}P_{-\frac{1}{2}+i\tau}(-\cos\theta_0).$$

Thus the solution of the problem is given by the integral

$$u = -\frac{q}{\sqrt{ar}}\int_0^\infty \frac{P_{-\frac{1}{2}+i\tau}(\cos\theta)}{P_{-\frac{1}{2}+i\tau}(\cos\theta_0)}P_{-\frac{1}{2}+i\tau}(-\cos\theta_0)\frac{\cos[\tau\log(r/a)]}{\cosh\pi\tau}d\tau. \tag{8.5.9}$$

It is not hard to see that this integral is absolutely and uniformly convergent for $r_1 \leqslant r \leqslant r_2$, $0 \leqslant \theta \leqslant \theta_0$, where $0 < r_1 < r_2 < \infty$. In fact, it follows from (8.5.3) that the integral in question is majorized by the integral[13]

$$\int_0^\infty P_{-\frac{1}{2}+i\tau}(-\cos\theta_0)\frac{d\tau}{\cosh\pi\tau} = \frac{1}{2}\cos\frac{\theta_0}{2}. \tag{8.5.10}$$

[13] To verify (8.5.10), set $\beta = \pi$ in (8.12.8).

Using this result, we can prove that formula (8.5.9) actually gives the solution of our problem.[14]

8.6. Solution of Laplace's Equation in Spheroidal Coordinates

We now turn to other systems of orthogonal coordinates permitting separation of variables in Laplace's equation, and leading to particular solutions which can be expressed in terms of spherical harmonics. We begin our discussion by examining two coordinate systems suitable for solving boundary value problems for spheroidal domains.[15] First we consider *prolate spheroidal coordinates* α, β, φ, related to the rectangular coordinates x, y, z by the formulas

$$x = c \sinh \alpha \sin \beta \cos \varphi, \quad y = c \sinh \alpha \sin \beta \sin \varphi, \quad z = c \cosh \alpha \cos \beta, \tag{8.6.1}$$

where

$$0 \leqslant \alpha < \infty, \quad 0 \leqslant \beta \leqslant \pi, \quad -\pi < \varphi \leqslant \pi,$$

and $c > 0$ is a scale factor.[16] Then every point of space is characterized by a unique triple of numbers α, β, φ. The corresponding triply orthogonal system of surfaces consists of the prolate spheroids $\alpha = \text{const}$ with foci at the points $(0, 0, \pm c)$, the double-sheeted hyperboloids of revolution $\beta = \text{const}$, which are confocal with the spheroids, and the planes $\varphi = \text{const}$ passing through the z-axis (see Figure 29). A simple calculation shows that the square of the element of arc length is

$$ds^2 = c^2 (\sinh^2 \alpha + \sin^2 \beta)(d\alpha^2 + d\beta^2) + c^2 \sinh^2 \alpha \sin^2 \beta \, d\varphi^2. \tag{8.6.2}$$

Therefore the metric coefficients are

$$h_\alpha = h_\beta = c\sqrt{\sinh^2 \alpha + \sin^2 \beta}, \qquad h = c \sinh \alpha \sin \beta,$$

and Laplace's equation takes the form [cf. (8.2.3)]

$$\nabla^2 u = \frac{1}{c^2 (\sinh^2 \alpha + \sin^2 \beta)} \left[\frac{1}{\sinh \alpha} \frac{\partial}{\partial \alpha} \left(\sinh \alpha \frac{\partial u}{\partial \alpha} \right) + \frac{1}{\sin \beta} \frac{\partial}{\partial \beta} \left(\sin \beta \frac{\partial u}{\partial \beta} \right) \right.$$
$$\left. + \left(\frac{1}{\sinh^2 \alpha} + \frac{1}{\sin^2 \beta} \right) \frac{\partial^2 u}{\partial \varphi^2} \right] = 0. \tag{8.6.3}$$

---

[14] In examining the convergence of integrals involving Legendre functions of complex degree $\nu = -\frac{1}{2} + i\tau$, it is useful to recall the asymptotic formulas proved in Problem 14, p. 202.

[15] The terms *spheroid* and *ellipsoid of revolution* are synonymous, and spheroidal coordinates might be called *degenerate ellipsoidal coordinates*, since cross sections of the coordinate surfaces normal to the z-axis are circles rather than ellipses (concerning ellipsoidal coordinates, see E. W. Hobson, *op. cit.*, Chap. 11).

[16] If a point has cylindrical coordinates r, φ and z, then $z + ir = c \cosh (\alpha + i\beta)$.

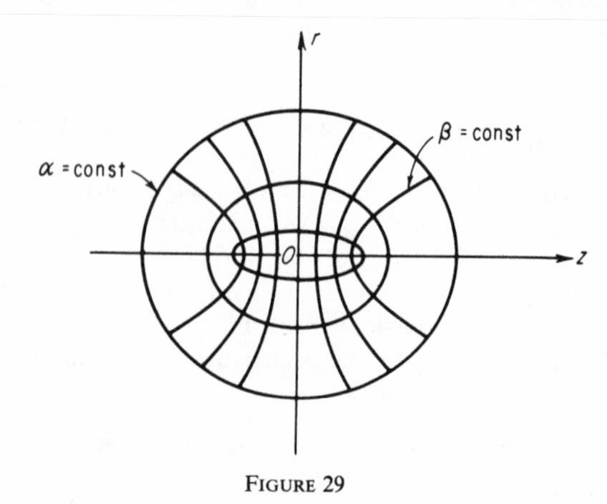

FIGURE 29

Now suppose we look for solutions of (8.6.3) which have the form

$$u = \mathrm{A}(\alpha)\mathrm{B}(\beta)\Phi(\varphi). \tag{8.6.4}$$

Then the variables separate, just as in Sec. 8.2, and the factors A, B, Φ satisfy the ordinary differential equations

$$\frac{d^2\Phi}{d\varphi^2} + \mu^2\Phi = 0, \tag{8.6.5}$$

$$\frac{1}{\sin\beta}\frac{d}{d\beta}\left(\sin\beta\frac{d\mathrm{B}}{d\beta}\right) + \left[\nu(\nu+1) - \frac{\mu^2}{\sin^2\beta}\right]\mathrm{B} = 0, \tag{8.6.6}$$

$$\frac{1}{\sinh\alpha}\frac{d}{d\alpha}\left(\sinh\alpha\frac{d\mathrm{A}}{d\alpha}\right) - \left[\nu(\nu+1) + \frac{\mu^2}{\sinh^2\alpha}\right]\mathrm{A} = 0, \tag{8.6.7}$$

where μ and ν are parameters whose choice is dictated by the concrete conditions of the problem. For example, in the rotationally symmetric case where u is independent of the variable φ, we set $\mu = 0$, $\Phi = 1$, while in the more general case where u depends on φ, we set $\mu = m$ ($m = 0, 1, 2, \ldots$), since u must be periodic in φ.

Next we consider *oblate spheroidal coordinates* α, β, φ, related to the rectangular coordinates x, y, z by the formulas

$$x = c\cosh\alpha\sin\beta\cos\varphi, \quad y = c\cosh\alpha\sin\beta\sin\varphi, \quad z = c\sinh\alpha\cos\beta, \tag{8.6.8}$$

where[17]

$$0 \leqslant \alpha < \infty, \quad 0 \leqslant \beta \leqslant \pi, \quad -\pi < \varphi \leqslant \pi.$$

---

[17] If a point has cylindrical coordinates r, φ and z, we now have $z + ir = \sinh(\alpha + i\beta)$.

In this case, the triply orthogonal system of surfaces consists of the oblate spheroids $\alpha = \text{const}$, the single-sheeted hyperboloids of revolution $\beta = \text{const}$ and the planes $\varphi = \text{const}$ (see Figure 30). The square of the element of arc length and Laplace's equation now take the form

$$ds^2 = c^2 (\cosh^2 \alpha - \sin^2 \beta)(d\alpha^2 + d\beta^2) + c^2 \cosh^2 \alpha \sin^2 \beta \, d\varphi^2, \qquad (8.6.9)$$

$$\nabla^2 u = \frac{1}{c^2 (\cosh^2 \alpha - \sin^2 \beta)} \left[\frac{1}{\cosh \alpha} \frac{\partial}{\partial \alpha} \left(\cosh \alpha \frac{\partial u}{\partial \alpha} \right) + \frac{1}{\sin \beta} \frac{\partial}{\partial \beta} \left(\sin \beta \frac{\partial u}{\partial \beta} \right) + \left(\frac{1}{\sin^2 \beta} - \frac{1}{\cosh^2 \alpha} \right) \frac{\partial^2 u}{\partial \varphi^2} \right] = 0. \qquad (8.6.10)$$

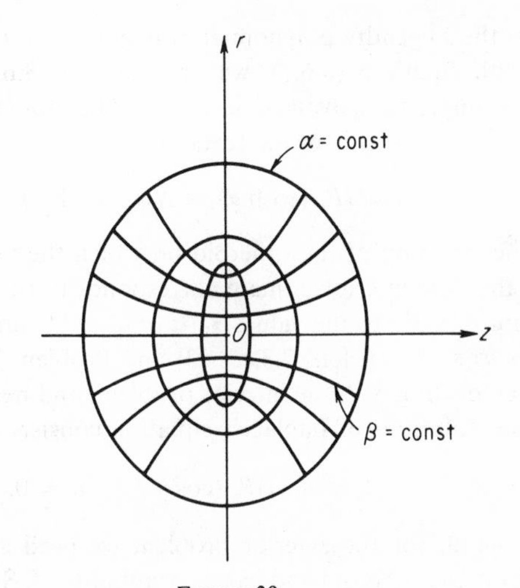

FIGURE 30

Separating variables, instead of (8.6.5–7) we find the following system of equations for determining the factors A, B and Φ:

$$\frac{d^2 \Phi}{d\varphi^2} + \mu^2 \Phi = 0, \qquad (8.6.11)$$

$$\frac{1}{\sin \beta} \frac{d}{d\beta} \left(\sin \beta \frac{dB}{d\beta} \right) + \left[\nu(\nu + 1) - \frac{\mu^2}{\sin^2 \beta} \right] B = 0, \qquad (8.6.12)$$

$$\frac{1}{\cosh \alpha} \frac{d}{d\alpha} \left(\cosh \alpha \frac{dA}{d\alpha} \right) - \left[\nu(\nu + 1) - \frac{\mu^2}{\cosh^2 \alpha} \right] A = 0. \qquad (8.6.13)$$

8.7. The Dirichlet Problem for a Spheroid

Using the particular solutions of Laplace's equation $\nabla^2 u = 0$ found in Sec. 8.6, we can construct functions harmonic in the interior or exterior of a spheroid, thereby solving the boundary value problems of potential theory for domains of this type. To keep things as simple as possible, we consider the Dirichlet problem, assuming that the boundary function f and the solution u are independent of the angle φ. We begin with the case of a prolate spheroid. The rotational symmetry of the problem corresponds to setting $\Phi = 1$ in (8.6.11) and $\mu = 0$ in (8.6.12–13). Then equation (8.6.12) reduces to the differential equation for the Legendre functions of argument $x = \cos \beta$ (cf. Sec. 8.3), whose only bounded solutions in the closed interval $[0, \pi]$ are of the form

$$\mathrm{B} = CP_n(\cos \beta), \qquad n = 0, 1, 2, \ldots, \tag{8.7.1}$$

where $P_n(x)$ is the Legendre polynomial of degree n [cf. (8.3.2).][18]

To deal with equation (8.6.7), we observe that (8.6.7) transforms into equation (8.6.6) under the substitution $\beta = i\alpha$. Therefore the general solution of (8.6.7) for $\mu = 0$, $\nu = n$ is of the form

$$\mathrm{A} = MP_n(\cosh \alpha) + NQ_n(\cosh \alpha). \tag{8.7.2}$$

If $\alpha = \alpha_0$ is the equation of the spheroid on which the boundary conditions are specified, then the interior domain corresponds to the values $0 \leqslant \alpha < \alpha_0$ and the exterior domain to the values $\alpha_0 < \alpha < \infty$.[19] Since $P_n(\cosh \alpha) \to 1$, $Q_n(\cosh \alpha) \to \infty$ as $\alpha \to 0$ [cf. (7.3.13, 23) and Problem 7, p. 201], we must set $N = 0$ when dealing with the interior problem, and hence the appropriate set of particular solutions of Laplace's equation consists of the functions

$$u = u_n = M_nP_n(\cosh \alpha)P_n(\cos \beta), \qquad n = 0, 1, 2, \ldots \tag{8.7.3}$$

On the other hand, for the exterior problem we need solutions which are harmonic outside the spheroid and vanish at infinity (cf. Sec. 8.5). According to (7.6.1, 3), this requires setting $M = 0$, so that the appropriate particular solutions of Laplace's equation are now of the form

$$u = u_n = N_nQ_n(\cosh \alpha)P_n(\cos \beta), \qquad n = 0, 1, 2, \ldots \tag{8.7.4}$$

Next we consider the case of an oblate spheroid. Since equations (8.6.6) and (8.6.12) are identical, the only difference between this case and the case of a prolate spheroid is that equation (8.6.7) is replaced by equation (8.6.13). Therefore we have the same admissible values of the parameter ν as before, i.e., $\nu = n$ $(n = 0, 1, 2, \ldots)$, and the factor $\mathrm{B}(\beta)$ is again given by (8.7.1). Since equation (8.6.13) transforms into equation (8.6.12) under the substitution $\beta = \frac{1}{2}\pi - i\alpha$, the general solution of (8.6.13) for the case $\mu = 0$, $\nu = n$ is of the form

$$\mathrm{A} = MP_n(i \sinh \alpha) + NQ_n(i \sinh \alpha), \tag{8.7.5}$$

[18] This assertion holds for both the interior and the exterior problem.

[19] This is true for either a prolate or an oblate spheroid.

corresponding to the following particular solutions of Laplace's equation:

$$u = u_n = [M_n P_n(i \sinh \alpha) + N_n Q_n(i \sinh \alpha)]P_n (\cos \beta). \tag{8.7.6}$$

We now show that N_n must be set equal to zero if the solutions (8.7.6) are to be harmonic inside the spheroid. The proof of this assertion is less trivial than in the case of the prolate spheroid, since both solutions $P_n(i \sinh \alpha)$ and $Q_n(i \sinh \alpha)$ are bounded in the whole interval $0 \leqslant \alpha < \alpha_0$. In fact, we must now examine the behavior of grad u near the singular curve of the transformation (8.6.8), i.e., the curve $\alpha = 0$, $\beta = \pi/2$ on which the Jacobian $\partial(x, y, z)/\partial(\alpha, \beta, \varphi)$ vanishes. It is an immediate consequence of (8.6.9) that

$$(\text{grad } u)^2 = \frac{1}{c^2 (\cosh^2 \alpha - \sin^2 \beta)} \left[\left(\frac{\partial u}{\partial \alpha}\right)^2 + \left(\frac{\partial u}{\partial \beta}\right)^2\right], \tag{8.7.7}$$

if we assume that u is independent of the angle φ. The denominator in the right-hand side of (8.7.7) vanishes on the curve $\alpha = 0$, $\beta = \pi/2$, and therefore a necessary condition for grad u to be finite is that the expression in brackets should also vanish for $\alpha = 0$, $\beta = \pi/2$, i.e., that $N_n = 0$, since (8.7.6) and (7.6.9–10) imply

$$\left[\left(\frac{\partial u}{\partial \alpha}\right)^2 + \left(\frac{\partial u}{\partial \beta}\right)^2\right]_{\alpha=0,\beta=\pi/2} = (-1)^{n-1} N_n^2.$$

Moreover, this condition is also sufficient. In fact, if

$$u = u_n = M_n P_n(i \sinh \alpha) P_n (\cos \beta), \qquad n = 0, 1, 2, \ldots, \tag{8.7.8}$$

then

$$\left(\frac{\partial u}{\partial \alpha}\right)^2 + \left(\frac{\partial u}{\partial \beta}\right)^2 = M_n^2 [P_n^2(i \sinh \alpha) P_n'^2 (\cos \beta) \sin^2 \beta - P_n'^2(i \sinh \alpha) P_n^2 (\cos \beta) \cosh^2 \alpha].$$

The expression in brackets is a polynomial in $\cos \beta$ which vanishes if $\cos \beta = \pm i \sinh \alpha$ and hence is divisible by $\cosh^2 \alpha - \sin^2 \beta$. It follows that grad u is well-behaved on the curve $\alpha = 0$, $\beta = \pi/2$, so that (8.7.8) gives the appropriate solutions of Laplace's equation in the interior of an oblate spheroid. In the case of the exterior problem, we must set $M_n = 0$ as before, which gives the solutions

$$u = u_n = N_n Q_n(i \sinh \alpha) P_n (\cos \beta), \qquad n = 0, 1, 2, \ldots \tag{8.7.9}$$

The Dirichlet problem for a spheroid can now be solved by superposition of the solutions (8.7.3–4) and (8.7.8–9). For example, consider the interior problem for a prolate spheroid, and suppose the boundary function $f = f(\beta)$ can be expanded in a series of Legendre polynomials

$$\begin{gathered} f(\beta) = \sum_{n=0}^{\infty} f_n P_n (\cos \beta), \qquad 0 \leqslant \beta \leqslant \pi, \\ f_n = (n + \tfrac{1}{2}) \int_0^{\pi} f(\beta) P_n (\cos \beta) \sin \beta \, d\beta, \end{gathered} \tag{8.7.10}$$

which is uniformly convergent in the closed interval $[0, \pi]$. Then, using Harnack's theorem on sequences of harmonic functions (mentioned on p. 208) we see that the series

$$u = \sum_{n=0}^{\infty} f_n \frac{P_n(\cosh \alpha)}{P_n(\cosh \alpha_0)} P_n(\cos \beta), \tag{8.7.11}$$

with terms of the form (8.7.3), converges uniformly for $0 \leqslant \alpha \leqslant \alpha_0$ to a harmonic function with boundary values

$$u|_{\alpha=\alpha_0} = f(\beta),$$

and hence solves the given boundary value problem.

Remark 1. The solutions of the Neumann problem and the mixed problem, involving the boundary conditions (6.3.1b) and (6.3.1c), can be obtained by similar methods.

Remark 2. In the case of the more general problem where $f = f(\beta, \varphi)$ is a function of both coordinates β and φ, it turns out that the appropriate set of particular solutions of Laplace's equation for prolate and oblate spheroid are

$$u = u_{mn} = [M_{mn} \cos m\varphi + N_{mn} \sin m\varphi] P_n^m(\cos \beta) \begin{matrix} P_n^m(\cosh \alpha) \\ Q_n^m(\cosh \alpha) \end{matrix}, \tag{8.7.12}$$

$$u = u_{mn} = [M_{mn} \cos m\varphi + N_{mn} \sin m\varphi] P_n^m(\cos \beta) \begin{matrix} P_n^m(i \sinh \alpha) \\ Q_n^m(i \sinh \alpha) \end{matrix}, \tag{8.7.13}$$

respectively, where $m = 0, 1, 2, \ldots$ and $n = m + 1, m + 2, \ldots$ The upper row in (8.7.12–13) corresponds to the interior problem and the lower row to the exterior problem.

8.8. The Gravitational Attraction of a Homogeneous Solid Spheroid

As a simple example of the results of the preceding two sections, we now calculate the gravitational potential of a homogeneous solid prolate spheroid of mass m and density ρ. Let the potentials inside and outside the spheroid be denoted by ψ_i and ψ_e, respectively. Then, as is well known,[20] the problem reduces to finding the solution of the equations

$$\nabla^2 \psi_i = -4\pi\rho, \qquad \nabla^2 \psi_e = 0, \tag{8.8.1}$$

which satisfy the boundary conditions

$$\psi_i|_\sigma = \psi_e|_\sigma, \qquad \left.\frac{\partial \psi_i}{\partial n}\right|_\sigma = \left.\frac{\partial \psi_e}{\partial n}\right|_\sigma, \qquad \psi_e|_\infty = 0, \tag{8.8.2}$$

where σ is the surface of the spheroid and $\partial/\partial n$ denotes the derivative with

[20] W. J. Sternberg and T. L. Smith, *op. cit.*, p. 134.

respect to the exterior normal to σ.[21] Solving this problem is equivalent to solving the equations

$$\nabla^2\psi^* = 0, \qquad \nabla^2\psi_e = 0,$$

if we represent ψ_i in the form of a sum

$$\psi_i = \psi_0 + \psi^*, \tag{8.8.3}$$

where ψ^* is harmonic inside the spheroid and ψ_i is a particular solution of Poisson's equation, e.g.,

$$\psi_0 = -\pi\rho(x^2 + y^2). \tag{8.8.4}$$

Using (8.6.1) to introduce spheroidal coordinates α, β, φ, and applying the superposition method to the particular solutions (8.7.3–4), we write the functions ψ^* and ψ_e in the form

$$\begin{aligned} \psi^* &= \sum_{n=0}^{\infty} M_n P_n(\cosh \alpha) P_n(\cos \beta), \\ \psi_e &= \sum_{n=0}^{\infty} N_n Q_n(\cosh \alpha) P_n(\cos \beta). \end{aligned} \tag{8.8.5}$$

To determine the coefficients M_n and N_n, we have the boundary conditions

$$\psi_i|_{\alpha=\alpha_0} = \psi_e|_{\alpha=\alpha_0}, \qquad \left.\frac{\partial\psi_i}{\partial\alpha}\right|_{\alpha=\alpha_0} = \left.\frac{\partial\psi_e}{\partial\alpha}\right|_{\alpha=\alpha_0}, \tag{8.8.6}$$

where α_0 is the value of the coordinate α corresponding to the surface of the spheroid.[22] Noting that

$$\psi_0 = -\pi\rho c^2 \sinh^2 \alpha \sin^2 \beta = -\frac{2\pi\rho c^2}{3} \sinh^2 \alpha\, [P_0(\cos \beta) - P_2(\cos \beta)], \tag{8.8.7}$$

and comparing coefficients in both sides of each of the equations (8.8.6), we find that

$$\begin{aligned} M_0 - \frac{2\pi\rho c^2}{3} \sinh^2 \alpha_0 &= N_0 Q_0(\cosh \alpha_0), \\ M_2 P_2(\cosh \alpha_0) + \frac{2\pi\rho c^2}{3} \sinh^2 \alpha_0 &= N_2 Q_2(\cosh \alpha_0), \\ -\frac{4\pi\rho c^2}{3} \cosh \alpha_0 &= N_0 Q_0'(\cosh \alpha_0), \\ M_2 P_2'(\cosh \alpha_0) + \frac{4\pi\rho c^2}{3} \cosh \alpha_0 &= N_2 Q_2'(\cosh \alpha_0), \end{aligned} \tag{8.8.8}$$

[21] The first of the equations (8.8.1) is known as *Poisson's equation*. As usual, we assume that ψ_i, ψ_e and their first and second derivatives with respect to x, y, z are continuous.

[22] If a is the semi-major axis and c the distance from the origin to the focus of the spheroid, then

$$\cosh \alpha_0 = \frac{a}{c}.$$

and

$$\begin{aligned} M_n P_n(\cosh \alpha_0) &= N_n Q_n(\cosh \alpha_0), \qquad n = 1, 3, 4, 5, \dots, \\ M_n P_n'(\cosh \alpha_0) &= N_n Q_n'(\cosh \alpha_0), \qquad n = 1, 3, 4, 5, \dots. \end{aligned} \tag{8.8.9}$$

It follows from (8.8.9) that $M_n = N_n = 0$ for all n different from 0 and 2. Therefore, using (8.8.8) to calculate the nonzero coefficients M_0, N_0, M_2, N_2, we can write the solution in closed form, susceptible to direct verification. After some simple calculations, during which we use (7.7.2) and the formula

$$m = \tfrac{4}{3}\pi\rho c^3 \cosh \alpha_0 \sinh^2 \alpha_0,$$

we arrive at the following expression for the potential outside the spheroid:

$$\psi_e = \frac{m}{c}\,[Q_0(\cosh \alpha) - Q_2(\cosh \alpha) P_2(\cos \beta)].$$

Similarly, we can easily find the potential inside the spheroid. Finally, using (7.9.1), we can express the potentials ψ_i and ψ_e in terms of elementary functions.

8.9. The Dirichlet Problem for a Hyperboloid of Revolution

The ability to separate variables in Laplace's equation written in spheroidal coordinates also allows us to solve boundary value problems for the domain bounded by a hyperboloid of revolution. If α, β, φ are the spheroidal coordinates described by (8.6.1), then the surface $\beta = \beta_0$ corresponds to a hyperboloid of revolution (see Figure 29). The Dirichlet problem for the case of axially symmetric (i.e., φ-independent) boundary conditions can be stated as follows: *Find the function* $u = u(\alpha, \beta)$ *such that* 1) u *is harmonic in the domain* $0 \leqslant \beta < \beta_0$ *and continuous in the closed domain* $0 \leqslant \beta \leqslant \beta_0$, *and* 2) u *satisfies the boundary condition* $u|_{\beta=\beta_0} = f(\alpha)$ *and the condition at infinity* $u|_{\alpha\to\infty} \to 0$ *uniformly in* β, *where* $f(\alpha)$ *is continuous in the interval* $0 \leqslant \alpha < \infty$ *and* $f(\alpha)|_{\alpha\to\infty} \to 0$.

As we now show, under certain conditions, the solution of this problem is given by a superposition of the following particular solutions of Laplace's equation:

$$u = u_\tau = M_\tau P_{-1/2+i\tau}(\cosh \alpha) P_{-1/2+i\tau}(\cos \beta), \qquad \tau \geqslant 0. \tag{8.9.1}$$

In fact, setting $\mu = 0$, $\nu = -\frac{1}{2} + i\tau$ in (8.6.6–7), we obtain

$$\begin{aligned} \mathrm{B} &= CP_{-1/2+i\tau}(\cos \beta) + DP_{-1/2+i\tau}(-\cos \beta), \\ \mathrm{A} &= MP_{-1/2+i\tau}(\cosh \alpha) + NP_{-1/2+i\tau}(-\cosh \alpha), \end{aligned}$$

and then the condition that the solutions be bounded on the axis of the hyperboloid, where either α or β vanishes, implies $D = N = 0$. Moreover, according to (7.6.3), we have $P_{-1/2+i\tau}(\cosh \alpha)|_{\alpha\to\infty} \to 0$, and hence $u_\tau|_{\alpha\to\infty} \to 0$, as required. The possibility of making a superposition of solutions (8.9.1) which satisfies the boundary conditions is based on the *Mehler-Fock theorem*, which states that[23]

$$f(x) = \int_0^\infty \tau \tanh \pi\tau P_{-\frac{1}{2}+i\tau}(x)\, d\tau \int_1^\infty f(\xi) P_{-\frac{1}{2}+i\tau}(\xi)\, d\xi, \qquad 1 < x < \infty \qquad (8.9.2)$$

at every continuity point of $f(x)$, provided that

1. The function $f(x)$, defined in the infinite interval $(1, \infty)$, is piecewise continuous and of bounded variation in every finite subinterval $[x_1, x_2]$, where $1 < x_1 < x_2 < \infty$;
2. The integrals

$$\int_1^a |f(x)|(x-1)^{-3/4}\, dx, \qquad \int_a^\infty |f(x)| x^{-1/2} \log x\, dx$$

are finite, for every $a > 1$.

Thus, if the boundary function $f(\alpha)$ satisfies appropriate conditions,[24] we can write

$$f(\alpha) = \int_0^\infty F(\tau) P_{-\frac{1}{2}+i\tau}(\cosh \alpha)\, d\tau, \qquad 0 \leqslant \alpha < \infty, \qquad (8.9.3)$$

where

$$F(\tau) = \tau \tanh \pi\tau \int_0^\infty f(\alpha) P_{-\frac{1}{2}+i\tau}(\cosh \alpha) \sinh \alpha\, d\alpha.$$

Then the integral

$$u = \int_0^\infty F(\tau) \frac{P_{-\frac{1}{2}+i\tau}(\cos \beta)}{P_{-\frac{1}{2}+i\tau}(\cos \beta_0)} P_{-\frac{1}{2}+i\tau}(\cosh \alpha)\, d\tau \qquad (8.9.4)$$

gives the solution of our problem, at least formally. For further details, including the solution of a problem of electrostatics, we refer the reader elsewhere.[25]

8.10. Solution of Laplace's Equation in Toroidal Coordinates

In addition to spherical and spheroidal coordinates, there are other coordinate systems whose use is intimately connected with Legendre functions.

---

[23] See N. N. Lebedev's dissertation (cited on p. 131), and V. A. Fock, *On the representation of an arbitrary function by an integral involving Legendre's functions with a complex index*, Doklady Akad. Nauk SSSR, **39**, 253 (1943). At discontinuity points, the integral in the right-hand side of (8.9.2) equals

$$\tfrac{1}{2}[f(x+0) + f(x-0)].$$

[24] E.g., if $f(\alpha)$ is continuous in $[0, A]$ for every finite A, and if $f(\alpha)$ falls off like $e^{-(\frac{1}{2}+\varepsilon)\alpha}$, $\varepsilon > 0$ as $\alpha \to \infty$.

[25] See N. N. Lebedev, *Solution of the Dirichlet problem for hyperboloids of revolution* (in Russian), Prikl. Mat. Mekh., **11**, 251 (1947).

First we consider *toroidal coordinates* α, β, φ, related to the rectangular coordinates x, y, z by the formulas

$$x = \frac{c \sinh \alpha \cos \varphi}{\cosh \alpha - \cos \beta}, \qquad y = \frac{c \sinh \alpha \sin \varphi}{\cosh \alpha - \cos \beta}, \qquad z = \frac{c \sin \beta}{\cosh \alpha - \cos \beta}, \tag{8.10.1}$$

where

$$0 \leqslant \alpha < \infty, \qquad -\pi < \beta \leqslant \pi, \qquad -\pi < \varphi \leqslant \pi,$$

and $c > 0$ is a scale factor.[26] This coordinate system is useful for solving boundary value problems involving the domain bounded by a torus, or the domain bounded by two intersecting spheres.[27] If a point has cylindrical coordinates r, φ and z, then

$$r = \frac{c \sinh \alpha}{\cosh \alpha - \cos \beta}, \qquad z = \frac{c \sin \beta}{\cosh \alpha - \cos \beta},$$

or more concisely,

$$z + ir = ic \coth \frac{\alpha + i\beta}{2}.$$

The corresponding triply orthogonal system of surfaces consists of the toroidal surfaces $\alpha = \text{const}$, described by the equation

$$(r - c \coth \alpha)^2 + z^2 = \left(\frac{c}{\sinh \alpha}\right)^2, \tag{8.10.2}$$

the spheres $\beta = \text{const}$, described by the equation

$$(z - c \cot \beta)^2 + r^2 = \left(\frac{c}{\sin \beta}\right)^2, \tag{8.10.3}$$

and the planes $\varphi = \text{const}$ (see Figure 31). It should be noted that all the spheres (8.10.3) intersect in the circle $r = c$, $z = 0$.

It follows from (8.10.1) that the square of the element of arc length is

$$ds^2 = \frac{c^2}{(\cosh \alpha - \cos \beta)^2}(d\alpha^2 + d\beta^2 + \sinh^2 \alpha \, d\varphi^2), \tag{8.10.4}$$

corresponding to the metric coefficients

$$h_\alpha = h_\beta = \frac{c}{\cosh \alpha - \cos \beta}, \qquad h_\varphi = \frac{c \sinh \alpha}{\cosh \alpha - \cos \beta}.$$

---

[26] It is clear from (8.10.1) that x, y and z are periodic in β and φ, with period 2π. Therefore we can choose $\beta_1 < \beta \leqslant \beta_1 + 2\pi$, $\varphi_1 < \varphi \leqslant \varphi_1 + 2\pi$ instead of $-\pi < \beta \leqslant \pi$, $-\pi < \varphi \leqslant \pi$ (which corresponds to the particular choice $\beta_1 = \varphi_1 = -\pi$), and it is sometimes convenient to do so (see Sec. 8.12).

[27] Later on, in Sec. 8.13, we will consider a closely related coordinate system, i.e., *bipolar coordinates*.

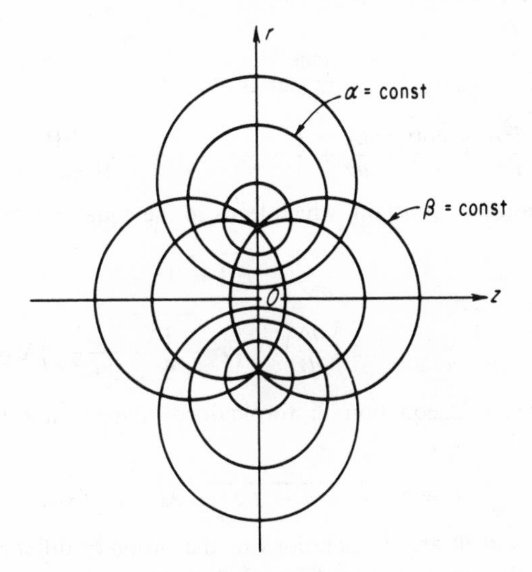

FIGURE 31

Therefore Laplace's equation in toroidal coordinates has the form

$$\frac{\partial}{\partial \alpha}\left(\frac{\sinh \alpha}{\cosh \alpha - \cos \beta}\frac{\partial u}{\partial \alpha}\right) + \frac{\partial}{\partial \beta}\left(\frac{\sinh \alpha}{\cosh \alpha - \cos \beta}\frac{\partial u}{\partial \beta}\right) + \frac{1}{(\cosh \alpha - \cos \beta)\sinh \alpha}\frac{\partial^2 u}{\partial \varphi^2} = 0. \tag{8.10.5}$$

Unlike the cases considered previously, we cannot separate variables in this equation. However, if we introduce a new unknown function v by making the substitution

$$u = \sqrt{2\cosh \alpha - 2\cos \beta}\, v, \tag{8.10.6}$$

then (8.10.5) goes into the equation

$$\frac{\partial^2 v}{\partial \alpha^2} + \frac{\partial^2 v}{\partial \beta^2} + \coth \alpha \frac{\partial v}{\partial \alpha} + \frac{1}{4}v + \frac{1}{\sinh^2 \alpha}\frac{\partial^2 v}{\partial \varphi^2} = 0, \tag{8.10.7}$$

which belongs to the class of equations permitting separation of variables. In fact, setting

$$v = \mathrm{A}(\alpha)\mathrm{B}(\beta)\Phi(\varphi), \tag{8.10.8}$$

we find that

$$\sinh^2 \alpha \frac{1}{\mathrm{A}}\frac{d^2\mathrm{A}}{d\alpha^2} + \frac{1}{\mathrm{B}}\frac{d^2\mathrm{B}}{d\beta^2} + \frac{\coth \alpha}{\mathrm{A}}\frac{d\mathrm{A}}{d\alpha} + \frac{1}{4} = -\frac{1}{\Phi}\frac{d^2\Phi}{d\varphi^2} = \mu^2,$$

where μ^2 is a constant. This implies

$$\frac{d^2\Phi}{d\varphi^2} + \mu^2\Phi = 0,$$
$$\frac{1}{A}\frac{d^2A}{d\alpha^2} + \frac{\coth\alpha}{A}\frac{dA}{d\alpha} + \frac{1}{4} - \frac{\mu^2}{\sinh^2\alpha} = -\frac{1}{B}\frac{d^2B}{d\beta^2} = \nu^2, \tag{8.10.9}$$

where ν^2 is another constant, which leads to the equations

$$\frac{d^2B}{d\beta^2} + \nu^2 B = 0, \tag{8.10.10}$$

$$\frac{1}{\sinh\alpha}\frac{d}{d\alpha}\left(\sinh\alpha\frac{dA}{d\alpha}\right) - \left(\nu^2 - \frac{1}{4} + \frac{\mu^2}{\sinh^2\alpha}\right)A = 0. \tag{8.10.11}$$

Thus Laplace's equation in toroidal coordinates has infinitely many solutions of the form

$$u = \sqrt{2\cosh\alpha - 2\cos\beta}\, A(\alpha)B(\beta)\Phi(\varphi), \tag{8.10.12}$$

where A, B and Φ are the solutions of the ordinary differential equations (8.10.9–11). By superposition of these solutions, we can solve various boundary value problems of mathematical physics for the domains mentioned at the beginning of this section. As usual, the case of rotational symmetry, where the function u is independent of the coordinate φ, corresponds to setting $\mu = 0$ and $\Phi = 1$. In this case, solving equations (8.10.10–11), we find that

$$u = \sqrt{2\cosh\alpha - 2\cos\beta}\,[AP_{\nu-\frac{1}{2}}(\cosh\alpha) + BQ_{\nu-\frac{1}{2}}(\cosh\alpha)] \times [C\cos\nu\beta + D\sin\nu\beta]. \tag{8.10.13}$$

8.11. The Dirichlet Problem for a Torus

To illustrate the application of toroidal coordinates, we now solve both the interior and exterior Dirichlet problems for the domain bounded by the toroidal surface $\alpha = \alpha_0$. To keep things simple, we consider the case of rotational symmetry, corresponding to $\mu = 0$, $\Phi = 1$. We also have the continuity conditions

$$u|_{\beta=-\pi} = u|_{\beta=\pi}, \qquad \left.\frac{\partial u}{\partial\beta}\right|_{\beta=-\pi} = \left.\frac{\partial u}{\partial\beta}\right|_{\beta=\pi}, \tag{8.11.1}$$

which are equivalent to the physical requirement that the solutions be periodic in the "cyclic" coordinate β. The conditions (8.11.1) are possible only if the parameter ν is an integer, which, without loss of generality, we can assume to be nonnegative, i.e., $\nu = n$ ($n = 0, 1, 2, \ldots$).

For the interior problem, we need solutions bounded in the domain

$\alpha_0 < \alpha \leqslant \infty$. Therefore, because of the behavior of $P_{n-\frac{1}{2}}(\cosh \alpha)$, $Q_{n-\frac{1}{2}}(\cosh \alpha)$ for large α, given by formulas (7.10.1, 8), we must set $A = 0$. On the other hand, for the exterior problem, which corresponds to the domain $0 \leqslant \alpha < \alpha_0$, we have to consider the behavior of $P_{n-\frac{1}{2}}(\cosh \alpha)$, $Q_{n-\frac{1}{2}}(\cosh \alpha)$ as $\alpha \to 0$, and then, according to (7.3.13, 23), we must set $B = 0$ if the solutions are to remain bounded. Thus the solutions of Laplace's equation suitable for solving the interior Dirichlet problem for a torus are

$$u = u_n = \sqrt{2 \cosh \alpha - 2 \cos \beta}\,[M_n \cos n\beta + N_n \sin n\beta] Q_{n-\frac{1}{2}}(\cosh \alpha), \qquad n = 0, 1, 2, \ldots, \tag{8.11.2}$$

while those suitable for solving the exterior problem are

$$u = u_n = \sqrt{2 \cosh \alpha - 2 \cos \beta}\,[M_n \cos n\beta + N_n \sin n\beta] P_{n-\frac{1}{2}}(\cosh \alpha), \qquad n = 0, 1, 2, \ldots \tag{8.11.3}$$

For this reason, $P_{n-\frac{1}{2}}(\cosh \alpha)$ and $Q_{n-\frac{1}{2}}(\cosh \alpha)$ are often called *toroidal functions.*

Example. *Find the electrostatic field due to a charged toroidal conductor at potential V.*

This problem reduces to solving the exterior Dirichlet problem with the boundary condition

$$\psi_{\alpha=\alpha_0} = V, \tag{8.11.4}$$

where ψ is the electrostatic potential. According to (8.10.2), the relation between the quantities c, α_0 and the geometric parameters α, l of the torus (see Figure 32) is given by

$$c \coth \alpha_0 = l, \qquad \frac{c}{\sinh \alpha_0} = a,$$

and hence

$$c = \sqrt{l^2 - a^2}, \qquad \cosh \alpha_0 = \frac{l}{a}.$$

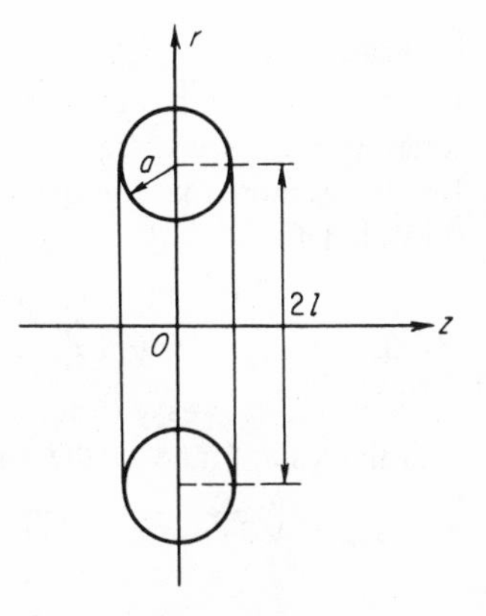

FIGURE 32

As shown above, we should look for a solution in the form of a series (8.11.2), where, because of the symmetry of the problem with respect to the plane $z = 0$, we must set $N_n = 0$, obtaining

$$u = \sqrt{2 \cosh \alpha - 2 \cos \beta} \sum_{n=0}^{\infty} M_n P_{n-\frac{1}{2}}(\cosh \alpha) \cos n\beta. \tag{8.11.5}$$

The boundary condition (8.11.4) will be satisfied if we determine the coefficients M_n from the relation

$$\frac{V}{\sqrt{2\cosh\alpha_0 - 2\cos\beta}} = \sum_{n=0}^{\infty} M_n P_{n-1/2}(\cosh\alpha_0)\cos n\beta, \qquad -\pi \leqslant \beta \leqslant \pi. \tag{8.11.6}$$

Expanding the left-hand side of (8.11.6) in a Fourier series in the interval $[-\pi, \pi]$ and using (7.10.10), we find that

$$\begin{aligned} M_n P_{n-1/2}(\cosh\alpha_0) &= \frac{2V}{\pi}\int_0^{\pi} \frac{\cos n\beta}{\sqrt{2\cosh\alpha_0 - 2\cos\beta}}\, d\beta \\ &= \frac{2V}{\pi} Q_{n-1/2}(\cosh\alpha_0), \qquad n = 1, 2, \ldots, \\ M_0 P_{-1/2}(\cosh\alpha_0) &= \frac{V}{\pi} Q_{-1/2}(\cosh\alpha_0), \end{aligned}$$

which leads to the following formal solution for ψ:

$$\psi = \frac{V}{\pi}\sqrt{2\cosh\alpha - 2\cos\beta}\left[\frac{P_{-1/2}(\cosh\alpha)}{P_{-1/2}(\cosh\alpha_0)} Q_{-1/2}(\cosh\alpha_0) + 2\sum_{n=1}^{\infty} \frac{P_{n-1/2}(\cosh\alpha)}{P_{n-1/2}(\cosh\alpha_0)} Q_{n-1/2}(\cosh\alpha_0)\cos n\beta\right]\cdot \tag{8.11.7}$$

By using the asymptotic representations of Sec. 7.11, it can be shown that the series (8.11.6) converges and actually gives the solution of our problem. Finally, we note that the charge density on the toroidal surface is given by the formula [cf. (6.6.10)]

$$\sigma = -\frac{1}{4\pi h_\alpha}\left.\frac{\partial u}{\partial \alpha}\right|_{\alpha=\alpha_0} = -\frac{1}{4\pi c}(\cosh\alpha_0 - \cos\beta)\left.\frac{\partial u}{\partial\alpha}\right|_{\alpha=\alpha_0}$$

Remark. It is easy to see that in the case where u is a function of all three coordinates α, β and φ, the appropriate solutions of Laplace's equation are

$$u = u_{mn} = \sqrt{2\cosh\alpha - 2\cos\beta}\,[M_{mn}\cos n\beta + N_{mn}\sin n\beta] \times Q^m_{n-1/2}(\cosh a)\begin{matrix}\cos m\varphi \\ \sin m\varphi\end{matrix}, \qquad m, n = 0, 1, 2, \ldots \tag{8.11.8}$$

for the interior problem, and

$$u = u_{mn} = \sqrt{2\cosh\alpha - \cos\beta}\,[M_{mn}\cos n\beta + N_{mn}\sin n\beta] \times P^m_{n-1/2}(\cosh\alpha)\begin{matrix}\cos m\varphi \\ \sin m\varphi\end{matrix}, \qquad m, n = 0, 1, 2, \ldots \tag{8.11.9}$$

for the exterior problem.

8.12. The Dirichlet Problem for a Domain Bounded by Two Intersecting Spheres

Toroidal coordinates can also be used to solve boundary value problems involving a domain bounded by two spheres S_1 and S_2 which intersect in a circle γ. Let x, y, z be a system of rectangular coordinates with origin at the center of γ, and let the z-axis pass through the center of the spheres (see Figure 33). Let α, β, φ be a system of toroidal coordinates related to x, y, z by the formulas (8.10.1), and choose the constant c equal to the radius of γ. Finally, let β_p be the angle between the plane $z = 0$ and the tangent plane to the sphere S_p $(p = 1, 2)$, drawn through any point of the circle γ, where $0 < \beta_1 < \beta_2 < 2\pi$. Then it follows from (8.10.3) that the equation of the sphere S_p in toroidal coordinates is $\beta = \beta_p$. Moreover, of the two domains bounded by the spheres, the interior domain D_i corresponds to the interval $\beta_1 < \beta < \beta_2$, while the exterior domain D_e corresponds to the interval $\beta_2 < \beta < \beta_1 + 2\pi$. In both D_i and D_e, the variable α ranges over the interval $0 \leqslant \alpha < \infty$, where points on the z-axis correspond to $\alpha = 0$ and points on the edge γ correspond to $\alpha = \infty$.[28]

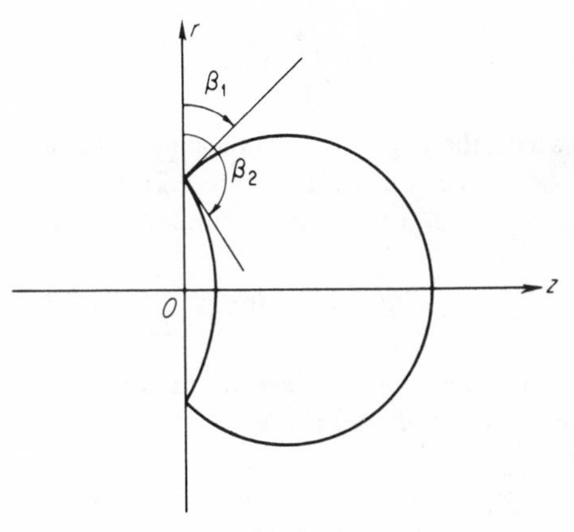

FIGURE 33

We now consider the Dirichlet problem for the domains D_i and D_e, confining ourselves to the rotationally symmetric case. Just as before, we start from the solutions (8.10.13), but unlike Sec. 8.11, there is no longer any need to restrict ν to be a nonnegative integer. In fact, as we shall soon see, the solution of our problem can be constructed by superposition of solutions of the form

$$u = u_\tau = \sqrt{2\cosh\alpha - 2\cos\beta}\,[M_\tau \cosh\tau\beta + N_\tau \sinh\tau\beta] \times P_{-\frac{1}{2}+i\tau}(\cosh\alpha), \qquad \tau \geqslant 0, \quad (8.12.1)$$

obtained from (8.10.13) by choosing $\nu = i\tau$ and setting $B = 0$.[29] We begin with the interior problem, and assume that the functions $f_p = f_p(\alpha)$ appearing in the boundary conditions

$$u\big|_{\beta=\beta_p} = f_p, \qquad p = 1, 2 \quad (8.12.2)$$

[28] Note also that $x = y = 0$, $z \to \pm\infty$ if $\alpha = 0$, $\beta \to 2\pi\pm$.

[29] This is necessary for the solution to be bounded on the z-axis.

are such that the functions

$$\varphi_p(\alpha) = \frac{f_p(\alpha)}{\sqrt{2\cosh\alpha - 2\cos\beta_p}}, \qquad p = 1, 2$$

can be represented as integrals of the form

$$\varphi_p(\alpha) = \int_0^\infty \Phi_p(\tau) P_{-\frac{1}{2}+i\tau}(\cosh\alpha)\, d\tau, \qquad 0 \leqslant \alpha < \infty, \tag{8.12.3}$$

where the expansion coefficients $\Phi_p(\tau)$ are independent of α. According to the Mehler-Fock theorem (8.9.2), such a representation is possible, and the functions $\Phi_p(\tau)$ can be calculated from the formula

$$\Phi_p(\tau) = \tau \tanh \pi\tau \int_0^\infty \varphi_p(\alpha) P_{-\frac{1}{2}+i\tau}(\cosh\alpha) \sinh\alpha\, d\alpha, \tag{8.12.4}$$

if the functions $f_p(\alpha)$ are continuous and of bounded variation in $[0, A]$ for every finite A, and if the integrals

$$\int_0^\infty \alpha |f_p(\alpha)|\, d\alpha, \qquad p = 1, 2 \tag{8.12.5}$$

are finite.

The last condition presupposes that the $f_p(\alpha)$ approach zero sufficiently rapidly as $\alpha \to \infty$, i.e., as the circular edge γ is approached. On the other hand,

$$\lim_{\alpha\to\infty} f_p = f_p(\infty) = u_\gamma,$$

where u_γ is the value taken by the solution u on the edge γ, and this value is usually not zero.[30] However, in most cases of practical importance, the modified functions

$$f_p^*(\alpha) = f_p(\alpha) - f_p(\infty), \qquad p = 1, 2$$

fall off sufficiently rapidly as $\alpha \to \infty$, and hence there exists an expansion

$$\varphi_p^*(\alpha) = \frac{f_p(\alpha) - f_p(\infty)}{\sqrt{2\cosh\alpha - 2\cos\beta_p}} = \int_0^\infty \Phi_p^*(\tau) P_{-\frac{1}{2}+i\tau}(\cosh\alpha)\, d\tau, \tag{8.12.6}$$

where $\Phi_p^*(\tau)$ is given by

$$\Phi_p^*(\tau) = \tau \tanh \pi\tau \int_0^\infty \varphi_p^*(\alpha) P_{-\frac{1}{2}+i\tau}(\cosh\alpha) \sinh\alpha\, d\alpha. \tag{8.12.7}$$

---

[30] Here we assume that the boundary function is continuous, but all our considerations can easily be extended to the case of piecewise continuity, where $\lim_{\alpha\to\infty} f_1$ may not equal $\lim_{\alpha\to\infty} f_2$.

Moreover, it is not hard to show that[31]

$$\frac{1}{\sqrt{2\cosh\alpha - 2\cos\beta_p}} = \int_0^\infty \frac{\cosh(\pi - \beta_p)\tau}{\cosh\pi\tau} P_{-\frac{1}{2}+i\tau}(\cosh\alpha)\,d\tau \qquad (8.12.8)$$

for $0 < \beta_p < 2\pi$. Multiplying (8.12.8) by $f_p(\infty)$ and adding the result to (8.12.6), we obtain an expansion for $\varphi_p(\alpha)$ of the required form (8.12.3), where

$$\Phi_p(\tau) = \Phi_p^*(\tau) + \frac{f_p(\infty)}{\cosh\pi\tau}\cosh(\pi - \beta_p)\tau. \qquad (8.12.9)$$

Now consider the integral

$$u = \sqrt{2\cosh\alpha - 2\cos\beta}\int_0^\infty \frac{\Phi_2\sinh(\beta - \beta_1)\tau + \Phi_1\sinh(\beta_2 - \beta)\tau}{\sinh(\beta_2 - \beta_1)\tau} \times P_{-\frac{1}{2}+i\tau}(\cosh\alpha)\,d\tau, \qquad (8.12.10)$$

made up of particular solutions of the form (8.12.1). We see at once that (8.12.10) satisfies the boundary conditions (8.12.2) and hence gives the solution of the interior Dirichlet problem. Similarly, the solution of the exterior Dirichlet problem can be written in the form

$$u = \sqrt{2\cosh\alpha - 2\cos\beta}\int_0^\infty \frac{\Phi_1\sinh(\beta - \beta_2)\tau + \Phi_2\sinh(2\pi + \beta_1 - \beta)\tau}{\sinh(2\pi + \beta_1 - \beta_2)\tau} \times P_{-\frac{1}{2}+i\tau}(\cosh\alpha)\,d\tau, \qquad (8.12.11)$$

where

$$f_1(\alpha) = u|_{\beta=\beta_1+2\pi}, \qquad f_2(\alpha) = u|_{\beta=\beta_2},$$

and the rest of the notation is the same as before.

Example. *Consider the "spherical bowl" or zone obtained by setting $\beta_1 = \beta_2 = \beta_0$ in Figure 33. Find the electrostatic field due to a thin charged conductor of this shape at potential V.*

This is just the exterior Dirichlet problem for determining the electrostatic potential ψ, in the special case where

$$\beta_1 = \beta_2 = \beta_0, \qquad f_1(\alpha) = f_2(\alpha) = V.$$

[31] Combining the formulas

$$\frac{1}{\sqrt{2\cosh x + 2\cosh\alpha}} = \frac{2}{\pi}\int_0^\infty \cos x\tau\,d\tau\int_0^\infty \frac{\cos\tau\theta}{\sqrt{2\cosh\theta + 2\cosh\alpha}}\,d\theta$$

[cf. (6.5.3–4)] and

$$P_{-\frac{1}{2}+i\tau}(\cosh\alpha) = \frac{2}{\pi}\cosh\pi\tau\int_0^\infty \frac{\cos\tau\theta}{\sqrt{2\cosh\theta + 2\cosh\alpha}}\,d\theta$$

[cf. (7.4.6)], we find that

$$\frac{1}{\sqrt{2\cosh x + 2\cosh\alpha}} = \int_0^\infty \frac{\cos x\tau}{\cosh\pi\tau} P_{-\frac{1}{2}+i\tau}(\cosh\alpha)\,d\tau,$$

which gives (8.12.8) after setting $x = i(\pi - \beta_p)$.

The functions Φ_1 and Φ_2 can be read off at once from (8.12.8):

$$\Phi_1 = \Phi_2 = V\frac{\cosh(\pi - \beta_0)\tau}{\cosh \pi\tau}.$$

Then formula (8.12.11) becomes

$$\psi = V\sqrt{2\cosh\alpha - 2\cos\beta}\int_0^\infty \frac{\cosh(\pi-\beta_0)\tau}{\cosh^2\pi\tau}\cosh(\pi+\beta_0-\beta)\tau \times P_{-\frac{1}{2}+i\tau}(\cosh\alpha)\,d\tau,\quad \beta_0 < \beta < \beta_0 + 2\pi. \tag{8.12.12}$$

Substituting

$$P_{-\frac{1}{2}+i\tau}(\cosh\alpha) = \frac{2}{\pi}\coth\pi\tau\int_\alpha^\infty \frac{\sin\theta\tau}{\sqrt{2\cosh\theta - 2\cosh\alpha}}\,d\theta$$

[cf. (7.4.7)] into (8.12.12), and integrating first with respect to τ and then with respect to θ, we find after some manipulation that the solution can be expressed in closed form in terms of elementary functions:[32]

$$\psi = \frac{V}{2}\left[1 + \sqrt{\frac{\cosh\alpha - \cos\beta}{\cosh\alpha - \cos(2\beta_0 - \beta)}}\right] + \frac{V}{\pi}\left[\arctan\frac{\sqrt{2}\cos\frac{\beta}{2}}{\sqrt{\cosh\alpha\cos\beta}} - \sqrt{\frac{\cosh\alpha - \cos\beta}{\cosh\alpha - \cos(2\beta_0-\beta)}}\arctan\frac{\sqrt{2}\cos\frac{2\beta_0 - \beta}{2}}{\sqrt{\cosh\alpha - \cos(2\beta_0 - \beta)}}\right]. \tag{8.12.13}$$

The fact that (8.12.13) satisfies the boundary conditions is immediately apparent.

8.13. Solution of Laplace's Equation in Bipolar Coordinates

There is still another coordinate system which leads to solutions of Laplace's equation involving Legendre functions, i.e., three-dimensional *bipolar coordinates* α, β, φ, related to the rectangular coordinates x, y, z by the formulas

$$x = \frac{c\sin\alpha\cos\varphi}{\cosh\beta - \cos\alpha},\qquad y = \frac{c\sin\alpha\sin\varphi}{\cosh\beta - \cos\alpha},\qquad z = \frac{c\sinh\beta}{\cosh\beta - \cos\alpha}, \tag{8.13.1}$$

---

[32] In integrating with respect to τ, use the formula

$$\int_0^\infty \frac{\cosh p\tau}{\sinh q\tau}\sin r\tau\,d\tau = \frac{\pi}{2q}\frac{\sinh\frac{\pi r}{q}}{\cosh\frac{\pi r}{q} + \cos\frac{\pi p}{q}},\qquad 0 \leqslant p < q.$$

where

$$0 \leqslant \alpha \leqslant \pi, \qquad -\infty < \beta < \infty, \qquad -\pi < \varphi \leqslant \pi,$$

and $c > 0$ is a scale factor. This system is closely related to the toroidal coordinates studied in Secs. 8.10–12, and is suitable for solving boundary value problems for the domain bounded by two nonintersecting spheres. If a point has cylindrical coordinates r, φ and z, then

$$r = \frac{c \sin \alpha}{\cosh \beta - \cos \alpha}, \qquad z = \frac{c \sinh \beta}{\cosh \beta - \cos \alpha},$$

or more concisely

$$z + ir = ic \cot \frac{\alpha + i\beta}{2}.$$

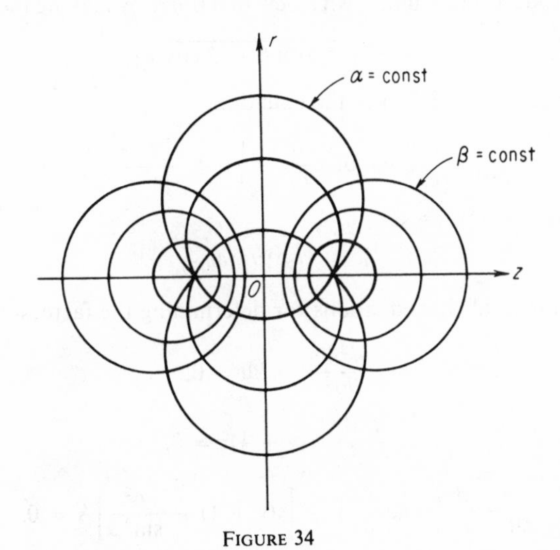

FIGURE 34

The corresponding triply orthogonal family of surfaces consists of the "spindle-shaped" surfaces $\alpha = \text{const}$, described by the equation

$$(r - c \cot \alpha)^2 + z^2 = \left(\frac{c}{\sin \alpha}\right)^2, \tag{8.13.2}$$

the spheres $\beta = \text{const}$, described by the equation

$$(z - c \coth \beta)^2 + r^2 = \left(\frac{c}{\sinh \beta}\right)^2, \tag{8.13.3}$$

and the planes $\varphi = \text{const}$ (see Figure 34). The points $r = 0$, $z = \pm c$ correspond to the values $\beta = \infty$, while $r = 0$, $z \to \pm\infty$ if $\alpha = 0$, $\beta \to 0\pm$.

It follows from (8.13.1) that the square of the element of arc length is

$$ds^2 = \frac{c^2}{(\cosh \beta - \cos \alpha)^2}(d\alpha^2 + d\beta^2 + \sin^2 \alpha \, d\varphi^2), \tag{8.13.4}$$

corresponding to the metric coefficients

$$h_\alpha = h_\beta = \frac{c}{\cosh \beta - \cos \alpha}, \qquad h_\varphi = \frac{c \sin \alpha}{\cosh \beta - \cos \alpha}.$$

Therefore Laplace's equation in bipolar coordinates has the form

$$\frac{\partial}{\partial \alpha}\left(\frac{\sin \alpha}{\cosh \beta - \cos \alpha}\frac{\partial u}{\partial \alpha}\right) + \frac{\partial}{\partial \beta}\left(\frac{\sin \alpha}{\cosh \beta - \cos \alpha}\frac{\partial u}{\partial \beta}\right) + \frac{1}{(\cosh \beta - \cos \alpha)\sin \alpha}\frac{\partial^2 u}{\partial \varphi^2} = 0. \tag{8.13.5}$$

Just as in the case of toroidal coordinates, we can separate variables, provided we first introduce a new unknown a new function v by making the substitution

$$u = \sqrt{2 \cosh \beta - 2 \cos \alpha}\, v, \tag{8.13.6}$$

which transforms (8.13.5) into the equation

$$\frac{\partial^2 v}{\partial \alpha^2} + \frac{\partial^2 v}{\partial \beta^2} + \cot \alpha \frac{\partial v}{\partial \alpha} - \frac{1}{4} v + \frac{1}{\sin^2 \alpha}\frac{\partial^2 v}{\partial \varphi^2} = 0. \tag{8.13.7}$$

To solve (8.13.7), we set

$$v = \mathrm{A}(\alpha)\mathrm{B}(\beta)\Phi(\varphi). \tag{8.13.8}$$

This gives the following equations for determining the factors A, B and Φ:

$$\frac{d^2\Phi}{d\varphi^2} + \mu^2\Phi = 0, \tag{8.13.9}$$

$$\frac{d^2\mathrm{B}}{d\beta^2} - (\nu + \tfrac{1}{2})^2 = 0, \tag{8.13.10}$$

$$\frac{1}{\sin \alpha}\frac{d}{d\alpha}\left(\sin \alpha \frac{d\mathrm{A}}{d\alpha}\right) + \left[\nu(\nu + 1) - \frac{\mu^2}{\sin^2 \alpha}\right]\mathrm{A} = 0. \tag{8.13.11}$$

The first two equations can be solved in terms of elementary functions, and the third in terms of Legendre functions. In particular, for the rotationally symmetric case, where the solution u is independent of φ, we find that

$$u = \sqrt{2 \cosh \beta - 2 \cos \alpha}\,[AP_\nu(\cos \alpha) + BQ_\nu(\cos \alpha)] \times [C \cosh (\nu + \tfrac{1}{2})\beta + D \sinh (\nu + \tfrac{1}{2})\beta]. \tag{8.13.12}$$

In problems involving the domain bounded by two nonintersecting spheres $\beta = \beta_1$ and $\beta = \beta_2$, the variable α ranges over the closed interval $[0, \pi]$, and hence to obtain solutions which are finite on the z-axis we must set $B = 0$ and $\nu = n$ $(n = 0, 1, 2, \ldots)$, as in Sec. 8.3. Thus, for this class of problems, the appropriate particular solutions of Laplace's equation are

$$u = u_n = \sqrt{2 \cosh \beta - 2 \cos \alpha}\,[M_n \cosh (n + \tfrac{1}{2})\beta + N_n \sinh (n + \tfrac{1}{2})\beta] \times P_n(\cos \alpha), \qquad n = 0, 1, 2, \ldots \tag{8.13.13}$$

On the other hand, in problems involving the domain bounded by the surface $\alpha = \alpha_0$, the appropriate particular solutions are obtained by choosing $\nu = -\frac{1}{2} + i\tau$ ($\tau \geqslant 0$), and are of the form

$$u = u_\tau = \sqrt{2\cosh\beta - 2\cos\alpha}\,[M_\tau \cos\tau\beta + N_\tau \sin\tau\beta] \times P_{-\frac{1}{2}+i\tau}(\pm\cos\alpha), \qquad \tau \geqslant 0, \tag{8.13.14}$$

where the plus sign corresponds to the exterior problem ($0 \leqslant \alpha < \alpha_0$) and the minus sign to the interior problem ($\alpha_0 < \alpha \leqslant \pi$).

Example. *Find the electrostatic field between two spherical conductors of radius a, whose centers are a distance 2l apart, if one conductor is at potential −V and the other is at potential +V.*

The spheres have equations $\beta = \pm\beta_0$ in bipolar coordinates, if we choose the quantities c, β_0 such that

$$c\coth\beta_0 = l, \qquad \frac{c}{\sinh\beta_0} = a,$$

i.e.,

$$c = \sqrt{l^2 - a^2}, \qquad \cosh\beta_0 = \frac{l}{a}$$

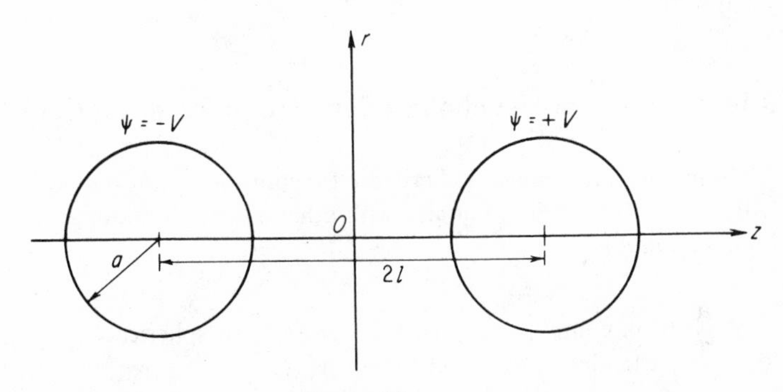

FIGURE 35

(see Figure 35). Then the problem reduces to finding a function ψ (where ψ is the electrostatic potential) which is harmonic in the domain $-\beta_0 < \beta < \beta_0$ and satisfies the boundary conditions

$$\psi|_{\beta=-\beta_0} = -V, \qquad \psi|_{\beta=\beta_0} = V.$$

Using (8.13.13) and noting that u is an odd function of β, we look for a solution of the form

$$\psi = \sqrt{2\cosh\beta - 2\cos\alpha}\sum_{n=0}^{\infty} M_n P_n(\cos\alpha)\sinh(n + \tfrac{1}{2})\beta. \tag{8.13.15}$$

The constants M_n can be determined from the condition

$$\frac{V}{\sqrt{2\cosh\beta_0 - 2\cos\alpha}} = \sum_{n=0}^{\infty} M_n P_n(\cos\alpha)\sinh(n+\tfrac{1}{2})\beta_0, \qquad 0 \leqslant \alpha \leqslant \pi.$$

Using (4.2.3) to expand the left-hand side in a series of Legendre polynomials, we obtain

$$\begin{aligned}\frac{V}{\sqrt{2\cosh\beta_0 - 2\cos\alpha}} &= \frac{Ve^{-\beta_0/2}}{\sqrt{1 - 2e^{-\beta_0}\cos\alpha + e^{-2\beta_0}}}\\ &= V\sum_{n=0}^{\infty} e^{-(n+\frac{1}{2})\beta_0} P_n(\cos\alpha),\end{aligned}$$

which implies

$$M_n \sinh(n+\tfrac{1}{2})\beta_0 = Ve^{-(n+\frac{1}{2})\beta_0}.$$

Thus the formal solution of the problem is given by the series

$$\psi = V\sqrt{2\cosh\beta - 2\cos\alpha}\sum_{n=0}^{\infty} e^{-(n+\frac{1}{2})\beta_0}\frac{\sinh(n+\frac{1}{2})\beta}{\sinh(n+\frac{1}{2})\beta_0}P_n(\cos\alpha). \quad (8.13.16)$$

The fact that (8.13.16) converges and satisfies the boundary conditions is easily verified.

8.14. Solution of Helmholtz's Equation in Spherical Coordinates

In mathematical physics, Legendre functions arise not only when dealing with Laplace's equation, but also with other equations, among which Helmholtz's equation

$$\nabla^2 u + k^2 u = 0 \quad (8.14.1)$$

is of particular importance. To solve (8.14.1) in spherical coordinates, we look for particular solutions of the form

$$u = R(r)\Theta(\theta)\Phi(\varphi),$$

just as in Sec. 8.2. The variables separate immediately, and we obtain the following differential equations for determining the factors R, Θ and Φ:

$$\frac{d^2\Phi}{d\varphi^2} + \mu^2\Phi = 0, \quad (8.14.2)$$

$$\frac{1}{\sin\theta}\frac{d}{d\theta}\left(\sin\theta\frac{d\Theta}{d\theta}\right) + \left[\nu(\nu+1) - \frac{\mu^2}{\sin^2\theta}\right]\Theta = 0, \quad (8.14.3)$$

$$\frac{d}{dr}\left(r^2\frac{dR}{dr}\right) + [k^2r^2 - \nu(\nu+1)]R = 0. \quad (8.14.4)$$

Here μ and ν are arbitrary real or complex parameters, but without loss of generality we can assume that $\operatorname{Re}\mu \geqslant 0$, $\operatorname{Re}\nu \geqslant -\frac{1}{2}$ (cf. footnote 2, p. 206).

Equations (8.14.2–3) coincide with equations (8.2.5–6), and can be solved in terms of elementary functions in the first case, and in terms of Legendre functions in the second case. Under the substitution

$$R = r^{-1/2}v,$$

equation (8.14.4) goes into

$$v'' + \frac{1}{r}v' + \left[k^2 - \frac{(\nu + \frac{1}{2})^2}{r^2}\right]v = 0. \tag{8.14.5}$$

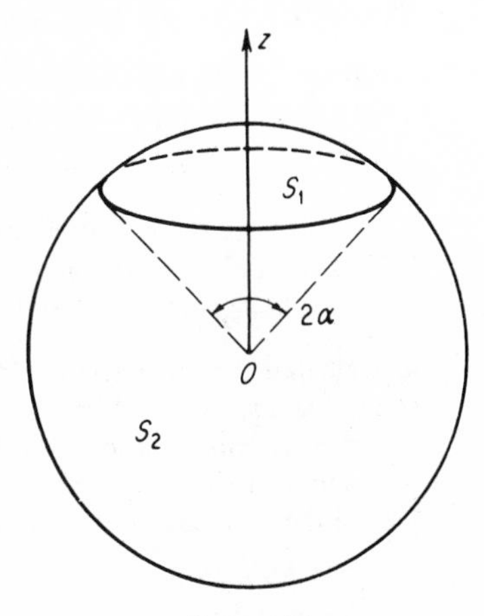

FIGURE 36

This is Bessel's equation of argument $z = kr$, whose general solution can be expressed in terms of cylinder functions. In particular, in the rotationally symmetric case, where u is independent of the coordinate φ, we have

$$u = r^{-1/2}[AJ_{\nu+1/2}(kr) + BH^{(2)}_{\nu+1/2}(kr)][CP_\nu(\cos\theta) + DQ_\nu(\cos\theta)], \tag{8.14.6}$$

where $J_\nu(z)$ is the Bessel function of the first kind and $H^{(2)}_{\nu+1/2}$ is the second Hankel function.[33] In problems where θ varies over the interval $[0, \pi]$, the boundedness requirement compels us to set $D = 0$ and $\nu = n$ $(n = 0, 1, 2, \ldots)$. By superposition of the particular solutions (8.14.6), we can solve many problems of mathematical physics, including the important problem of diffraction of electromagnetic waves by the earth's surface.[34]

PROBLEMS

1. Let the surface of a sphere of radius a be divided into two regions S_1 and S_2 as shown in Figure 36. Find the stationary distribution of temperature u in the sphere if S_1 is held at temperature u_0, while S_2 is held at temperature zero.

Ans.

$$u(r, \theta) = \frac{u_0}{2}\left\{1 - \cos\alpha - \sum_{n=1}^{\infty}[P_{n+1}(\cos\alpha) - P_{n-1}(\cos\alpha)]\left(\frac{r}{a}\right)^n P_n(\cos\theta)\right\}.$$

---

[33] This form of the solution is convenient for problems involving steady-state oscillations, when the time dependence is described by the factor $e^{i\omega t}$. If the time dependence is described by $e^{-i\omega t}$, we replace $H^{(2)}_{\nu+1/2}(kr)$ by $H^{(1)}_{\nu+1/2}(kr)$.

[34] G. A. Grinberg, *op. cit.*, Chap. 23.

2. Find the potential ψ of the electromagnetic field inside a sphere of radius a if one hemisphere (corresponding to $0 \leqslant \theta < \pi/2$) is held at potential V, while the other hemisphere (corresponding to $\pi/2 < \theta \leqslant \pi$) is held at potential zero (cf. footnote 17, p. 160).

Ans.

$$\psi(r, \theta) = \frac{V}{a}\left[1 + \sum_{n=0}^{\infty} \frac{4n+3}{2n+2} P_{2n}(0) \left(\frac{r}{a}\right)^{2n+1} P_{2n+1}(\cos\theta)\right].$$

3. Find the stationary distribution of temperature u in a prolate spheroid if half of its surface (corresponding to $z > 0$) is held at temperature u_0, while the other half (corresponding to $z < 0$) is held at temperature zero.

4. Calculate the gravitational potentials ψ_i, ψ_e (see Sec. 8.8) of a homogeneous solid oblate spheroid. Introducing spherical coordinates r, θ and φ, derive an asymptotic representation of ψ_e for small c, where c is the distance from the origin to the focus and verify that $\psi_e \to m/r$ as $c \to 0$, the result to be expected. Derive the corresponding asymptotic formula for the prolate spheroid.

Hint. Note that

$$\cosh\alpha = \frac{r}{2c}\left[\sqrt{1 + \frac{2c}{r}\cos\theta + \frac{c^2}{r^2}} + \sqrt{1 - \frac{2c}{r}\cos\theta + \frac{c^2}{r^2}}\right],$$

$$\cos\beta = \frac{r}{2c}\left[\sqrt{1 + \frac{2c}{r}\cos\theta + \frac{c^2}{r^2}} - \sqrt{1 - \frac{2c}{r}\cos\theta + \frac{c^2}{r^2}}\right].$$

Ans. For the prolate spheroid,

$$\psi_e|_{c\to 0} \approx m\left[\frac{1}{r} + \frac{c^2}{5r^3} P_2(\cos\theta)\right].$$

5. Find the electrostatic potential ψ inside a hollow prolate spheroid with semiaxes a and b, which has a point charge q at its center and whose surface is held at potential zero.

Hint. Write ψ as the sum of the potential ψ_0 due to the source and the potential u due to the induced charges. Use the formula [35]

$$\int_{-1}^{1} \frac{P_{2n}(x)}{\sqrt{\sinh^2\alpha + x^2}}\, dx = 2P_{2n}(0)Q_{2n}(\cosh\alpha).$$

Ans.

$$\psi(\alpha, \beta) = \frac{q}{\sqrt{a^2 - b^2}}\left[\frac{1}{\sqrt{\sinh^2\alpha + \cos^2\beta}} - \sum_{n=0}^{\infty}(4n+1)P_{2n}(0)\frac{Q_{2n}(\cosh\alpha_0)}{P_{2n}(\cosh\alpha_0)} P_{2n}(\cosh\alpha)P_{2n}(\cos\beta)\right],$$

where $\tanh\alpha_0 = b/a$.

---

[35] See Problem 16, formula (ii), p. 202.

6. Calculate the surface charge density σ on a conducting disk of radius a due to a point charge q a distance l from the disk along its axis of symmetry (see Figure 37).

Hint. Note that the disk is a limiting case of an oblate spheroid. Use the formula[36]

$$\int_{-1}^{1} \frac{P_{2n}(x)}{\sqrt{\cosh^2 \alpha - x^2}}\, dx = 2iP_{2n}(0)Q_{2n}(i \sinh \alpha).$$

Ans.

$$\sigma = -\frac{q}{4\pi a^2}\left[\left(\frac{a}{l}\right)^2\left(1 + \frac{a^2 \sin^2 \beta}{l^2}\right)^{-3/2} + \frac{2i}{\sqrt{\pi}\cos\beta}\sum_{n=0}^{\infty}(-1)^n \frac{(4n+1)n!}{\Gamma(n+\frac{1}{2})} Q_{2n}\left(\frac{il}{a}\right)P_{2n}(\cos\beta)\right],$$

where $\beta = \arcsin(r/a)$ and r is the distance from the center of the disk to an arbitrary point on its surface.

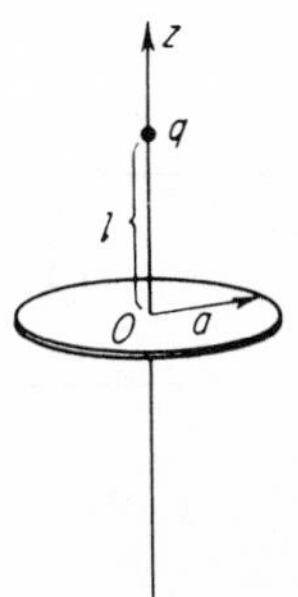

FIGURE 37

7. Suppose a constant electric field E_0 acts along the axis of symmetry of a grounded conducting torus. What is the electrostatic potential ψ along this axis?

Hint. Use formula (7.10.10), after integrating by parts.

Ans.

$$\psi|_{r=0} = -E_0 z + \frac{8}{\pi}E_0\sqrt{l^2 - a^2}\sin\frac{\beta}{2}\sum_{n=1}^{\infty}\frac{nQ_{n-\frac{1}{2}}(\cosh\alpha_0)}{P_{n-\frac{1}{2}}(\cosh\alpha_0)}\sin n\beta,$$

where $\cosh \alpha_0 = l/a$, and the notation is the same as in Sec. 8.11.

8. Solve the preceding problem, assuming instead that the external field is due to a point charge q at the center of the torus.

9. Find the electrostatic potential ψ outside a conductor at potential V, which has the form of the "spindle-shaped" surface mentioned on p. 231 in connection with bipolar coordinates.

Hint. Cf. (8.13.14) and (7.4.6).

Ans.

$$\psi(\alpha, \beta) = V\sqrt{2\cosh\beta - 2\cos\alpha} \times \int_0^{\infty}\frac{\cos\beta\tau}{\cosh\pi\tau}\frac{P_{-\frac{1}{2}+i\tau}(-\cos\alpha_0)}{P_{-\frac{1}{2}+i\tau}(\cos\alpha_0)}P_{-\frac{1}{2}+i\tau}(\cos\alpha)\, d\tau.$$

---

[36] See Problem 16, formula (i), p. 202.

9

HYPERGEOMETRIC FUNCTIONS

9.1. The Hypergeometric Series and Its Analytic Continuation

By the *hypergeometric series* (already introduced in Sec. 7.2) is meant the power series

$$\sum_{k=0}^{\infty} \frac{(\alpha)_k(\beta)_k}{(\gamma)_k k!} z^k, \tag{9.1.1}$$

where z is a complex variable, α, β and γ are parameters which can take arbitrary real or complex values (provided that $\gamma \neq 0, -1, -2, \ldots$), and the symbol $(\lambda)_k$ denotes the quantity

$$(\lambda)_0 = 1, \qquad (\lambda)_k = \frac{\Gamma(\lambda + k)}{\Gamma(\lambda)} = \lambda(\lambda + 1)\cdots(\lambda + k - 1), \qquad k = 1, 2, \ldots$$

If either α or β is zero or a negative integer, the series terminates after a finite number of terms, and its sum is then a polynomial in z. Except for this case, the radius of convergence of the hypergeometric series is 1, as is easily seen by using the ratio test.[1]

The sum of the series (9.1.1), i.e., the function

$$F(\alpha, \beta; \gamma; z) = \sum_{k=0}^{\infty} \frac{(\alpha)_k(\beta)_k}{(\gamma)_k k!} z^k, \qquad |z| < 1, \tag{9.1.2}$$

---

[1] Writing

$$u_k = \frac{(\alpha)_k(\beta)_k}{(\gamma)_k k!} z^k,$$

we have

$$\left|\frac{u_{k+1}}{u_k}\right| = \left|\frac{(\alpha + k)(\beta + k)}{(\gamma + k)(1 + k)} z\right| \to |z|$$

as $k \to \infty$, so that the hypergeometric series converges for $|z| < 1$ and diverges for $|z| > 1$.

is called the *hypergeometric function*, but this definition is only suitable when z lies inside the unit circle. We now show that there exists a complex function which is analytic in the z-plane cut along the segment $[1, \infty]$ and coincides with $F(\alpha, \beta; \gamma; z)$ for $|z| < 1$. This function is the analytic continuation of $F(\alpha, \beta; \gamma; z)$ into the cut plane, and will be denoted by the same symbol. To carry out this analytic continuation, we first assume that $\operatorname{Re}\gamma > \operatorname{Re}\beta > 0$ and use the integral representation

$$\frac{(\beta)_k}{(\gamma)_k} = \frac{\Gamma(\gamma)}{\Gamma(\beta)\Gamma(\gamma-\beta)} \int_0^1 t^{\beta-1+k}(1-t)^{\gamma-\beta-1}\,dt, \qquad k = 0, 1, 2, \ldots, \tag{9.1.3}$$

implied by the formulas of Sec. 1.5. Substitution of (9.1.3) into (9.1.2) gives

$$\begin{aligned} F(\alpha, \beta; \gamma; z) &= \frac{\Gamma(\gamma)}{\Gamma(\beta)\Gamma(\gamma-\beta)} \sum_{k=0}^{\infty} \frac{(\alpha)_k}{k!} z^k \int_0^1 t^{\beta-1+k}(1-t)^{\gamma-\beta-1}\,dt \\ &= \frac{\Gamma(\gamma)}{\Gamma(\beta)\Gamma(\gamma-\beta)} \int_0^1 t^{\beta-1}(1-t)^{\gamma-\beta-1}\,dt \sum_{k=0}^{\infty} \frac{(\alpha)_k}{k!}(zt)^k, \end{aligned}$$

where, as usual, reversing the order of summation and integration is justified by an absolute convergence argument.[2] According to the binomial expansion (cf. footnote 17, p. 121),

$$\sum_{k=0}^{\infty} \frac{(\alpha)_k}{k!}(zt)^k = (1-tz)^{-\alpha}, \qquad 0 \leqslant t \leqslant 1, \quad |z| < 1,$$

and hence $F(\alpha, \beta; \gamma; z)$ has the representation

$$F(\alpha, \beta; \gamma; z) = \frac{\Gamma(\gamma)}{\Gamma(\beta)\Gamma(\gamma-\beta)} \int_0^1 t^{\beta-1}(1-t)^{\gamma-\beta-1}(1-tz)^{-\alpha}\,dt,$$
$$\operatorname{Re}\gamma > \operatorname{Re}\beta > 0, \quad |z| < 1. \tag{9.1.4}$$

The next step is to show that the integral in (9.1.4) has meaning and represents an analytic function of z in the plane cut along $[1, \infty]$. If z belongs to the closed domain

$$\rho \leqslant |z-1| \leqslant R, \qquad |\arg(1-z)| \leqslant \pi - \delta, \tag{9.1.5}$$

where $R > 0$ is arbitrarily large and $\rho > 0$, $\delta > 0$ are arbitrarily small, and if $0 < t < 1$, then the integrand

$$t^{\beta-1}(1-t)^{\gamma-\beta-1}(1-tz)^{-\alpha}$$

is continuous in t for every z and analytic in z for every t, and we need only

---

[2] In fact, if $\operatorname{Re}\gamma > \operatorname{Re}\beta > 0$ and $|z| < 1$, then

$$\begin{aligned} \sum_{k=0}^{\infty} \frac{|(\alpha)_k|}{k!}|z|^k \int_0^1 |t^{\beta-1+k}(1-t)^{\gamma-\beta-1}|\,dt &\leqslant \sum_{k=0}^{\infty} \frac{(|\alpha|)_k}{k!}|z|^k \int_0^1 t^{\operatorname{Re}\beta-1+k}(1-t)^{\operatorname{Re}\gamma-\operatorname{Re}\beta-1}\,dt \\ &= \frac{\Gamma(\operatorname{Re}\beta)\Gamma(\operatorname{Re}(\gamma-\beta))}{\Gamma(\operatorname{Re}\gamma)} F(|\alpha|, \operatorname{Re}\beta; \operatorname{Re}\gamma; |z|). \end{aligned}$$

show that the integral is uniformly convergent in the indicated region.[3] But this follows at once from the estimate

$$|t^{\beta-1}(1-t)^{\gamma-\beta-1}(1-tz)^{-\alpha}| \leqslant Mt^{\operatorname{Re}\beta-1}(1-t)^{\operatorname{Re}\gamma-\operatorname{Re}\beta-1}$$

where M is the maximum value of the continuous function $|(1-tz)|^{-\alpha}$ for t in $[0, 1]$ and z in the domain (9.1.5), and from the fact that the integral

$$M\int_0^1 t^{\operatorname{Re}\beta-1}(1-t)^{\operatorname{Re}\gamma-\operatorname{Re}\beta-1}\,dt$$

converges for $\operatorname{Re}\gamma > \operatorname{Re}\beta > 0$. Therefore the condition $|z| < 1$ can be dropped in (9.1.4), and the desired analytic continuation of the hypergeometric function is given by the formula

$$F(\alpha, \beta; \gamma; z) = \frac{\Gamma(\gamma)}{\Gamma(\beta)\Gamma(\gamma-\beta)}\int_0^1 t^{\beta-1}(1-t)^{\gamma-\beta-1}(1-tz)^{-\alpha}\,dt,$$
$$\operatorname{Re}\gamma > \operatorname{Re}\beta > 0, \quad |\arg(1-z)| < \pi. \qquad (9.1.6)$$

In the general case where the parameters have arbitrary values, the analytic continuation of $F(\alpha, \beta; \gamma; z)$ into the plane cut along $[1, \infty]$ can be written as a contour integral obtained by using residue theory to sum the series (9.1.2).[4] A more elementary method of carrying out the analytic continuation, which, however, does not lead to a general analytic expression for the hypergeometric function in explicit form, involves the use of the recurrence relation[5]

$$\begin{aligned}\gamma(\gamma+1)F(\alpha, \beta; \gamma; z) &= \gamma(\gamma-\alpha+1)F(\alpha, \beta+1; \gamma+2; z)\\ &\quad + \alpha[\gamma-(\gamma-\beta)z]F(\alpha+1, \beta+1; \gamma+2; z).\end{aligned} \qquad (9.1.7)$$

By repeated application of this identity, we can represent the function $F(\alpha, \beta; \gamma; z)$ with arbitrary parameters ($\gamma \neq 0, -1, -2, \ldots$) as a sum

$$F(\alpha, \beta; \gamma; z) = \sum_{s=0}^{p} a_{sp}(\alpha, \beta; \gamma; z)F(\alpha+s, \beta+p; \gamma+2p; z), \qquad (9.1.8)$$

where p is a positive integer and the $a_{sp}(\alpha, \beta; \gamma; z)$ are polynomials in z. If we

---

[3] E. C. Titchmarsh, *op. cit.*, pp. 99–100.

[4] E. T. Whittaker and G. N. Watson, *op. cit.*, p. 288.

[5] To verify (9.1.7), we substitute from (9.1.2), noting that the coefficient of z^k in the right-hand side of (9.1.7) becomes

$$\gamma(\gamma-\alpha+1)\frac{(\alpha)_k(\beta+1)_k}{(\gamma+2)_k k!} + \alpha\gamma\frac{(\alpha+1)_k(\beta+1)_k}{(\gamma+2)_k k!} - \alpha(\gamma-\beta)\frac{(\alpha+1)_{k-1}(\beta+1)_{k-1}}{(\gamma+2)_{k-1}(k-1)!}$$
$$= \frac{(\alpha)_k(\beta)_k}{(\gamma+2)_k k!}\left[\gamma(\gamma-\alpha+1)\frac{\beta+k}{\beta} + \alpha\gamma\frac{\alpha+k}{\alpha}\frac{\beta+k}{\beta} - \alpha(\gamma-\beta)\frac{(\gamma+k+1)k}{\alpha\beta}\right]$$
$$= \frac{(\alpha)_k(\beta)_k}{(\gamma+2)_k k!}(\gamma+k)(\gamma+k+1) \equiv \gamma(\gamma+1)\frac{(\alpha)_k(\beta)_k}{(\gamma)_k k!},$$

choose p so large that $\operatorname{Re}\beta > -p$, $\operatorname{Re}(\gamma - \beta) > -p$, then we can use formula (9.1.6) to make the analytic continuation of each of the functions $F(\alpha + s, \beta + p; \gamma + 2p; z)$ appearing in the right-hand side of (9.1.8). Substituting the corresponding expressions into (9.1.8), we obtain the desired analytic continuation of $F(\alpha, \beta; \gamma; z)$, since the resulting function is analytic in the plane cut along $[1, \infty]$ and coincides with (9.1.2) for $|z| < 1$.

The hypergeometric function $F(\alpha, \beta; \gamma; z)$ plays an important role in mathematical analysis and its applications. Introduction of this function allows us to solve many interesting problems, such as conformal mapping of triangular domains bounded by line segments or circular arcs, various problems of quantum mechanics, etc. Moreover, as will be seen in Sec. 9.8, a number of special functions can be expressed in terms of the hypergeometric function, so that the theory of these functions can be regarded as a special case of the general theory developed in this chapter (cf. footnote 20, p. 176).

9.2. Elementary Properties of the Hypergeometric Function

In this section we consider some properties of the hypergeometric function which are immediate consequences of its definition by the series (9.1.2).[6] First of all, observing that the terms of the series do not change if the parameters α and β are permuted, we obtain the *symmetry property*

$$F(\alpha, \beta; \gamma; z) = F(\beta, \alpha; \gamma; z). \tag{9.2.1}$$

Next, differentiating (9.2.1) with respect to z, we find that

$$\begin{aligned}\frac{d}{dz} F(\alpha, \beta; \gamma; z) &= \sum_{k=1}^{\infty} \frac{(\alpha)_k(\beta)_k}{(\gamma)_k(k-1)!} z^{k-1} = \sum_{k=0}^{\infty} \frac{(\alpha)_{k+1}(\beta)_{k+1}}{(\gamma)_{k+1}k!} z^k \\ &= \frac{\alpha\beta}{\gamma} \sum_{k=0}^{\infty} \frac{(\alpha+1)_k(\beta+1)_k}{(\gamma+1)_k k!} z^k = \frac{\alpha\beta}{\gamma} F(\alpha+1, \beta+1; \gamma+1; z),\end{aligned}$$

and hence[7]

$$\frac{d}{dz} F(\alpha, \beta; \gamma; z) = \frac{\alpha\beta}{\gamma} F(\alpha + 1, \beta + 1; \gamma + 1; z). \tag{9.2.2}$$

Repeated application of (9.2.2) leads to the formula

$$\frac{d^m}{dz^m} F(\alpha, \beta; \gamma; z) = \frac{(\alpha)_m(\beta)_m}{(\gamma)_m} F(\alpha + m, \beta + m; \gamma + m; z), \qquad m = 1, 2, \ldots \tag{9.2.3}$$

---

[6] It follows from the principle of analytic continuation that all the formulas proved here, under the assumption that $|z| < 1$, remain valid in the whole domain of definition of $F(\alpha, \beta; \gamma; z)$.

[7] Cf. formula (7.12.25), p. 197.

From now on, to simplify the notation, we write

$$F(\alpha, \beta; \gamma; z) \equiv F, \qquad F(\alpha \pm 1, \beta; \gamma; z) \equiv F(\alpha \pm 1),$$
$$F(\alpha, \beta \pm 1; \gamma; z) \equiv F(\beta \pm 1), \qquad F(\alpha, \beta; \gamma \pm 1; z) \equiv F(\gamma \pm 1).$$

Then the functions $F(\alpha \pm 1)$, $F(\beta \pm 1)$ and $F(\gamma \pm 1)$ are said to be *contiguous* to F. The function F and any two functions contiguous to F are connected by recurrence relations whose coefficients are linear functions of the variable z.[8] Among the relations of this type we cite the formulas

$$(\gamma - \alpha - \beta)F + \alpha(1 - z)F(\alpha + 1) - (\gamma - \beta)F(\beta - 1) = 0, \qquad (9.2.4)$$
$$(\gamma - \alpha - 1)F + \alpha F(\alpha + 1) - (\gamma - 1)F(\gamma - 1) = 0, \qquad (9.2.5)$$
$$\gamma(1 - z)F - \gamma F(\alpha - 1) + (\gamma - \beta)zF(\gamma + 1) = 0, \qquad (9.2.6)$$

which can be verified by direct substitution of the series (9.1.2). For example, substituting (9.1.2) into (9.2.4), we obtain

$$(\gamma - \alpha - \beta)F + \alpha(1 - z)F(\alpha + 1) - (\gamma - \beta)F(\beta - 1)$$
$$= \sum_{k=1}^{\infty} \left[(\gamma - \alpha - \beta)\frac{(\alpha)_k(\beta)_k}{(\gamma)_k k!} + \alpha\frac{(\alpha + 1)_k(\beta)_k}{(\gamma)_k k!} \right.$$
$$\left. -(\gamma - \beta)\frac{(\alpha)_k(\beta - 1)_k}{(\gamma)_k k!} - \alpha\frac{(\alpha + 1)_{k-1}(\beta)_{k-1}}{(\gamma)_{k-1}(k - 1)!} \right] z^k$$
$$= \sum_{k=1}^{\infty} \frac{(\alpha)_k(\beta)_{k-1}}{(\gamma)_k k!} [(\gamma - \alpha - \beta)(\beta + k - 1) + (\alpha + k)(\beta + k - 1)$$
$$- (\gamma - \beta)(\beta - 1) - (\gamma + k - 1)k]z^k \equiv 0,$$

and similarly for (9.2.5–6). Three other formulas are an immediate consequence of (9.2.4–6) and the symmetry condition (9.2.1):

$$(\gamma - \alpha - \beta)F + \beta(1 - z)F(\beta + 1) - (\gamma - \alpha)F(\alpha - 1) = 0, \qquad (9.2.7)$$
$$(\gamma - \beta - 1)F + \beta F(\beta + 1) - (\gamma - 1)F(\gamma - 1) = 0, \qquad (9.2.8)$$
$$\gamma(1 - z)F - \gamma F(\beta - 1) + (\gamma - \alpha)zF(\gamma + 1) = 0. \qquad (9.2.9)$$

The rest of the recurrence relations can be obtained from (9.2.4–9) by eliminating a common contiguous function from an appropriate pair of formulas. For example, combining (9.2.5) and (9.2.8), or (9.2.6) and (9.2.9), we obtain

$$(\alpha - \beta)F - \alpha F(\alpha + 1) + \beta F(\beta + 1) = 0, \qquad (9.2.10)$$
$$(\alpha - \beta)(1 - z)F + (\gamma - \alpha)F(\alpha - 1) - (\gamma - \beta)F(\beta - 1) = 0, \qquad (9.2.11)$$

and so on.[9]

[8] Obviously, the total number of such relations is

$$\binom{6}{2} = 15.$$

[9] The list of all fifteen recurrence relations involving F and its contiguous functions is given in the Bateman Manuscript Project, *Higher Transcendental Functions, Vol. 1*, p. 103.

Besides the recurrence relations just given, there exist similar relations between the function $F(\alpha, \beta; \gamma; z)$ and any pair of functions of the form $F(\alpha + l, \beta + m; \gamma + n; z)$, where l, m and n are arbitrary integers. Some simple relations of this type are[10]

$$F(\alpha, \beta; \gamma; z) - F(\alpha, \beta; \gamma - 1; z) = -\frac{\alpha\beta z}{\gamma(\gamma - 1)} F(\alpha + 1, \beta + 1; \gamma + 1; z), \quad (9.2.12)$$

$$F(\alpha, \beta + 1; \gamma; z) - F(\alpha, \beta; \gamma; z) = \frac{\alpha z}{\gamma} F(\alpha + 1, \beta + 1; \gamma + 1; z), \quad (9.2.13)$$

$$F(\alpha, \beta + 1; \gamma + 1; z) - F(\alpha, \beta; \gamma; z) = \frac{\alpha(\gamma - \beta)z}{\gamma(\gamma + 1)} F(\alpha + 1, \beta + 1; \gamma + 2; z), \quad (9.2.14)$$

$$F(\alpha - 1, \beta + 1; \gamma; z) - F(\alpha, \beta; \gamma; z) = \frac{(\alpha - \beta - 1)z}{\gamma} F(\alpha, \beta + 1; \gamma + 1; z). \quad (9.2.15)$$

Formulas (9.2.12–15) are proved by direct substitution of (9.1.2), or by repeated use of the relations between $F(\alpha, \beta; \gamma; z)$ and its contiguous functions.

Finally, we recall from Sec. 7.2 that the hypergeometric function $u = F(\alpha, \beta; \gamma; z)$ is a solution of the hypergeometric equation

$$z(1 - z)u'' + [\gamma - (\alpha + \beta + 1)z]u' - \alpha\beta u = 0, \quad (9.2.16)$$

which is analytic in a neighborhood of the point $z = 0$.

9.3. Evaluation of $\lim_{z \to 1-} F(\alpha, \beta; \gamma; z)$ for $\operatorname{Re}(\gamma - \alpha - \beta) > 0$

In developing the theory of the hypergeometric function, it is important to know the limit as $z \to 1-$ of the function (9.1.2), where the parameters satisfy the condition $\operatorname{Re}(\gamma - \alpha - \beta) > 0$.[11] Suppose that besides this condition, $\operatorname{Re} \gamma > \operatorname{Re} \beta > 0$ as well. Then the desired result can be obtained by passing to the limit behind the integral sign in (9.1.6), which gives

$$\lim_{z \to 1-} F(\alpha, \beta; \gamma; z) = \frac{\Gamma(\gamma)}{\Gamma(\beta)\Gamma(\gamma - \beta)} \int_0^1 t^{\beta - 1} (1 - t)^{\gamma - \alpha - \beta - 1} \, dt,$$

[10] Formula (9.1.7) is also a relation of this type.

[11] It can be shown that if this condition is not satisfied, then, with certain exceptions, the sum of the hypergeometric series becomes infinite as $z \to 1-$.

or, in view of (1.5.2, 6),

$$\lim_{z\to 1-} F(\alpha, \beta; \gamma; z) = \frac{\Gamma(\gamma)\Gamma(\gamma-\alpha-\beta)}{\Gamma(\gamma-\alpha)\Gamma(\gamma-\beta)}, \tag{9.3.1}$$

where, for the time being, we assume that

$$\operatorname{Re}(\gamma-\alpha-\beta) > 0, \qquad \operatorname{Re}\gamma > \operatorname{Re}\beta > 0. \tag{9.3.2}$$

To justify the passage to the limit, it is sufficient to prove that the conditions (9.3.2) imply that the integral (9.1.6) is uniformly convergent for $0 \leqslant z \leqslant 1$. To this end, we note that

$$1 - t \leqslant |1 - tz| \leqslant 1$$

for $0 \leqslant z \leqslant 1$, $0 \leqslant t \leqslant 1$, and hence

$$|t^{\beta-1}(1-t)^{\gamma-\beta-1}(1-tz)^{-\alpha}| \leqslant t^{\operatorname{Re}\beta-1}(1-t)^{\lambda-1}, \tag{9.3.3}$$

where

$$\lambda = \begin{cases} \operatorname{Re}(\gamma-\alpha-\beta) & \text{if } \operatorname{Re}\alpha > 0, \\ \operatorname{Re}(\gamma-\beta) & \text{if } \operatorname{Re}\alpha < 0. \end{cases}$$

The estimate (9.3.3) shows that the integral (9.1.6) is uniformly convergent for $0 \leqslant z \leqslant 1$, since the integral

$$\int_0^1 t^{\operatorname{Re}\beta-1}(1-t)^{\lambda-1}\,dt,$$

which majorizes (9.1.6), is convergent if the conditions (9.3.2) hold.

We now show that the second of the conditions (9.3.2) is not essential. Suppose that instead of (9.3.2), the parameters of the hypergeometric functions satisfy the weaker inequalities

$$\operatorname{Re}(\gamma-\alpha-\beta) > 0, \quad \operatorname{Re}(\gamma-\beta) > -1, \quad \operatorname{Re}\beta > -1.$$

Then the restrictions under which we proved (9.3.1) are satisfied by each of the hypergeometric functions in the right-hand side of the recurrence relation (9.1.7). It follows that

$$\begin{aligned} \lim_{z\to 1-} F(\alpha, \beta; \gamma; z) &= \frac{\gamma-\alpha+1}{\gamma+1}\,\frac{\Gamma(\gamma+2)\Gamma(\gamma-\alpha-\beta+1)}{\Gamma(\gamma-\alpha+2)\Gamma(\gamma-\beta+1)} \\ &\quad + \frac{\alpha\beta}{\gamma(\gamma+1)}\,\frac{\Gamma(\gamma+2)\Gamma(\gamma-\alpha-\beta)}{\Gamma(\gamma-\alpha+1)\Gamma(\gamma-\beta+1)} \\ &\equiv \frac{\Gamma(\gamma)\Gamma(\gamma-\alpha-\beta)}{\Gamma(\gamma-\alpha)\Gamma(\gamma-\beta)}, \end{aligned}$$

which is just the previous result. Repeating this argument, we can prove by induction that

$$\lim_{z\to 1-} F(\alpha, \beta; \gamma; z) = \frac{\Gamma(\gamma)\Gamma(\gamma-\alpha-\beta)}{\Gamma(\gamma-\alpha)\Gamma(\gamma-\beta)}, \tag{9.3.4}$$

provided only that $\operatorname{Re}(\gamma - \alpha - \beta) > 0$. Formula (9.3.4) plays an important role in the derivation of various relations satisfied by the hypergeometric function.

9.4 $F(\alpha, \beta; \gamma; z)$ as a Function of its Parameters

In this section we show that the function

$$f(\alpha, \beta; \gamma; z) = \frac{1}{\Gamma(\gamma)} F(\alpha, \beta; \gamma; z) \tag{9.4.1}$$

is an entire function of α, β and γ, for fixed z. If $|z| < 1$, the proof is an immediate consequence of the expansion

$$f(\alpha, \beta; \gamma; z) = \sum_{k=0}^{\infty} \frac{(\alpha)_k(\beta)_k}{\Gamma(\gamma + k)k!} z^k, \qquad |z| < 1, \tag{9.4.2}$$

obtained by substituting (9.1.2) into (9.4.1). In fact, since the terms of the series (9.4.2) are entire functions of α, β, γ, and since the series is uniformly convergent in the region $|\alpha| \leqslant A$, $|\beta| \leqslant B$, $|\gamma| \leqslant C$ (where A, B and C are arbitrarily large),[12] it follows that $f(\alpha, \beta; \gamma; z)$ is an entire function of its parameters.

Now let z be an arbitrary point in the complex plane cut along $[1, \infty]$, and consider the formulas

$$f(\alpha, \beta; \gamma; z) = \frac{1}{\Gamma(\beta)\Gamma(\gamma - \beta)} \int_0^1 t^{\beta-1}(1 - t)^{\gamma-\beta-1}(1 - tz)^{-\alpha}\, dt, \qquad \operatorname{Re}\gamma > \operatorname{Re}\beta > 0, \quad |\arg(1 - z)| < \pi, \tag{9.4.3}$$

$$\begin{aligned} f(\alpha, \beta; \gamma; z) &= \gamma(\gamma - \alpha + 1) f(\alpha, \beta + 1; \gamma + 2; z) \\ &\quad + \alpha[\gamma - (\gamma - \beta)z] f(\alpha + 1, \beta + 1; \gamma + 2; z), \end{aligned} \tag{9.4.4}$$

which are the analogues of (9.1.6) and (9.1.7). Since the integrand in the right-hand side of (9.4.3) is an entire function of the parameters α, β, γ for any t in $(0, 1)$, and since the integral is uniformly convergent in the region

$$|\alpha| \leqslant A, \quad \delta \leqslant \operatorname{Re}\beta \leqslant B, \quad \delta \leqslant \operatorname{Re}(\gamma - \beta) \leqslant C,$$

---

[12] Use the criterion given in footnote 4, p. 102, noting that if

$$u_k = \frac{(\alpha)_k(\beta)_k}{\Gamma(\gamma + k)k!} z^k,$$

then

$$\left|\frac{u_{k+1}}{u_k}\right| = \left|\frac{(\alpha + k)(\beta + k)}{(\gamma + k)(1 + k)} z\right| \leqslant \frac{(A + k)(B + k)}{(k - C)(1 + k)} |z| \leqslant q < 1$$

for $|z| < 1$ and sufficiently large k.

where $\delta > 0$ is arbitrarily small, it follows that $f(\alpha, \beta; \gamma; z)$ is an analytic function of its parameters in the region

$$|\alpha| < \infty, \quad \operatorname{Re} \beta > 0, \quad \operatorname{Re} (\gamma - \beta) > 0.$$

By repeated application of the recurrence relation (9.4.4), we can represent the function $f(\alpha, \beta; \gamma; z)$ as a sum

$$f(\alpha, \beta; \gamma; z) = \sum_{s=0}^{p} b_{sp}(\alpha, \beta; \gamma; z) f(\alpha + s, \beta + p; \gamma + 2p; z), \tag{9.4.5}$$

where the $b_{sp}(\alpha, \beta; \gamma; z)$ are polynomials in α, β, γ and z, and p is a positive integer. As just shown, each term of this sum is an analytic function in the region $|\alpha| < \infty$, $\operatorname{Re} \beta > -p$, $\operatorname{Re} (\gamma - \beta) > -p$, and hence $f(\alpha, \beta; \gamma; z)$ is an entire function of its parameters. It follows that for fixed z in the plane cut along $[1, \infty]$, the hypergeometric function $F(\alpha, \beta; \gamma; z)$ is an entire function of α and β, and a meromorphic function of γ, with simple poles at the points $\gamma = 0, -1, -2, \ldots$

9.5. Linear Transformations of the Hypergeometric Function

Consider the class of all fractional linear transformations

$$z' = \frac{az + b}{cz + d}$$

carrying the points $z = 0, 1, \infty$ into the points $z' = 0, 1, \infty$ chosen in any order. It is easy to see that besides the identity transformation $z' = z$, this class consists of the following five transformations:

$$z' = \frac{z}{z - 1}, \quad z' = 1 - z, \quad z' = \frac{1}{1 - z}, \quad z' = \frac{1}{z}, \quad z' = \frac{z - 1}{z}.$$

We now derive various linear relations connecting the hypergeometric functions with variables z and z'. Relations of this kind are among the most important in the theory of the hypergeometric function, and are known as *linear transformations* of the hypergeometric function. In particular, these formulas enable us to make the analytic continuation of $F(\alpha, \beta; \gamma; z)$ into any part of the plane cut along $[1, \infty]$.[13]

We begin by deriving a relation which is useful in the case where one requires the analytic continuation of the hypergeometric function into the half-plane $\operatorname{Re} z < \frac{1}{2}$. Suppose z belongs to the plane cut along $[1, \infty]$, and assume for the time being that $\operatorname{Re} \gamma > \operatorname{Re} \beta > 0$. Then, using the integral

[13] The theoretical possibility of such an analytic continuation has already been proved in Sec. 9.1.

representation (9.1.6), and introducing the new variable of integration $s = 1 - t$, we find that

$$F(\alpha, \beta; \gamma; z) = \frac{\Gamma(\gamma)}{\Gamma(\beta)\Gamma(\gamma - \beta)} \int_0^1 s^{\gamma-\beta-1}(1 - s)^{\beta-1}(1 - z + sz)^{-\alpha}\, ds$$

$$= (1 - z)^{-\alpha} \frac{\Gamma(\gamma)}{\Gamma(\beta')\Gamma(\gamma - \beta')} \int_0^1 s^{\beta'-1}(1 - s)^{\gamma-\beta'-1}(1 - sz')^{-\alpha} ds,$$

where

$$\beta' = \gamma - \beta, \qquad z' = \frac{z}{z - 1},$$

and our assumptions imply that $\operatorname{Re} \gamma > \operatorname{Re} \beta' > 0$, while z' belongs to the plane cut along $[1, \infty]$.[14] According to (9.1.6), the expression on the right is just

$$(1 - z)^{-\alpha} F(\alpha, \beta'; \gamma; z'),$$

and hence

$$F(\alpha, \beta; \gamma; z) = (1 - z)^{-\alpha} F\left(\alpha, \gamma - \beta; \gamma; \frac{z}{z - 1}\right), \qquad |\arg(1 - z)| < \pi. \tag{9.5.1}$$

Formula (9.5.1) was proved under the temporary assumption that $\operatorname{Re} \gamma > \operatorname{Re} \beta > 0$, but, as we know from Sec. 9.4, after dividing by $\Gamma(\gamma)$, both sides become entire functions of β and γ.[15] Therefore, by the principle of analytic continuation, (9.5.1) remains valid for arbitrary β and γ, with the exception of the values $\gamma = 0, -1, -2, \ldots$ for which $F(\alpha, \beta; \gamma; z)$ is not defined. Moreover, if $\operatorname{Re} z < \frac{1}{2}$, then

$$\left|\frac{z}{z - 1}\right| < 1,$$

and the hypergeometric function in the right-hand side of (9.5.1) can be replaced by the sum of the hypergeometric series, i.e., (9.5.1) gives the analytic continuation of $F(\alpha, \beta; \gamma; z)$ into the half-plane $\operatorname{Re} z < \frac{1}{2}$.

Permuting α and β in (9.5.1), and using the symmetry property (9.2.1), we arrive at the relation

$$F(\alpha, \beta; \gamma; z) = (1 - z)^{-\beta} F\left(\gamma - \alpha, \beta; \gamma; \frac{z}{z - 1}\right), \qquad |\arg(1 - z)| < \pi, \tag{9.5.2}$$

which can also be used to make the analytic continuation of the hypergeometric function into the half-plane $\operatorname{Re} z < \frac{1}{2}$. To obtain another important

[14] Note that under the transformation $z' = z/(z - 1)$, the plane cut along $[1, \infty]$ goes into itself.

[15] The expression $F[f(\alpha, \beta, \gamma, \ldots), g(\alpha, \beta, \gamma, \ldots), \ldots]$ is an entire function of $\alpha, \beta, \gamma, \ldots$ if $F, f, g, \ldots$ are entire functions of their arguments.

result, we perform the transformations (9.5.1) and (9.5.2) consecutively, obtaining

$$F(\alpha, \beta; \gamma; z) = (1 - z)^{-\alpha}\left(1 - \frac{z}{z-1}\right)^{-(\gamma-\beta)} F(\gamma - \alpha, \ \gamma - \beta; \gamma; z),$$
$$|\arg(1 - z)| < \pi,$$

or

$$F(\alpha, \beta; \gamma; z) = (1 - z)^{\gamma-\alpha-\beta} F(\gamma - \alpha, \gamma - \beta; \gamma; z),$$
$$|\arg(1 - z)| < \pi. \tag{9.5.3}$$

To derive a relation between the hypergeometric function with variable z and the hypergeometric function with variable $z' = 1 - z$, we use a general method from the theory of linear differential equations. First we note that the general solution of the hypergeometric equation

$$z(1 - z)u'' + [\gamma - (\alpha + \beta + 1)z]u' - \alpha\beta u = 0 \tag{9.5.4}$$

can be written in the form[16]

$$u = A_1 F(\alpha, \beta; \gamma; z) + A_2 z^{1-\gamma} F(1 - \gamma + \alpha, 1 - \gamma + \beta; 2 - \gamma; z),$$
$$|\arg(1 - z)| < \pi, \quad |\arg z| < \pi, \quad \gamma \neq 0, \pm 1, \pm 2, \ldots \tag{9.5.5}$$

Under the transformation $z' = 1 - z$, the domain $|\arg(1 - z)| < \pi$, $|\arg z| < \pi$ goes into the domain $|\arg(1 - z')| < \pi$, $|\arg z'| < \pi$, and equation (9.5.4) goes into the hypergeometric equation with parameters $\alpha' = \alpha$, $\beta' = \beta$, $\gamma' = 1 + \alpha + \beta - \gamma$. Therefore the expression

$$u = B_1 F(\alpha, \beta; 1 + \alpha + \beta - \gamma; 1 - z) + B_2(1 - z)^{\gamma-\alpha-\beta}$$
$$\times F(\gamma - \alpha, \gamma - \beta; 1 - \alpha - \beta + \gamma; 1 - z), \tag{9.5.6}$$
$$|\arg(1 - z)| < \pi, \quad |\arg z| < \pi, \quad \alpha + \beta - \gamma \neq 0, \pm 1, \pm 2, \ldots$$

is also a general solution of equation (9.5.4). In particular, this implies the existence of a linear relation of the form

$$F(\alpha, \beta; \gamma; z) = C_1 F(\alpha, \beta; 1 + \alpha + \beta - \gamma; 1 - z)$$
$$+ C_2(1 - z)^{\gamma-\alpha-\beta} F(\gamma - \alpha, \gamma - \beta; 1 - \alpha - \beta + \gamma; 1 - z),$$
$$\alpha + \beta - \gamma \neq 0, \pm 1, \pm 2, \ldots$$

To determine the constants C_1 and C_2, we assume temporarily that $\mathrm{Re}\,(\alpha + \beta) < \mathrm{Re}\,\gamma < 1$, and then take the limit of the last equality, first as $z \to 1-$ and then as $z \to 0+$. Using (9.3.4), we obtain

$$C_1 = \frac{\Gamma(\gamma)\Gamma(\gamma - \alpha - \beta)}{\Gamma(\gamma - \alpha)\Gamma(\gamma - \beta)},$$

$$C_1 \frac{\Gamma(1 + \alpha + \beta - \gamma)\Gamma(1 - \gamma)}{\Gamma(1 + \alpha - \gamma)\Gamma(1 + \beta - \gamma)} + C_2 \frac{\Gamma(1 - \alpha - \beta + \gamma)\Gamma(1 - \gamma)}{\Gamma(1 - \alpha)\Gamma(1 - \beta)} = 1.$$

[16] See Sec. 7.2, noting that by the principle of analytic continuation, formula (7.2.6) remains valid in the whole domain $|\arg(1 - z)| < \pi$, $|\arg z| < \pi$.

It follows that

$$C_2 = \frac{\Gamma(\gamma)\Gamma(\alpha + \beta - \gamma)}{\Gamma(\alpha)\Gamma(\beta)},$$

after some simple calculations involving the identity (1.2.2). Therefore the required formula is

$$\begin{aligned} F(\alpha, \beta; \gamma; z) &= \frac{\Gamma(\gamma)\Gamma(\gamma - \alpha - \beta)}{\Gamma(\gamma - \alpha)\Gamma(\gamma - \beta)} F(\alpha, \beta; 1 + \alpha + \beta - \gamma; 1 - z) \\ &\quad + (1 - z)^{\gamma - \alpha - \beta} \frac{\Gamma(\gamma)\Gamma(\alpha + \beta - \gamma)}{\Gamma(\alpha)\Gamma(\beta)} \\ &\quad \times F(\gamma - \alpha, \gamma - \beta; 1 - \alpha - \beta + \gamma; 1 - z), \end{aligned} \tag{9.5.7}$$

$$|\arg z| < \pi, \quad |\arg (1 - z)| < \pi, \quad \alpha + \beta - \gamma \neq 0, \pm 1, \pm 2, \ldots$$

To get rid of the superfluous restrictions imposed on the parameters α, β and γ, we note that after multiplication by $\sin \pi(\gamma - \alpha - \beta)/\Gamma(\gamma)$, both sides of (9.5.7) are entire functions of the parameters.[17] Therefore, according to the principle of analytic continuation, the relation (9.5.7) is valid for all values of the parameters except those for which $\alpha + \beta - \gamma = 0, \pm 1, \pm 2, \ldots$ Formula (9.5.7) gives the analytic continuation of the hypergeometric function into the domain $|z - 1| < 1$, $|\arg (1 - z)| < \pi$.

The remaining relations between the hypergeometric functions with variables z and z' can be obtained by combining the formulas just derived. For example, consecutive application of (9.5.1) and (9.5.7) leads to the relation[18]

$$\begin{aligned} F(\alpha, \beta; \gamma; z) &= (1 - z)^{-\alpha} \frac{\Gamma(\gamma)\Gamma(\beta - \alpha)}{\Gamma(\gamma - \alpha)\Gamma(\beta)} F\left(\alpha, \gamma - \beta; 1 + \alpha - \beta; \frac{1}{1 - z}\right) \\ &\quad + (1 - z)^{-\beta} \frac{\Gamma(\gamma)\Gamma(\alpha - \beta)}{\Gamma(\gamma - \beta)\Gamma(\alpha)} F\left(\gamma - \alpha, \beta; 1 - \alpha + \beta; \frac{1}{1 - z}\right) \end{aligned}$$

$$|\arg (-z)| < \pi, \quad |\arg (1 - z)| < \pi, \quad \alpha - \beta \neq 0, \pm 1, \pm 2, \ldots \tag{9.5.8}$$

which enables us to make the analytic continuation of $F(\alpha, \beta; \gamma; z)$ into the domain $|z - 1| > 1$, $|\arg (1 - z)| < \pi$. Then, combining (9.5.8) with (9.5.1–2), we obtain

$$\begin{aligned} F(\alpha, \beta; \gamma; z) &= (-z)^{-\alpha} \frac{\Gamma(\gamma)\Gamma(\beta - \alpha)}{\Gamma(\gamma - \alpha)\Gamma(\beta)} F\left(\alpha, 1 + \alpha - \gamma; 1 + \alpha - \beta; \frac{1}{z}\right) \\ &\quad + (-z)^{-\beta} \frac{\Gamma(\gamma)\Gamma(\alpha - \beta)}{\Gamma(\gamma - \beta)\Gamma(\alpha)} F\left(\beta, 1 + \beta - \gamma; 1 + \beta - \alpha; \frac{1}{z}\right), \end{aligned}$$

$$|\arg (-z)| < \pi, \quad |\arg (1 - z)| < \pi, \quad \alpha - \beta \neq 0, \pm 1, \pm 2, \ldots, \tag{9.5.9}$$

---

[17] Here we again make use of (1.2.2).

[18] Note that under the transformation $z' = z/(z - 1)$, the domain $|\arg (-z)| < \pi$, $|\arg (1 - z)| < \pi$ goes into the domain $|\arg z'| < \pi$, $|\arg (1 - z')| < \pi$, which guarantees that (9.5.1) and (9.5.7) can be applied consecutively.

which gives the analytic continuation of $F(\alpha, \beta; \gamma; z)$ into the domain $|z| > 1$, $|\arg (1 - z)| < \pi$. Finally, consecutive application of (9.5.7) and (9.5.1) gives

$$F(\alpha, \beta; \gamma; z) = z^{-\alpha} \frac{\Gamma(\gamma)\Gamma(\gamma - \alpha - \beta)}{\Gamma(\gamma - \alpha)\Gamma(\gamma - \beta)} F\left(\alpha, 1 + \alpha - \gamma; 1 + \alpha + \beta - \gamma; \frac{z-1}{z}\right)$$

$$+ z^{\alpha-\gamma}(1 - z)^{\gamma-\alpha-\beta} \frac{\Gamma(\gamma)\Gamma(\alpha + \beta - \gamma)}{\Gamma(\alpha)\Gamma(\beta)}$$

$$\times F\left(\gamma - \alpha, 1 - \alpha; 1 + \gamma - \alpha - \beta; \frac{z-1}{z}\right),$$

$$|\arg z| < \pi, \quad |\arg (1 - z)| < \pi, \quad \alpha + \beta - \gamma \neq 0, \pm 1, \pm 2, \ldots, \tag{9.5.10}$$

which can be used to make the analytic continuation of $F(\alpha, \beta; \gamma; z)$ into the domain $\operatorname{Re} z > \frac{1}{2}$, $|\arg (1 - z)| < \pi$.

The problem of the analytic continuation of the hypergeometric function into the z-plane cut along $[1, \infty]$ is solved by using formulas (9.5.1–3) and (9.5.7–10). Some exceptional cases, where these formulas are not applicable, will be considered in Sec. 9.7.

9.6. Quadratic Transformations of the Hypergeometric Function

The relations between hypergeometric functions derived in the preceding section are valid for arbitrary values of the parameters α, β, γ (apart from certain exceptional values). One can also consider relations where the parameters satisfy certain constraints; although less general, relations of this type are also useful in making various transformations and carrying out analytic continuation. Among such relations, the most interesting involve hypergeometric functions with two arbitrary parameters. As will be seen below, they also contain expressions like

$$\frac{1 + \sqrt{1 - z}}{2}, \quad \frac{1 - \sqrt{1 - z}}{1 + \sqrt{1 - z}}, \quad \frac{-4z}{(1 - z)^2}, \ldots,$$

and hence are called *quadratic transformations of the hypergeometric function.*

As an example of a formula belonging to this class, consider the relation

$$F(\alpha, \beta; \alpha + \beta + \tfrac{1}{2}; z) = F\left(2\alpha, 2\beta; \alpha + \beta + \tfrac{1}{2}; \frac{1 - \sqrt{1 - z}}{2}\right), \tag{9.6.1}$$

$$|\arg (1 - z)| < \pi, \quad \alpha + \beta + \tfrac{1}{2} \neq 0, -1, -2, \ldots,$$

which can be proved as follows: The left-hand side is a solution of the hypergeometric equation (9.5.4) with parameter $\gamma = \alpha + \beta + \frac{1}{2}$, which is analytic in the domain $|\arg (1 - z)| < \pi$. Under the substitution[19]

$$z' = \tfrac{1}{2}(1 - \sqrt{1 - z}),$$

[19] By $\sqrt{1 - z}$ is meant the branch which is positive for real z in the interval (0, 1).

this equation goes into an equation of the same form with parameters

$$\alpha' = 2\alpha, \quad \beta' = 2\beta, \quad \gamma' = \alpha + \beta + \tfrac{1}{2},$$

and the domain $|\arg(1 - z)| < \pi$ goes into the domain $\operatorname{Re} z' < \frac{1}{2}$, which is part of the domain $|\arg(1 - z')| < \pi$. But according to (7.2.6), the hypergeometric equation cannot have two linearly independent solutions which are analytic in a neighborhood of the point $z = 0$, and hence there must exist a relation of the form

$$F(\alpha, \beta; \alpha + \beta + \tfrac{1}{2}; z) = AF\left(2\alpha, 2\beta; \alpha + \beta + \tfrac{1}{2}; \frac{1 - \sqrt{1 - z}}{2}\right),$$

where A is a constant. Setting $z = 0$, we find that $A = 1$, thereby proving (9.6.1).

A large number of other relations of the same type can be deduced by applying the linear transformations of Sec. 9.5 to formula (9.6.1) and changing the independent variable or the parameters. For example, using (9.5.3) and (9.5.1) to transform the right-hand side of (9.6.1), we find that

$$F(\alpha, \beta; \alpha + \beta + \tfrac{1}{2}; z) = \left(\frac{1 + \sqrt{1 - z}}{2}\right)^{-2\alpha} F\left(2\alpha, \alpha - \beta + \tfrac{1}{2}; \alpha + \beta + \tfrac{1}{2}; \frac{\sqrt{1 - z} - 1}{\sqrt{1 - z} + 1}\right),$$

$$|\arg(1 - z)| < \pi, \quad \alpha + \beta + \tfrac{1}{2} \neq 0, -1, -2, \ldots, \qquad (9.6.2)$$

$$F(\alpha, \beta; \alpha + \beta + \tfrac{1}{2}; z) = \left(\frac{1 + \sqrt{1 - z}}{2}\right)^{1/2 - \alpha - \beta} F\left(\alpha - \beta + \tfrac{1}{2}, \beta - \alpha + \tfrac{1}{2}; \alpha + \beta + \tfrac{1}{2}; \frac{1 - \sqrt{1 - z}}{2}\right),$$

$$|\arg(1 - z)| < \pi, \quad \alpha + \beta + \tfrac{1}{2} \neq 0, -1, -2, \ldots \qquad (9.6.3)$$

Using (9.5.1) to transform the left-hand sides of (9.6.1) and (9.6.2), and then making the substitution

$$\frac{z}{z - 1} \to z, \qquad \alpha + \beta + \tfrac{1}{2} \to \gamma,$$

we obtain two other useful relations:

$$F(\alpha, \alpha + \tfrac{1}{2}; \gamma; z) = (1 - z)^{-\alpha} F\left(2\alpha, 2\gamma - 2\alpha - 1; \gamma; \frac{1}{2} - \frac{1}{2\sqrt{1 - z}}\right),$$

$$|\arg(1 - z)| < \pi, \qquad (9.6.4)$$

$$F(\alpha, \alpha + \tfrac{1}{2}; \gamma; z) = \left(\frac{1 + \sqrt{1 - z}}{2}\right)^{-2\alpha} F\left(2\alpha, 2\alpha - \gamma + 1; \gamma; \frac{1 - \sqrt{1 - z}}{1 + \sqrt{1 - z}}\right),$$

$$|\arg(1 - z)| < \pi. \qquad (9.6.5)$$

Finally, using (9.5.3) to transform the left-hand sides of (9.6.1) and (9.6.2), and then making the substitution

$$\alpha \to \alpha - \tfrac{1}{2}, \qquad \beta \to \beta - \tfrac{1}{2},$$

we arrive at the relations

$$F(\alpha, \beta; \alpha + \beta - \tfrac{1}{2}; z) = \frac{1}{\sqrt{1-z}} F\left(2\alpha - 1, 2\beta - 1; \alpha + \beta - \tfrac{1}{2}; \frac{1 - \sqrt{1-z}}{2}\right),$$

$$|\arg(1-z)| < \pi, \quad \alpha + \beta - \tfrac{1}{2} \neq 0, -1, -2, \ldots, \tag{9.6.6}$$

$$F(\alpha, \beta; \alpha + \beta - \tfrac{1}{2}; z) = \frac{1}{\sqrt{1-z}} \left(\frac{1 + \sqrt{1-z}}{2}\right)^{1-2\alpha} \times F\left(2\alpha - 1, \alpha - \beta + \tfrac{1}{2}; \alpha + \beta - \tfrac{1}{2}; \frac{\sqrt{1-z} - 1}{\sqrt{1-z} + 1}\right),$$

$$|\arg(1-z)| < \pi, \quad \alpha + \beta - \tfrac{1}{2} \neq 0, -1, -2, \ldots \tag{9.6.7}$$

It is interesting to note that formulas (9.6.2, 5, 7) continue the corresponding hypergeometric functions into the plane cut along $[1, \infty]$. In fact,

$$\left|\frac{1 - \sqrt{1-z}}{1 + \sqrt{1-z}}\right| < 1$$

if $|\arg(1-z)| < \pi$, and hence the hypergeometric function in the right-hand side of each of these formulas can be replaced by the sum of the corresponding hypergeometric series.

Further results can be obtained by taking inverses of the formulas just derived. For example, inversion of (9.6.1–3) gives[20]

$$F\{\alpha, \beta; \tfrac{1}{2}(\alpha + \beta + 1\}; z) = F\{\tfrac{1}{2}\alpha, \tfrac{1}{2}\beta; \tfrac{1}{2}(\alpha + \beta + 1); 4z(1-z)\},$$

$$\operatorname{Re} z < \tfrac{1}{2}, \quad \tfrac{1}{2}(\alpha + \beta + 1) \neq 0, -1, -2, \ldots, \tag{9.6.8}$$

$$F(\alpha, \beta; \alpha - \beta + 1; z) = (1-z)^{-\alpha} F\left\{\tfrac{1}{2}\alpha, \tfrac{1}{2}(\alpha + 1) - \beta; \alpha - \beta + 1; -\frac{4z}{(1-z)^2}\right\},$$

$$|z| < 1, \quad \alpha - \beta + 1 \neq 0, -1, -2, \ldots, \tag{9.6.9}$$

$$F(\alpha, 1 - \alpha; \gamma; z) = (1-z)^{\gamma-1} F\{\tfrac{1}{2}(\gamma - \alpha), \tfrac{1}{2}(\gamma + \alpha - 1); \gamma; 4z(1-z)\},$$

$$\operatorname{Re} z < \tfrac{1}{2}. \tag{9.6.10}$$

---

[20] In particular, (9.6.8) is obtained from (9.6.1) by making the substitution

$$2\alpha \to \alpha, \quad 2\beta \to \beta, \quad \frac{1 - \sqrt{1-z}}{2} \to z.$$

Moreover, combining these formulas with the linear transformations given in Sec. 9.5, we can obtain still another group of formulas. For example, applying the transformation (9.5.7) to the right-hand side of (9.6.8) and making the substitution

$$\alpha \to 2\alpha, \quad \beta \to 2\beta, \quad z \to \frac{1-z}{2},$$

we find that[21]

$$\begin{aligned} F\left(2\alpha, 2\beta; \alpha+\beta+\tfrac{1}{2}; \frac{1-z}{2}\right) &= \frac{\Gamma(\alpha+\beta+\frac{1}{2})\Gamma(\frac{1}{2})}{\Gamma(\alpha+\frac{1}{2})\Gamma(\beta+\frac{1}{2})} F(\alpha, \beta; \tfrac{1}{2}; z^2) \\ &+ z\frac{\Gamma(\alpha+\beta+\frac{1}{2})\Gamma(-\frac{1}{2})}{\Gamma(\alpha)\Gamma(\beta)} F(\alpha+\tfrac{1}{2}, \beta+\tfrac{1}{2}; \tfrac{3}{2}; z^2), \\ &|\arg(1 \pm z)| < \pi, \quad \alpha+\beta+\tfrac{1}{2} \neq 0, -1, -2, \ldots \end{aligned} \tag{9.6.11}$$

Formula (9.6.11) plays an important role in the theory of spherical harmonics. For example, the relation (7.6.9) is an immediate consequence of (9.6.11).

We conclude this section by deriving a few formulas of a more complicated nature. The first result is

$$\begin{aligned} &F(\alpha, \beta; 2\beta; z) \\ &= \left(\frac{1+\sqrt{1-z}}{2}\right)^{-2\alpha} F\left\{\alpha, \alpha-\beta+\tfrac{1}{2}; \beta+\tfrac{1}{2}; \left(\frac{1-\sqrt{1-z}}{1+\sqrt{1-z}}\right)^2\right\}, \\ &\qquad |\arg(1-z)| < \pi, \quad 2\beta \neq -1, -3, -5, \ldots, \end{aligned} \tag{9.6.12}$$

which is proved in the same way as (9.6.1), by noting that under the change of variables

$$z' = \left(\frac{1-\sqrt{1-z}}{1+\sqrt{1-z}}\right)^2, \qquad u = \left(\frac{1+\sqrt{1-z}}{2}\right)^{-2\alpha} v,$$

equation (9.5.4) goes into the hypergeometric equation with the new parameters

$$\alpha' = \alpha, \qquad \beta' = \alpha - \beta + \tfrac{1}{2}, \qquad \gamma' = \beta + \tfrac{1}{2}.$$

Since the verification of this fact is quite tedious, we supply some intermediate

---

[21] In the course of the derivation, it is convenient to assume temporarily that $|\arg(1-z)| < \pi$, $\operatorname{Re} z > 0$. The result can then be extended the whole domain $|\arg(1 \pm z)| < \pi$ by using the principle of analytic continuation.

steps which will serve to keep the reader on the right track during the course of the calculation:

$$z = 1 - \left(\frac{\sqrt{z'} - 1}{\sqrt{z'} + 1}\right)^2, \qquad \frac{dz'}{dz} = -\frac{z'(\sqrt{z'} + 1)^3}{2(\sqrt{z'} - 1)},$$

$$u = (\sqrt{z'} + 1)^{2\alpha} v, \tag{9.6.13}$$

$$\frac{du}{dz} = \frac{dz'}{dz}\frac{du}{dz'} = -\frac{(\sqrt{z'} + 1)^{2\alpha+2}}{2(\sqrt{z'} - 1)}\left[\alpha v + \sqrt{z'}(\sqrt{z'} + 1)\frac{dv}{dz'}\right], \tag{9.6.14}$$

$$\begin{aligned} z(1 - z)\frac{d^2u}{dz^2} &= \sqrt{z'}(\sqrt{z'} + 1)^{2\alpha}\left\{\left[\alpha + 1 - \frac{1}{2}\frac{\sqrt{z'} + 1}{\sqrt{z'} - 1}\right]\left[\alpha v + \sqrt{z'}(\sqrt{z'} + 1)\frac{dv}{dz'}\right]\right. \\ &\quad \left. + \sqrt{z'}(\sqrt{z'} + 1)\left[\left(\alpha + 1 + \frac{1}{2\sqrt{z'}}\right)\frac{dv}{dz'} + \sqrt{z'}(\sqrt{z'} + 1)\frac{d^2v}{dz'^2}\right]\right\}. \end{aligned} \tag{9.6.15}$$

After using (9.6.13–15) to write the hypergeometric equation satisfied by u, we multiply the result by

$$\frac{1 - \sqrt{z'}}{\sqrt{z'}(1 + \sqrt{z'})},$$

obtaining

$$\begin{aligned} &\frac{1 - \sqrt{z'}}{1 + \sqrt{z'}}\left[\alpha + 1 - \frac{1}{2}\frac{\sqrt{z'} + 1}{\sqrt{z'} - 1}\right]\left[\alpha v + \sqrt{z'}(\sqrt{z'} + 1)\frac{dv}{dz'}\right] \\ &\quad + \sqrt{z'}(1 - \sqrt{z'})\left[\left(\alpha + 1 + \frac{1}{2\sqrt{z'}}\right)\frac{dv}{dz'} + \sqrt{z'}(\sqrt{z'} + 1)\frac{d^2v}{dz'^2}\right] \\ &\quad + \frac{\sqrt{z'} + 1}{\sqrt{z'}}\left[\beta - (\alpha + \beta + 1)\frac{2\sqrt{z'}}{(\sqrt{z'} + 1)^2}\right]\left[\alpha v + \sqrt{z'}(\sqrt{z'} + 1)\frac{dv}{dz'}\right] \\ &\quad - \frac{\alpha\beta(1 - \sqrt{z'})}{\sqrt{z'}(1 + \sqrt{z'})} v = 0, \end{aligned}$$

which can now be reduced quite easily to the hypergeometric equation

$$z'(1 - z')\frac{d^2v}{dz'^2} + [(\beta + \tfrac{1}{2}) - (2\alpha - \beta + \tfrac{3}{2})z']\frac{dv}{dz'} - \alpha(\alpha - \beta + \tfrac{1}{2})v = 0,$$

satisfied by v. Making the substitution

$$\frac{1 - \sqrt{1 - z}}{1 + \sqrt{1 - z}} \to z$$

in (9.6.12), we obtain the formula

$$F\left\{\alpha, \beta; 2\beta; \frac{4z}{(1+z)^2}\right\} = (1+z)^{2\alpha} F(\alpha, \alpha - \beta + \tfrac{1}{2}; \beta + \tfrac{1}{2}; z^2),$$

$$|z| < 1, \quad 2\beta \neq -1, -3, -5, \ldots \qquad (9.6.16)$$

Our final result is

$$F(\alpha, \beta; 2\beta; z) = \left(1 - \frac{z}{2}\right)^{-\alpha} F\left\{\tfrac{1}{2}\alpha, \tfrac{1}{2}(\alpha+1); \beta + \tfrac{1}{2}; \left(\frac{z}{2-z}\right)^2\right\},$$

$$|\arg(1 - z)| < \pi, \quad 2\beta \neq -1, -3, -5, \ldots, \qquad (9.6.17)$$

which can be derived as follows: Applying the transformation (9.5.1) to the right-hand side of (9.6.9) and replacing β by $\alpha - \beta + \frac{1}{2}$, we obtain

$$F(\alpha, \alpha - \beta + \tfrac{1}{2}; \beta + \tfrac{1}{2}; z) = (1+z)^{-\alpha} F\left\{\tfrac{1}{2}\alpha, \tfrac{1}{2}(\alpha+1); \beta + \tfrac{1}{2}; \frac{4z}{(1+z)^2}\right\},$$

$$|z| < 1, \quad 2\beta \neq -1, -3, -5, \ldots \qquad (9.6.18)$$

Then, comparing (9.6.13) and (9.6.15), we find that

$$F\left\{\alpha, \beta; 2\beta; \frac{4z}{(1+z)^2}\right\} = \frac{(1+z)^{2\alpha}}{(1+z^2)^{\alpha}} F\left\{\tfrac{1}{2}\alpha, \tfrac{1}{2}(\alpha+1); \beta + \tfrac{1}{2}; \frac{4z^2}{(1+z^2)^2}\right\},$$

and the desired result is obtained by making the substitution

$$\frac{4z}{(1+z)^2} \to z,$$

which implies

$$\frac{1+z^2}{(1+z)^2} \to \frac{2-z}{2}, \qquad \frac{4z^2}{(1+z^2)^2} \to \left(\frac{z}{2-z}\right)^2.$$

The theory of quadratic transformations of the hypergeometric function was developed by Gauss, Kummer and Goursat, and also from a more general point of view in Riemann's investigations of a class of differential equations including the hypergeometric equation as a special case.[22] We refer the reader to these sources for a more detailed treatment of the subject.[23]

---

[22] See E. Goursat, *Sur l'équation différentielle linéaire, qui admet pour intégrale la série hypergéometrique*, Ann. Sci. École Norm. Sup. (2), **10**, 3 (1881). The relevant references by Gauss, Kummer and Riemann are given on p. 296 of the book by Whittaker and Watson (*op. cit.*).

[23] See also the Bateman Manuscript Project, *Higher Transcendental Functions, Vol. 1*, p. 110 ff., for an extensive list of quadratic transformations of the hypergeometric function.

9.7. Formulas for Analytic Continuation of $F(\alpha, \beta; \gamma; z)$ in Exceptional Cases

The formulas derived in Sec. 9.5 allow us to obtain the analytic continuation of the hypergeometric function into any part of the z-plane cut along $[1, \infty]$. However, some of these formulas are no longer meaningful for certain values of the parameters, and must therefore be modified in a way we now indicate. The general approach is to start from the formulas of Sec. 9.5 and then carry out appropriate passages to the limit.

For example, suppose we want to find the analytic continuation of the function $F(\alpha, \beta; \gamma; z)$ into the domain $|z - 1| < 1$, $|\arg(1 - z)| < \pi$. If $\alpha + \beta - \gamma \neq 0, \pm 1, \pm 2, \ldots$, we can use (9.5.7), but this formula is not applicable if $\gamma = \alpha + \beta \pm n$ $(n = 0, 1, 2, \ldots)$. To derive a formula allowing us to carry out the analytic continuation in the latter case, we replace the hypergeometric functions in the right-hand side of (9.5.7) by the corresponding series, and use (1.2.2) to transform the result, obtaining

$$\begin{aligned}\frac{1}{\Gamma(\gamma)} F(\alpha, \beta; \gamma; z) &= \frac{\pi}{\sin \pi(\gamma - \alpha - \beta)}\left[\frac{1}{\Gamma(\gamma - \alpha)\Gamma(\gamma - \beta)} \sum_{k=0}^{\infty} \frac{(\alpha)_k(\beta)_k}{\Gamma(1 + \alpha + \beta - \gamma + k)k!}(1 - z)^k\right.\\ &\qquad \left.- \frac{1}{\Gamma(\alpha)\Gamma(\beta)} \sum_{k=0}^{\infty} \frac{(\gamma - \alpha)_k(\gamma - \beta)_k}{\Gamma(1 - \alpha - \beta + \gamma + k)} \frac{(1 - z)^{k+\gamma-\alpha-\beta}}{k!}\right]\\ &= \frac{\pi}{\sin \pi(\gamma - \alpha - \beta)}(g_1 - g_2). \end{aligned} \tag{9.7.1}$$

It is easily verified that

$$\lim_{\gamma \to \alpha+\beta+n} g_1 = \lim_{\gamma \to \alpha+\beta+n} g_2 = \frac{1}{\Gamma(\alpha)\Gamma(\beta)} \sum_{k=0}^{\infty} \frac{(\alpha + n)_k(\beta + n)_k}{(n + k)!k!}(1 - z)^{k+n},$$

and hence the right-hand side of (9.7.1) becomes indeterminate for $\gamma = \alpha + \beta + n$. Using L'Hospital's rule to eliminate this indeterminacy, we have

$$\frac{1}{\Gamma(\alpha + \beta + n)} F(\alpha, \beta; \alpha + \beta + n; z) = (-1)^n\left[\left.\frac{\partial g_1}{\partial \gamma}\right|_{\gamma=\alpha+\beta+n} - \left.\frac{\partial g_2}{\partial \gamma}\right|_{\gamma=\alpha+\beta+n}\right]. \tag{9.7.2}$$

After some calculations resembling those made in Sec. 5.5, we find that[24]

$$\begin{aligned}\left.\frac{\partial g_1}{\partial \gamma}\right|_{\gamma=\alpha+\beta+n} &= \frac{1}{\Gamma(\alpha + n)\Gamma(\beta + n)} \sum_{k=0}^{n-1} \frac{(-1)^{n-k}(n - k - 1)!(\alpha)_k(\beta)_k}{k!}(1 - z)^k\\ &\quad + \frac{1}{\Gamma(\alpha)\Gamma(\beta)} \sum_{k=0}^{\infty} \frac{(\alpha + n)_k(\beta + n)_k}{(n + k)!k!}\\ &\quad \times [\psi(k + 1) - \psi(\alpha + n) - \psi(\beta + n)](1 - z)^{k+n}, \end{aligned} \tag{9.7.3}$$

---

[24] In differentiating g_2, we use the formula

$$\frac{d}{d\lambda}(\lambda)_k = (\lambda)_k[\psi(\lambda + k) - \psi(\lambda)].$$

From now on, we assume that $\alpha, \beta \neq 0, -1, -2, \ldots$

$$\left.\frac{\partial g_2}{\partial \gamma}\right|_{\gamma=\alpha+\beta+n} = \frac{1}{\Gamma(\alpha)\Gamma(\beta)} \sum_{k=0}^{\infty} \frac{(\alpha+n)_k(\beta+n)_k}{(n+k)!k!}$$
$$\times [\psi(\alpha+n+k) - \psi(\alpha+n) + \psi(\beta+n+k)$$
$$- \psi(\beta+n) - \psi(1+n+k) + \log(1-z)](1-z)^{k+n}, \quad (9.7.4)$$

where $\psi(z) = \Gamma'(z)/\Gamma(z)$ is the logarithmic derivative of the gamma function. Substituting (9.7.3–4) into (9.7.2), we obtain

$$F(\alpha, \beta; \alpha+\beta+n; z)$$
$$= \frac{\Gamma(\alpha+\beta+n)}{\Gamma(\alpha+n)\Gamma(\beta+n)} \sum_{k=0}^{n-1} \frac{(-1)^k(n-k-1)!(\alpha)_k(\beta)_k}{k!}(1-z)^k$$
$$+ \frac{(-1)^n\Gamma(\alpha+\beta+n)}{\Gamma(\alpha)\Gamma(\beta)} \sum_{k=0}^{\infty} \frac{(\alpha+n)_k(\beta+n)_k}{(n+k)!k!}[\psi(k+1)+\psi(n+k+1)$$
$$-\psi(\alpha+n+k)-\psi(\beta+n+k)-\log(1-z)](1-z)^{n+k},$$
$$|z-1|<1,\ |\arg(1-z)|<\pi, \quad n=0,1,2,\ldots, \quad \alpha, \beta \neq 0, -1, -2, \ldots \quad (9.7.5)$$

As usual, the meaningless sum

$$\sum_{k=0}^{-1} \cdots,$$

which appears when $n = 0$, is set equal to zero.

Formula (9.7.5) is no longer applicable if α or β equals $0, -1, -2, \ldots$, but then $F(\alpha, \beta; \alpha+\beta+n; z)$ reduces to a polynomial, and there is no need for analytic continuation. Moreover, the case $\gamma = \alpha+\beta-n$ reduces to that just considered by using the transformation (9.5.3), which becomes

$$F(\alpha, \beta; \alpha+\beta-n; z) = (1-z)^{-n}F(\alpha', \beta'; \alpha'+\beta'+n; z) \quad (9.7.6)$$

if $\alpha' = \alpha - n$, $\beta' = \beta - n$.

Similar considerations apply to the other formulas of Secs. 9.5–6. To give another example, we derive a formula suitable for making the analytic continuation of $F(\alpha, \beta; \gamma; z)$ into the domain $|z| > 1$, $|\arg(-z)| < \pi$ in the case where $\alpha - \beta = 0, \pm 1, \pm 2, \ldots$ Here we have to pass to the limit $\beta \to \alpha \pm n$ $(n = 0, 1, 2, \ldots)$ in (9.5.9). A calculation like that given above leads to the following formula (for the case $\beta = \alpha + n$):[25]

$$F(\alpha, \alpha+n; \gamma; z)$$
$$= \frac{\Gamma(\gamma)(-z)^{-\alpha}}{\Gamma(\gamma-\alpha)\Gamma(\alpha+n)} \sum_{k=0}^{n-1} \frac{(n-k-1)!(\alpha)_k(1-\gamma+\alpha)_k}{k!}(-z)^{-k}$$
$$+ \frac{\Gamma(\gamma)(-z)^{-\alpha}}{\Gamma(\alpha)\Gamma(\gamma-\alpha-n)} \sum_{k=0}^{\infty} \frac{(\alpha+n)_k(1+\alpha-\gamma+n)_k}{(n+k)!k!}$$
$$\times [\psi(k+1) + \psi(n+k+1) - \psi(\alpha+n+k)$$
$$- \psi(\gamma-\alpha-n-k) + \log(-z)]z^{-n-k},$$
$$|z| > 1,\ |\arg(1-z)| < \pi, \quad n = 0, 1, 2, \ldots, \quad \alpha \neq 0, -1, -2, \ldots,$$
$$\gamma - \alpha \neq 0, \pm 1, \pm 2, \ldots, \quad \gamma \neq 0, -1, -2, \ldots \quad (9.7.7)$$

[25] In the last step of the calculation, use formula (1.3.4).

We now examine the cases where formula (9.7.7) is not applicable. If $\alpha = 0, -1, -2, \ldots$, the function $F(\alpha, \alpha + n; \gamma; z)$ reduces to a polynomial, and there is no need for analytic continuation. According to (9.5.3),

$$F(\alpha, \alpha + n; \gamma; z) = (1 - z)^{\gamma - 2\alpha - n} F(\gamma - \alpha, \gamma - \alpha - n; \gamma; z), \quad (9.7.8)$$

and therefore $F(\alpha, \alpha + n; \gamma; z)$ reduces to an algebraic function if $\gamma - \alpha = 0, -1, -2, \ldots$ or $\gamma - \alpha = 1, 2, \ldots, n$, and analytic continuation is again unnecessary. If $\gamma - \alpha = n + 1, n + 2, \ldots$ and $\alpha \neq 0, \pm 1, \pm 2, \ldots$, then the hypergeometric function in the right-hand side of (9.7.8) satisfies the conditions allowing it to be continued by using formula (9.7.7). If $\gamma - \alpha = n + 1, n + 2, \ldots$ and $\alpha = 1, 2, \ldots$, the hypergeometric function can be represented by an integral of the type (9.1.6) with a rational integrand, i.e., $F(\alpha, \alpha + n; \gamma; z)$ can be expressed in finite form in terms of rational functions. Finally, we note that the case $\beta = \alpha - n$ reduces to that just considered if we again use the transformation (9.5.3).

9.8. Representation of Various Functions in Terms of the Hypergeometric Function

As we now show, various familiar functions of mathematical analysis are special cases of the hypergeometric function $F(\alpha, \beta; \gamma; z)$, corresponding to suitable choices of the parameters α, β, γ and the variable z:[26]

1. *Elementary functions.* The hypergeometric function $F(\alpha, \beta; \gamma; z)$ reduces to a polynomial if $\alpha = 0, -1, -2, \ldots$ or $\beta = 0, -1, -2, \ldots$ For example,

$$F(\alpha, 0; \gamma; z) = 1, \qquad F(\alpha, -2; \gamma; z) = 1 - 2\frac{\alpha}{\gamma} z + \frac{\alpha(\alpha + 1)}{\gamma(\gamma + 1)} z^2,$$

and so on. The transformation

$$F(\alpha, \beta; \gamma; z) = (1 - z)^{\gamma - \alpha - \beta} F(\gamma - \alpha, \gamma - \beta; \gamma; z), \quad |\arg (1 - z)| < \pi$$

[cf. (9.5.3)] shows that $F(\alpha, \beta; \gamma; z)$ reduces to an algebraic function if $\gamma - \alpha = 0, -1, -2, \ldots$ or $\gamma - \beta = 0, -1, -2, \ldots$ In particular,

$$F(\alpha, \beta; \beta; z) = (1 - z)^{-\alpha}, \qquad |\arg (1 - z)| < \pi \quad (9.8.1)$$

for any value of β, and

$$(1 - z)^{\nu} = F(-\nu, 1; 1; z), \qquad (1 - z)^{-1/2} = F(\tfrac{1}{2}, 1; 1; z),$$
$$z^n = F(-n, 1; 1; 1 - z), \qquad n = 0, 1, 2, \ldots \quad (9.8.2)$$

---

[26] Further examples are given in the Bateman Manuscript Project, *Higher Transcendental Functions, Vol. 1*, pp. 89, 101.

Other representations of this type can be derived from the formulas of Sec. 9.6. Thus, setting $\beta = \alpha + \frac{1}{2}$ in (9.6.2) and (9.6.7), we obtain

$$F(\alpha, \alpha + \tfrac{1}{2}; 2\alpha + 1; z) = \left(\frac{1 + \sqrt{1 - z}}{2}\right)^{-2\alpha}, \qquad |\arg (1 - z)| < \pi,$$

$$F(\alpha, \alpha + \tfrac{1}{2}; 2\alpha; z) = \frac{1}{\sqrt{1 - z}}\left(\frac{1 + \sqrt{1 - z}}{2}\right)^{1-2\alpha}, \ |\arg (1 - z)| < \pi. \tag{9.8.3}$$

By starting from the series expansion

$$\log (1 - z) = -\sum_{k=0}^{\infty} \frac{z^{k+1}}{k + 1} = -z \sum_{k=0}^{\infty} \frac{(1)_k(1)_k}{(2)_k k!} z^k, \qquad |z| < 1$$

of the logarithm, we find that

$$\log (1 - z) = -zF(1, 1; 2; z), \qquad |\arg (1 - z)| < \pi. \tag{9.8.4}$$

Similarly, we deduce the following formulas for the inverse trigonometric functions:

$$\begin{aligned} \text{arc tan } z &= zF(\tfrac{1}{2}, 1; \tfrac{3}{2}; -z^2), \qquad |\arg (1 \pm zi)| < \pi, \\ \text{arc sin } z &= zF(\tfrac{1}{2}, \tfrac{1}{2}; \tfrac{3}{2}; z^2), \qquad |\arg (1 \pm z)| < \pi. \end{aligned} \tag{9.8.5}$$

2. *Elliptic integrals.* The complete elliptic integrals

$$K(z) = \int_0^{\pi/2} (1 - z^2 \sin^2 \varphi)^{-1/2}\, d\varphi, \quad E(z) = \int_0^{\pi/2} (1 - z^2 \sin^2 \varphi)^{1/2}\, d\varphi$$

of the first and second kinds [cf. (7.10.11)], where z is a complex variable belonging to the domain $|\arg (1 \pm z)| < \pi$, can also be represented in terms of the hypergeometric function. Assuming temporarily that $|z| < 1$ and using the binomial expansion, we find that

$$K(z) = \sum_{k=0}^{\infty} \frac{(\frac{1}{2})_k}{k!} z^{2k} \int_0^{\pi/2} \sin^{2k} \varphi \, d\varphi = \frac{\pi}{2} \sum_{k=0}^{\infty} \frac{(\frac{1}{2})_k(\frac{1}{2})_k}{(1)_k k!} z^{2k},$$

which implies

$$K(z) = \frac{\pi}{2} F(\tfrac{1}{2}, \tfrac{1}{2}; 1; z^2), \qquad |\arg (1 \pm z)| < \pi. \tag{9.8.6}$$

Similarly, we have the following representation of the elliptic integral of the second kind:

$$E(z) = \frac{\pi}{2} F(-\tfrac{1}{2}, \tfrac{1}{2}; 1; z^2), \qquad |\arg (1 \pm z)| < \pi. \tag{9.8.7}$$

Starting from these formulas, one can develop the theory of elliptic integrals, regarded as functions of the modulus z.

3. *Spherical harmonics.* One of the most important classes of functions which can be expressed in terms of the hypergeometric function consists of the spherical harmonics studied in Chapter 7. In fact, formulas (7.12.27) and (7.12.29) immediately imply the following representations of the associated Legendre functions:

$$P_\nu^m(z) = \frac{\Gamma(\nu + m + 1)}{\Gamma(\nu - m + 1)} \frac{(z^2 - 1)^{m/2}}{2^m \Gamma(m + 1)} \times F\left(m - \nu, m + \nu + 1; m + 1; \frac{1 - z}{2}\right),$$

$$|\arg(z \pm 1)| < \pi, \qquad m = 0, 1, 2, \ldots, \tag{9.8.8}$$

$$Q_\nu^m(z) = \frac{(-1)^m \sqrt{\pi}\,\Gamma(\nu + m + 1)}{2^{\nu+1}\Gamma(\nu + \frac{3}{2}) z^{\nu+m+1}} (z^2 - 1)^{m/2} \times F\left(\frac{m + \nu + 2}{2}, \frac{m + \nu + 1}{2}; \nu + \frac{3}{2}; \frac{1}{z^2}\right),$$

$$|\arg z| < \pi, \qquad |\arg(z \pm 1)| < \pi, \qquad m = 0, 1, 2, \ldots \tag{9.8.9}$$

In particular, the Legendre polynomials (see Sec. 4.2) are given by the formula

$$P_n(z) = F\left(-n, n + 1; 1; \frac{1 - z}{2}\right), \qquad n = 0, 1, 2, \ldots \tag{9.8.10}$$

By regarding (9.8.8–10) as definitions and using the general theory of the hypergeometric function, it is a simple matter to develop the theory of spherical harmonics. This approach is especially convenient for deriving the relations of Sec. 7.6 and their generalizations to the case of arbitrary m.

9.9 The Confluent Hypergeometric Function

Besides the hypergeometric function $F(\alpha, \beta; \gamma; z)$, an important role is played in the theory of special functions by a related function

$$\Phi(\alpha, \gamma; z) = \sum_{k=0}^{\infty} \frac{(\alpha)_k}{(\gamma)_k} \frac{z^k}{k!}, \qquad |z| < \infty, \quad \gamma \neq 0, -1, -2, \ldots, \tag{9.9.1}$$

known as the *confluent hypergeometric function.* Here z is a complex variable, α and γ are parameters which can take arbitrary real or complex values (except that $\gamma \neq 0, -1, -2, \ldots$), and, as always,

$$(\lambda)_0 = 1, \qquad (\lambda)_k = \frac{\Gamma(\lambda + k)}{\Gamma(\lambda)} = \lambda(\lambda + 1)\cdots(\lambda + k - 1), \qquad k = 1, 2, \ldots$$

As indicated, the series (9.9.1) converges for all finite z,[27] and therefore represents an entire function of z.

If we set

$$\varphi(\alpha, \gamma; z) = \frac{1}{\Gamma(\gamma)} \Phi(\alpha, \gamma; z) = \sum_{k=0}^{\infty} \frac{(\alpha)_k}{\Gamma(\gamma + k)} \frac{z^k}{k!}, \tag{9.9.2}$$

then $\varphi(\alpha, \gamma; z)$ is an entire function of α and γ, for fixed z. In fact, the terms of the series (9.9.2) are entire functions of α and γ, and the series is uniformly convergent in the region $|\alpha| \leqslant A$, $|\gamma| \leqslant C$ (where A and C are arbitrarily large).[28] Therefore, for fixed z, $\Phi(\alpha, \gamma; z)$ is an entire function of α and a meromorphic function of γ, with simple poles at the points $\gamma = 0, -1, -2, \ldots$

A comparison of (9.1.2) and (9.1.3) shows at once that

$$\Phi(\alpha, \gamma; z) = \lim_{\beta \to \infty} F\left(\alpha, \beta; \gamma; \frac{z}{\beta}\right). \tag{9.9.3}$$

The function $\Phi(\alpha, \gamma; z)$ is very frequently encountered in analysis, mainly because of the fact that a large number of special functions can be obtained from $\Phi(\alpha, \gamma; z)$ by making suitable choices of the parameters α, γ and the variable z (see Sec. 9.13). This makes it possible to develop the general theory of these functions in a simple and compact form.

The definition of the confluent hypergeometric function immediately implies the identities

$$\frac{d}{dz} \Phi(\alpha, \gamma; z) = \frac{\alpha}{\gamma} \Phi(\alpha + 1, \gamma + 1; z), \tag{9.9.4}$$

$$\frac{d^m}{dz^m} \Phi(\alpha, \gamma; z) = \frac{(\alpha)_m}{(\gamma)_m} \Phi(\alpha + m, \gamma + m; z), \quad m = 1, 2, \ldots, \tag{9.9.5}$$

---

[27] Use the ratio test, noting that if

$$u_k = \frac{(\alpha)_k}{(\gamma)_k} \frac{z^k}{k!},$$

then

$$\left|\frac{u_{k+1}}{u_k}\right| = \left|\frac{\alpha + k}{(\gamma + k)(1 + k)} z\right| \to 0,$$

as $k \to \infty$.

[28] Use the criterion given in footnote 4, p. 102, noting that if

$$v_k = \frac{(\alpha)_k}{\Gamma(\gamma + k)} \frac{z^k}{k!},$$

then

$$\left|\frac{v_{k+1}}{v_k}\right| = \left|\frac{\alpha + k}{(\gamma + k)(1 + k)} z\right| \leqslant \frac{A + k}{(k - C)(1 + k)} |z| \leqslant q < 1,$$

for sufficiently large k.

and the recurrence relations

$$(\gamma - \alpha - 1)\Phi + \alpha\Phi(\alpha + 1) - (\gamma - 1)\Phi(\gamma - 1) = 0, \tag{9.9.6}$$

$$\gamma\Phi - \gamma\Phi(\alpha - 1) - z\Phi(\gamma + 1) = 0, \tag{9.9.7}$$

$$(\alpha - 1 + z)\Phi + (\gamma - \alpha)\Phi(\alpha - 1) - (\gamma - 1)\Phi(\gamma - 1) = 0, \tag{9.9.8}$$

$$\gamma(\alpha + z)\Phi - \alpha\gamma\Phi(\alpha + 1) - (\gamma - \alpha)z\Phi(\gamma + 1) = 0, \tag{9.9.9}$$

$$(\gamma - \alpha)\Phi(\alpha - 1) + (2\alpha - \gamma + z)\Phi - \alpha\Phi(\alpha + 1) = 0, \tag{9.9.10}$$

$$\gamma(\gamma - 1)\Phi(\gamma - 1) - \gamma(\gamma - 1 + z)\Phi + (\gamma - \alpha)z\Phi(\gamma + 1) = 0, \tag{9.9.11}$$

connecting the function $\Phi \equiv \Phi(\alpha, \gamma; z)$ with any two *contiguous* functions $\Phi(\alpha \pm 1) \equiv \Phi(\alpha \pm 1, \gamma; z)$ and $\Phi(\gamma \pm 1) \equiv \Phi(\alpha, \gamma \pm 1; z)$. Formulas (9.9.6–7) can be verified by direct substitution of the series (9.9.1), and then the other recurrence relations can be obtained by simple transformations of (9.9.6–7).

Besides the recurrence relations just given, there exist similar relations between the function $\Phi(\alpha, \gamma; z)$ and any pair of functions of the form $\Phi(\alpha + m, \gamma + n; z)$, where m and n are arbitrary integers. Two simple relations of this kind are[29]

$$\Phi(\alpha, \gamma; z) = \Phi(\alpha + 1, \gamma; z) - \frac{z}{\gamma}\Phi(\alpha + 1, \gamma + 1; z), \tag{9.9.12}$$

$$\Phi(\alpha, \gamma; z) = \frac{\gamma - \alpha}{\gamma}\Phi(\alpha, \gamma + 1; z) + \frac{\alpha}{\gamma}\Phi(\alpha + 1, \gamma + 1; z), \tag{9.9.13}$$

as can be verified by direct substitution of (9.9.1), or by repeated use of the relations between $\Phi(\alpha, \gamma; z)$ and its contiguous functions.

9.10. The Differential Equation for the Confluent Hypergeometric Function and Its Solutions. The Confluent Hypergeometric Function of the Second Kind

It is easy to see that the confluent hypergeometric function is a particular solution of the linear differential equation

$$zu'' + (\gamma - z)u' - \alpha u = 0, \tag{9.10.1}$$

where $\gamma \neq 0, -1, -2, \ldots$ In fact, denoting the left-hand side of this equation by $l(u)$, and setting $u = u_1 = \Phi(\alpha, \gamma; z)$, we have

$$l(u_1) = \sum_{k=2}^{\infty} \frac{k(k-1)(\alpha)_k}{(\gamma)_k k!} z^{k-1} + (\gamma - z)\sum_{k=1}^{\infty} \frac{(\alpha)_k k}{(\gamma)_k k!} z^{k-1} - \alpha \sum_{k=0}^{\infty} \frac{(\alpha)_k}{(\gamma)_k k!} z^k$$

$$= \left[\gamma\frac{(\alpha)_1}{(\gamma)_1} - \alpha\right] + \sum_{k=1}^{\infty} \frac{(\alpha)_k z^k}{(\gamma)_k k!}\left[k\frac{\alpha + k}{\gamma + k} + \gamma\frac{\alpha + k}{\gamma + k} - k - \alpha\right] \equiv 0.$$

---

[29] Note the similarity between formulas (9.9.6–13) and formulas (9.2.4–15).

To obtain a second linearly independent solution of (9.10.1), we assume that $|\arg z| < \pi$ and make the substitution $u = z^{1-\gamma}v$. Then equation (9.10.1) goes into an equation of the same form, i.e.,

$$zv'' + (\gamma' - z)v' - \alpha' v = 0,$$

with new parameters $\alpha' = 1 + \alpha - \gamma, \gamma' = 2 - \gamma$. It follows that the function

$$u = u_2 = z^{1-\gamma}\Phi(1 + \alpha - \gamma, 2 - \gamma; z)$$

is also a solution of (9.10.1) if $\gamma \neq 2, 3, \ldots$. Thus, if $\gamma \neq 0, \pm 1, \pm 2, \ldots$, both solutions u_1, u_2 are meaningful and are linearly independent of each other,[30] so that the general solution of (9.10.1) can be written in the form

$$u = A\Phi(\alpha, \gamma; z) + Bz^{1-\gamma}\Phi(1 + \alpha - \gamma; 2 - \gamma; z),$$
$$|\arg z| < \pi, \quad \gamma \neq 0, \pm 1, \pm 2, \ldots \tag{9.10.2}$$

With a view to obtaining an expression for the general solution of (9.10.1) which is suitable for arbitrary $\gamma \neq 0, -1, -2, \ldots$ [see (9.10.11) below], we introduce a new function

$$\Psi(\alpha, \gamma; z) = \frac{\Gamma(1 - \gamma)}{\Gamma(1 + \alpha - \gamma)}\Phi(\alpha, \gamma; z) + \frac{\Gamma(\gamma - 1)}{\Gamma(\alpha)}z^{1-\gamma}\Phi(1 + \alpha - \gamma, 2 - \gamma; z),$$
$$|\arg z| < \pi, \quad \gamma \neq 0, \pm 1, \pm 2, \ldots, \tag{9.10.3}$$

called the *confluent hypergeometric function of the second kind.* Formula (9.10.3) defines the function $\Psi(\alpha, \gamma; z)$ for arbitrary nonintegral γ, and moreover, as we now show, the right-hand side of (9.10.3) approaches a definite limit as $\gamma \to n + 1$ $(n = 0, 1, 2, \ldots)$. Replacing the Φ functions in (9.10.3) by the appropriate series, and using formula (1.2.2) from the theory of the gamma function, we obtain

$$\begin{aligned}\Psi(\alpha, \gamma; z) &= \frac{\pi}{\sin \pi\gamma}\left[\frac{1}{\Gamma(1 + \alpha - \gamma)}\sum_{k=0}^{\infty}\frac{(\alpha)_k}{\Gamma(\gamma + k)}\frac{z^k}{k!}\right.\\ &\quad\left. - \frac{1}{\Gamma(\alpha)}\sum_{k=0}^{\infty}\frac{(4\alpha - \gamma)}{\Gamma(2 - \gamma + k)}\frac{z^{k+1-\gamma}}{k!}\right] = \frac{\pi}{\sin \pi\gamma}(g_1 - g_2).\end{aligned} \tag{9.10.4}$$

Since

$$\lim_{\gamma \to n+1} g_1 = \frac{1}{\Gamma(\alpha - n)}\sum_{k=0}^{\infty}\frac{(\alpha)_k}{\Gamma(k + n + 1)}\frac{z^k}{k!} = \frac{1}{\Gamma(\alpha - n)}\sum_{k=0}^{\infty}\frac{(\alpha)_k}{(n + k)!}\frac{z^k}{k!},$$

$$\begin{aligned}\lim_{\gamma \to n+1} g_2 &= \frac{1}{\Gamma(\alpha)}\sum_{k=n}^{\infty}\frac{(\alpha - n)_k}{\Gamma(k - n + 1)}\frac{z^{k-n}}{k!}\\ &= \frac{1}{\Gamma(\alpha)}\sum_{k=0}^{\infty}\frac{(\alpha - n)_{k+n}}{\Gamma(k + 1)}\frac{z^k}{(n + k)!} = \frac{1}{\Gamma(\alpha - n)}\sum_{k=0}^{\infty}\frac{(\alpha)_k}{(n + k)!}\frac{z^k}{k!},\end{aligned}$$

---

[30] Note that $u_1 \equiv u_2$ if $\gamma = 1$.

the right-hand side of (9.10.4) becomes indeterminate as $\gamma \to n + 1$, and approaches a limit whose value can be found by using L'Hospital's rule, i.e.,

$$\Psi(\alpha, n + 1; z) = \lim_{\gamma \to n+1} \Psi(\alpha, \gamma; z) = (-1)^{n+1}\left[\frac{\partial g_1}{\partial \gamma}\bigg|_{\gamma = n+1} - \frac{\partial g_2}{\partial \gamma}\bigg|_{\gamma = n+1}\right],$$
$$|\arg z| < \pi, \quad n = 0, 1, 2, \ldots \qquad (9.10.5)$$

Calculations like those made in Sec. 5.5 show that[31]

$$\frac{\partial g_1}{\partial \gamma}\bigg|_{\gamma = n+1} = \frac{1}{\Gamma(\alpha - n)} \sum_{k=0}^{\infty} \frac{(\alpha)_k z^k}{(n + k)!k!} [\psi(\alpha - n) - \psi(n + k + 1)],$$

$$\frac{\partial g_2}{\partial \gamma}\bigg|_{\gamma = n+1} = \frac{1}{\Gamma(\alpha - n)} \sum_{k=0}^{\infty} \frac{(\alpha)_k z^k}{(n + k)!k!}$$
$$\times [\psi(1 + k) - \psi(\alpha + k) + \psi(\alpha - n) - \log z]$$
$$+ \frac{1}{\Gamma(\alpha)} \sum_{k=0}^{n-1} \frac{(-1)^{n-k}(n - k - 1)!(\alpha - n)_k}{k!} z^{k-n},$$

which leads to the following series expansion:

$$\Psi(\alpha, n + 1; z)$$
$$= \frac{(-1)^{n+1}}{\Gamma(\alpha - n)} \sum_{k=0}^{\infty} \frac{(\alpha)_k z^k}{(n + k)!k!} [\psi(\alpha + k) - \psi(1 + k) - \psi(n + 1 + k) + \log z]$$
$$+ \frac{1}{\Gamma(\alpha)} \sum_{k=0}^{n-1} \frac{(-1)^k (n - k - 1)!(\alpha - n)_k}{k!} z^{k-n}, \qquad (9.10.6)$$
$$|\arg z| < \pi, \quad n = 0, 1, 2, \ldots, \quad \alpha \neq 0, -1, -2, \ldots$$

Here $\psi(z) = \Gamma'(z)/\Gamma(z)$ is the logarithmic derivative of the gamma function, and the meaningless sum

$$\sum_{k=0}^{-1} \ldots,$$

which appears when $n = 0$, is set equal to zero.

If $\alpha = -m$ $(m = 0, 1, 2, \ldots)$, passage to the limit $\gamma \to n + 1$ $(n = 0, 1, 2, \ldots)$ in (9.10.3) leads to the expression[32]

$$\Psi(-m, n + 1; z) = (-1)^m \frac{(m + n)!}{n!} \Phi(-m, n + 1; z), \qquad (9.10.7)$$
$$m = 0, 1, 2, \ldots, \quad n = 0, 1, 2, \ldots$$

---

[31] In differentiating g_2, we use the formula

$$\frac{d}{d\lambda}(\lambda)_k = (\lambda)_k[\psi(\lambda + k) - \psi(\lambda)].$$

From now on, we assume that $\alpha \neq 0, -1, -2, \ldots$

[32] Here we again use formula (1.2.2).

Moreover, it is an immediate consequence of (9.10.3) that the confluent hypergeometric function of the second kind satisfies the relation

$$\Psi(\alpha, \gamma; z) = z^{1-\gamma}\Psi(1 + \alpha - \gamma, 2 - \gamma; z), \qquad |\arg z| < \pi. \tag{9.10.8}$$

Using this formula, we can define the function $\Psi(\alpha, \gamma; z)$ for $\gamma = 0, -1, -2, \ldots$, obtaining

$$\Psi(\alpha, 1 - n; z) = \lim_{\gamma \to 1-n} \Psi(\alpha, \gamma; z) = z^n \Psi(\alpha + n, n + 1; z), \qquad |\arg z| < \pi, \quad n = 1, 2, \ldots \tag{9.10.9}$$

Thus we see that $\Psi(\alpha, \gamma; z)$ is meaningful for arbitrary values of the parameters α and γ. It follows from the definition (9.10.3) and the properties of $\Phi(\alpha, \gamma; z)$ that $\Psi(\alpha, \gamma; z)$ is an analytic function of z in the plane cut along $[-\infty, 0]$, and an entire function of α and γ.

Next we show that $\Psi(\alpha, \gamma; z)$ is a solution of the differential equation (9.10.1). For $\gamma \neq 0, \pm 1, \pm 2, \ldots$, this is an immediate consequence of (9.10.3), and for integral γ, the result follows from the principle of analytic continuation (cf. footnote 12, p. 167). For $\alpha \neq 0, -1, -2, \ldots$, the solutions $\Phi(\alpha, \gamma; z)$ and $\Psi(\alpha, \gamma; z)$ are linearly independent, as can easily be verified by calculating the Wronskian[33]

$$W\{\Phi(\alpha, \gamma, z), \Psi(\alpha, \gamma; z)\} = -\frac{\Gamma(\gamma)}{\Gamma(\alpha)} z^{-\gamma}e^z, \qquad |\arg z| < \pi, \quad \gamma \neq 0, -1, -2, \ldots, \tag{9.10.10}$$

and then the general solution of (9.10.1) can be written in the form

$$u = A\Phi(\alpha, \gamma; z) + B\Psi(\alpha, \gamma; z), \qquad |\arg z| < \pi, \quad \alpha, \gamma \neq 0, -1, -2, \ldots \tag{9.10.11}$$

The function $\Psi(\alpha, \gamma; z)$ has a number of properties analogous to those of $\Phi(\alpha, \gamma; z)$. For example, we have the differentiation formulas

$$\begin{aligned} \frac{d}{dz}\Psi(\alpha, \gamma; z) &= -\alpha\Psi(\alpha + 1, \gamma + 1; z), \\ \frac{d^m}{dz^m}\Psi(\alpha, \gamma; z) &= (-1)^m(\alpha)_m\Psi(\alpha + m, \gamma + m; z), \qquad m = 1, 2, \ldots, \end{aligned} \tag{9.10.12}$$

the recurrence relations

$$\Psi - \alpha\Psi(\alpha + 1) - \Psi(\gamma - 1) = 0, \tag{9.10.13}$$

$$(\gamma - \alpha)\Psi + \Psi(\alpha - 1) - z\Psi(\gamma + 1) = 0, \tag{9.10.14}$$

---

[33] Equation (9.10.1) implies

$$W\{\Phi, \Psi\} = Cz^{-\gamma}e^z.$$

Comparing both sides of this identity as $z \to 0$, we find that

$$C = -\frac{\Gamma(\gamma)}{\Gamma(\alpha)}.$$

$$(\alpha - 1 + z)\Psi - \Psi(\alpha - 1) + (\alpha - \gamma + 1)\Psi(\gamma - 1) = 0, \tag{9.10.15}$$

$$(\alpha + z)\Psi + \alpha(\gamma - \alpha - 1)\Psi(\alpha + 1) - z\Psi(\gamma + 1) = 0, \tag{9.10.16}$$

$$\Psi(\alpha - 1) - (2\alpha - \gamma + z)\Psi + \alpha(\alpha - \gamma + 1)\Psi(\alpha + 1) = 0, \tag{9.10.17}$$

$$(\gamma - \alpha - 1)\Psi(\gamma - 1) - (\gamma - 1 + z)\Psi + z\Psi(\gamma + 1) = 0, \tag{9.10.18}$$

$$\Psi \equiv \Psi(\alpha, \gamma; z), \quad \Psi(\alpha \pm 1) \equiv \Psi(\alpha \pm 1, \gamma; z), \quad \Psi(\gamma \pm 1) \equiv \Psi(\alpha, \gamma \pm 1; z)$$

and so on, whose validity follows from the definition of the Ψ function and the corresponding properties of the Φ function.

9.11. Integral Representations of the Confluent Hypergeometric Functions

The functions $\Phi(\alpha, \gamma; z)$ and $\Psi(\alpha, \gamma; z)$ have simple integral representations which play an important role in the theory and applications of confluent hypergeometric functions. Here we consider only the basic representations in terms of integrals evaluated along an interval of the real axis, referring the reader elsewhere for more general representations in terms of contour integrals.[34]

The simplest integral representation of the function $\Phi(\alpha, \gamma; z)$ can be obtained by summing the series (9.9.1) with the help of formula (9.1.2):

$$\frac{(\alpha)_k}{(\gamma)_k} = \frac{\Gamma(\gamma)}{\Gamma(\alpha)\Gamma(\gamma - \alpha)} \int_0^1 t^{\alpha - 1 + k}(1 - t)^{\gamma - \alpha - 1}\, dt,$$

$$\operatorname{Re} \gamma > \operatorname{Re} \alpha > 0, \quad k = 0, 1, 2, \ldots$$

This gives

$$\Phi(\alpha, \gamma; z) = \frac{\Gamma(\gamma)}{\Gamma(\alpha)\Gamma(\gamma - \alpha)} \sum_{k=0}^{\infty} \frac{z^k}{k!} \int_0^1 t^{\alpha - 1 + k}(1 - t)^{\gamma - \alpha - 1}\, dt$$

$$= \frac{\Gamma(\gamma)}{\Gamma(\alpha)\Gamma(\gamma - \alpha)} \int_0^1 t^{\alpha - 1}(1 - t)^{\gamma - \alpha - 1} dt \sum_{k=0}^{\infty} \frac{(zt)^k}{k!},$$

or

$$\Phi(\alpha, \gamma; z) = \frac{\Gamma(\gamma)}{\Gamma(\alpha)\Gamma(\gamma - \alpha)} \int_0^1 e^{zt} t^{\alpha - 1}(1 - t)^{\gamma - \alpha - 1}\, dt, \qquad \operatorname{Re} \gamma > \operatorname{Re} \alpha > 0, \tag{9.11.1}$$

where reversing the order of integration and summation is justified by the usual absolute convergence argument (cf. footnote 2, p. 239).

---

[34] See the Bateman Manuscript Project, *Higher Transcendental Functions, Vol. 1*, pp. 256, 271 ff.

We can use the integral representation (9.11.1) to deduce an important relation satisfied by the function $\Phi(\alpha, \gamma; z)$. Assuming temporarily that $\operatorname{Re}\gamma > \operatorname{Re}\alpha > 0$, we make the change of variable $t = 1 - s$. Then (9.11.1) becomes

$$\Phi(\alpha, \gamma; z) = \frac{\Gamma(\gamma)}{\Gamma(\alpha)\Gamma(\gamma - \alpha)} e^z \int_0^1 e^{-zs} s^{\gamma-\alpha-1}(1 - s)^{\alpha-1}\, ds,$$

which implies

$$\Phi(\alpha, \gamma; z) = e^z \Phi(\gamma - \alpha, \gamma; z), \tag{9.11.2}$$

since $\operatorname{Re}\gamma > \operatorname{Re}(\gamma - \alpha)$. The relation (9.11.2) was proved under the assumption that $\operatorname{Re}\gamma > \operatorname{Re}\alpha > 0$, but after dividing by $\Gamma(\gamma)$, both sides become entire functions of α and γ. Therefore, according to the principle of analytic continuation, (9.11.2) remains valid for arbitrary α and γ, provided that $\gamma \neq 0, -1, -2, \ldots$

To obtain an integral representation of $\Psi(\alpha, \gamma; z)$, we first note that the function u, defined by

$$u = \frac{1}{\Gamma(\alpha)} \int_0^\infty e^{-zt} t^{\alpha-1}(1 + t)^{\gamma-\alpha-1}\, dt, \qquad \operatorname{Re}\alpha > 0, \quad \operatorname{Re} z > 0, \tag{9.11.3}$$

is a solution of the differential equation (9.10.1). In fact, denoting the left-hand side of (9.11.3) by $l(u)$, we have[35]

$$\begin{aligned} l(u) &= \frac{1}{\Gamma(\alpha)} \int_0^\infty e^{-zt} t^{\alpha-1}(1 + t)^{\gamma-\alpha-1}[zt^2 - (\gamma - z)t - \alpha]\, dt \\ &= -\frac{1}{\Gamma(\alpha)} \int_0^\infty \frac{d}{dt}[e^{-zt} t^{\alpha}(1 + t)^{\gamma-\alpha}]\, dt = -\frac{1}{\Gamma(\alpha)} e^{-zt} t^{\alpha}(1 + t)^{\gamma-\alpha}\Big|_{t=0}^{t=\infty} \equiv 0. \end{aligned}$$

According to (9.10.2), the solution u can be written in the form

$$u = A\Phi(\alpha, \gamma; z) + Bz^{1-\gamma}\Phi(1 + \alpha - \gamma, 2 - \gamma; z), \qquad |\arg z| < \pi, \quad \gamma \neq 0, \pm 1, \pm 2, \ldots \tag{9.11.4}$$

Assuming temporarily that $0 < \operatorname{Re}\gamma < 1$ and $z > 0$, we take the limit of (9.11.3) as $z \to 0+$. This gives

$$A = \lim_{z \to 0+} u = \frac{1}{\Gamma(\alpha)} \int_0^\infty t^{\alpha-1}(1 + t)^{\gamma-\alpha-1}\, dt = \frac{\Gamma(1 - \gamma)}{\Gamma(1 + \alpha - \gamma)},$$

where we have used formulas (1.5.3) and (1.5.6) from the theory of the gamma function, and the passage to the limit behind the integral sign is easily

[35] With our restrictions on α and z, the differentiation behind the integral sign is justified.

justified. Moreover, differentiating (9.11.4) with respect to z, multiplying by z^γ and then taking the limit as $z \to 0+$, we obtain

$$B = \frac{1}{1-\gamma}\lim_{z\to 0+} z^\gamma u' = \frac{1}{\gamma - 1}\frac{1}{\Gamma(\alpha)}\lim_{z\to 0+} z^\gamma \int_0^\infty e^{-zt}t^\alpha(1+t)^{\gamma-\alpha-1}\,dt$$

$$= \frac{1}{(\gamma-1)\Gamma(\alpha)}\lim_{z\to 0+}\int_0^\infty e^{-s}s^\alpha(s+z)^{\gamma-\alpha-1}\,ds$$

$$= \frac{1}{(\gamma-1)\Gamma(\alpha)}\int_0^\infty e^{-s}s^{\gamma-1}\,ds = \frac{\Gamma(\gamma-1)}{\Gamma(\alpha)}.$$

It follows that

$$u = \frac{\Gamma(1-\gamma)}{\Gamma(1+\alpha-\gamma)}\Phi(\alpha,\gamma;z) + \frac{\Gamma(\gamma-1)}{\Gamma(\alpha)}z^{1-\gamma}\Phi(1+\alpha-\gamma, 2-\gamma; z) \equiv \Psi(\alpha,\gamma;z). \quad (9.11.5)$$

Since both sides are entire functions of the parameter γ and analytic functions of the variable z in the half-plane $\operatorname{Re} z > 0$ (see Sec. 9.10), the temporary restrictions imposed on γ and z can be dropped, and we arrive at the integral representation

$$\Psi(\alpha,\gamma;z) = \frac{1}{\Gamma(\alpha)}\int_0^\infty e^{-zt}t^{\alpha-1}(1+t)^{\gamma-\alpha-1}\,dt, \quad \operatorname{Re}\alpha > 0, \quad \operatorname{Re} z > 0. \quad (9.11.6)$$

Some other integral representations of the functions $\Phi(\alpha,\gamma;z)$ and $\Psi(\alpha,\gamma;z)$ are given in Problems 11–13, p. 278.

9.12. Asymptotic Representations of the Confluent Hypergeometric Functions for Large $|z|$

We begin by deriving the asymptotic representation of $\Psi(\alpha,\gamma;z)$ for large $|z|$, which turns out to be simpler than the corresponding representation of $\Phi(\alpha,\gamma;z)$. Suppose that

$$\operatorname{Re}\alpha > 0, \qquad |\arg z| < \frac{\pi}{2} - \delta,$$

where $\delta > 0$ is arbitrarily small. According to (5.11.2),

$$(1+t)^{\gamma-\alpha-1} = \sum_{k=0}^{n}\frac{(-1)^k(1+\alpha-\gamma)_k}{k!}t^k + \frac{(-1)^{n+1}(1+\alpha-\gamma)_n}{n!}t^{n+1}\int_0^1 (1-s)^n(1+st)^{\gamma-\alpha-n-2}\,ds.$$

Substituting this expansion into the integral representation (9.11.6) and integrating term by term, we obtain[36]

$$\Psi(\alpha, \gamma; z) = z^{-\alpha}\left[\sum_{k=0}^{n} \frac{(-1)^k(\alpha)_k(1+\alpha-\gamma)_k}{k!} z^{-k} + r_n(z)\right],$$

where

$$r_n(z) = \frac{(-1)^{n+1}(1+\alpha-\gamma)_n z^{\alpha}}{n!\Gamma(\alpha)} \int_0^\infty e^{-zt}t^{n+\alpha}\, dt \int_0^1 (1-s)^n(1+st)^{\gamma-\alpha-n-2}\, ds.$$

Estimating $|r_n(z)|$ we find that

$$|r_n(z)| \leqslant \left|\frac{(1+\alpha-\gamma)_n}{n!\Gamma(\alpha)} z^{\alpha}\right| \int_0^\infty e^{-|z|t \sin\delta} t^{n+\operatorname{Re}\alpha}\, dt$$
$$\times \int_0^1 (1-s)^n(1+st)^{\operatorname{Re}(\gamma-\alpha)-n-2}\, ds.$$

If we choose n so large that $\operatorname{Re}(\gamma-\alpha) - n - 2 \leqslant 0$, then

$$(1+st)^{\operatorname{Re}(\gamma-\alpha)-n-2} \leqslant 1,$$

and hence[37]

$$|r_n(z)| \leqslant \left|\frac{(1+\alpha-\gamma)_n}{(n+1)!\Gamma(\alpha)}\right| \frac{\Gamma(n+\operatorname{Re}\alpha+1)|z|^{\operatorname{Re}\alpha}e^{\pi|\operatorname{Im}\alpha|}}{(|z|\sin\delta)^{n+\operatorname{Re}\alpha+1}} = O(|z|^{-n-1}).$$

It follows that

$$\Psi(\alpha, \gamma; z) = z^{-\alpha}\left[\sum_{k=0}^{n} \frac{(-1)^k(\alpha)_k(1+\alpha-\gamma)_k}{k!} z^{-k} + O(|z|^{-n-1})\right],$$

$$\operatorname{Re}\alpha > 0, \quad |\arg z| \leqslant \frac{\pi}{2} - \delta, \quad n \geqslant \operatorname{Re}(\gamma-\alpha) - 2 \qquad (9.12.1)$$

for large $|z|$.

We now show that the conditions under which this formula has been proved can be considerably weakened. First we note that even if $\operatorname{Re}(\gamma-\alpha) - n - 2 > 0$, an integer $m > n$ can always be found such that $\operatorname{Re}(\gamma-\alpha) - m - 2 \leqslant 0$. Since the expansion (9.12.1) certainly holds with n replaced by m, we have

$$\sum_{k=0}^{m} \cdots + O(|z|^{-m-1}) = \sum_{k=0}^{n} \cdots + \sum_{k=n+1}^{m} \cdots + O(|z|^{-m-1})$$
$$= \sum_{k=0}^{n} \cdots + O(|z|^{-n-1})$$

---

[36] According to (1.5.1),

$$\frac{1}{\Gamma(\alpha)}\int_0^\infty e^{-zt}t^{\alpha+k-1}\, dt = (\alpha)_k z^{-\alpha-k}, \qquad \operatorname{Re}\alpha > 0, \quad \operatorname{Re} z > 0, \qquad k = 0, 1, 2, \ldots$$

[37] For complex a and b we have

$$|a^b| = |a|^{\operatorname{Re} b}e^{-\operatorname{Im} b \cdot \arg a} \leqslant |a|^{\operatorname{Re} b}e^{\pi|\operatorname{Im} b|}.$$

which again gives (9.12.1). Therefore the condition imposed on n can be dropped, and (9.12.1) is valid for arbitrary n.

Next we get rid of the restriction imposed on the parameter α. Suppose α satisfies the weaker condition $\operatorname{Re} \alpha > -1$. Then $\operatorname{Re}(\alpha + 1) > 0$, and formula (9.12.1) can be applied to each of the hypergeometric functions in the right-hand side of the identity

$$\Psi(\alpha, \gamma; z) = z\Psi(\alpha + 1, \gamma + 1; z) + (1 + \alpha - \gamma)\Psi(\alpha + 1, \gamma; z), \quad (9.12.2)$$

obtained by replacing α by $\alpha + 1$ in (9.10.14). Carrying out the necessary calculations, we again arrive at the asymptotic representation (9.12.1), but this time with the condition $\operatorname{Re} \alpha > -1$. Repeating this argument, we see that (9.12.1) holds for arbitrary values of α. Moreover, by slightly modifying the method used to prove (9.12.1), we can replace the condition $|\arg z| \leqslant \frac{1}{2}\pi - \delta$ by the weaker condition $|\arg z| \leqslant \pi - \delta$.[38] Thus, finally, we arrive at the following asymptotic representation of $\Psi(\alpha, \gamma; z)$ for large $|z|$:

$$\Psi(\alpha, \gamma; z) = z^{-\alpha}\left[\sum_{k=0}^{n} \frac{(-1)^k(\alpha)_k(1 + \alpha - \gamma)_k}{k!} z^{-k} + O(|z|^{-n-1})\right],$$
$$|\arg z| \leqslant \pi - \delta. \quad (9.12.3)$$

The corresponding asymptotic representation of the function $\Phi(\alpha, \gamma; z)$ can be deduced from (9.12.3) and the relation

$$\Phi(\alpha, \gamma; z) = \frac{\Gamma(\gamma)}{\Gamma(\gamma - \alpha)} e^{\pm\alpha\pi i}\,\Psi(\alpha, \gamma; z) + \frac{\Gamma(\gamma)}{\Gamma(\alpha)} e^{\pm(\alpha-\gamma)\pi i} e^{z}\,\Psi(\gamma - \alpha, \gamma; -z),$$
$$|\arg z| < \pi, \quad -z = ze^{\mp\pi i}, \quad \gamma \neq 0, -1, -2, \ldots, \quad (9.12.4)$$

which is the inverse of (9.10.3), where the plus sign is chosen if $\operatorname{Im} z > 0$ and the minus sign if $\operatorname{Im} z < 0$. To prove (9.12.4), we assume that $\gamma \neq 0, \pm 1, \pm 2, \ldots$ and use (9.10.3):

$$\Psi(\alpha, \gamma; z) = \frac{\Gamma(1 - \gamma)}{\Gamma(\alpha - \gamma + 1)}\,\Phi(\alpha, \gamma; z) + \frac{\Gamma(\gamma - 1)}{\Gamma(\alpha)} z^{1-\gamma}\Phi(1 + \alpha - \gamma, 2 - \gamma; z). \quad (9.12.5)$$

Replacing α by $\gamma - \alpha$ and z by $-z = ze^{\mp\pi i}$, we obtain

$$e^{z}\,\Psi(\gamma - \alpha, \gamma; -z) = \frac{\Gamma(1 - \gamma)}{\Gamma(1 - \alpha)}\,\Phi(\alpha, \gamma; z) - \frac{\Gamma(\gamma - 1)}{\Gamma(\gamma - \alpha)} z^{1-\gamma} e^{\pm\gamma\pi i}\Phi(1 + \alpha - \gamma, 2 - \gamma; z), \quad (9.12.6)$$

---

[38] Instead of (9.11.6), use the integral representation

$$\Psi(\alpha, \gamma; z) = \frac{1}{\Gamma(\alpha)} \int_0^{\infty \cdot e^{i\theta}} e^{-zt} t^{\alpha-1}(1 + t)^{\gamma-\alpha-1}\,dt, \qquad \operatorname{Re} \alpha > 0,$$

where

$$\theta = \begin{cases} \dfrac{\pi}{2} & \text{if} \quad -(\pi - \delta) \leqslant \arg z \leqslant -\left(\dfrac{\pi}{2} - \delta\right), \\ -\dfrac{\pi}{2} & \text{if} \quad \dfrac{\pi}{2} - \delta \leqslant \arg z \leqslant \pi - \delta. \end{cases}$$

where we have used (9.11.2). Eliminating $\Phi(1 + \alpha - \gamma, 2 - \gamma; z)$ from (9.12.5–6), we arrive at (9.12.4) after some simple calculations, where the validity of the result for positive integral values of γ follows from the principle of analytic continuation. Substituting (9.12.3) into (9.12.4), we find the desired asymptotic representation of $\Phi(\alpha, \gamma; z)$ for large $|z|$:

$$\Phi(\alpha, \gamma; z)$$

$$= \frac{\Gamma(\gamma)}{\Gamma(\gamma - \alpha)} e^{\pm \alpha\pi i} z^{-\alpha} \left[\sum_{k=0}^{n} \frac{(-1)^k (\alpha)_k (1 + \alpha - \gamma)_k}{k!} z^{-k} + O(|z|^{-n-1}) \right]$$

$$+ \frac{\Gamma(\gamma)}{\Gamma(\alpha)} e^z z^{-(\gamma-\alpha)} \left[\sum_{k=0}^{n} \frac{(\gamma - \alpha)_k (1 - \alpha)_k}{k!} z^{-k} + O(|z|^{-n-1}) \right],$$

$$|\arg z| \leqslant \pi - \delta, \qquad \gamma \neq 0, -1, -2, \ldots \qquad (9.12.7)$$

As before, the plus sign corresponds to $\operatorname{Im} z > 0$ and the minus sign to $\operatorname{Im} z < 0$. If $|\arg z| \leqslant \frac{1}{2}\pi - \delta$, the first term is small compared to the second, and (9.12.7) takes the form

$$\Phi(\alpha, \gamma; z) = \frac{\Gamma(\gamma)}{\Gamma(\alpha)} e^z z^{-(\gamma-\alpha)} \left[\sum_{k=0}^{n} \frac{(\gamma - \alpha)_k (1 - \alpha)_k}{k!} z^{-k} + O(|z|^{-n-1}) \right],$$

$$|\arg z| \leqslant \frac{\pi}{2} - \delta, \quad \alpha, \gamma \neq 0, -1, -2, \ldots \qquad (9.12.8)$$

9.13. Representation of Various Functions in Terms of the Confluent Hypergeometric Functions

As we now show, various familiar functions of mathematical analysis are special cases of the confluent hypergeometric functions $\Phi(\alpha, \gamma; z)$ and $\Psi(\alpha, \gamma; z)$, corresponding to suitable choices of the parameters α, γ and the variable z. Particular attention will be devoted to the special functions introduced in Chapters 2–5.

1. *Elementary functions.* Some typical relations involving elementary functions are

$$\Phi(\alpha, \alpha; z) = \sum_{k=0}^{\infty} \frac{z^k}{k!} = e^z,$$

$$\Phi(1, 2; z) = \sum_{k=0}^{\infty} \frac{z^k}{(k+1)!} = \frac{e^z - 1}{z},$$

$$\Phi(-2, 1; z) = 1 - 2z + \tfrac{1}{2}z^2.$$

2. *Error functions.* It follows from (2.1.5) and (2.1.2) that the error function has the expansion

$$\operatorname{Erf} z = \sum_{k=0}^{\infty} \frac{(-1)^k z^{2k+1}}{k!(2k+1)} = z \sum_{k=0}^{\infty} \frac{(\frac{1}{2})_k}{(\frac{3}{2})_k} \frac{(-z^2)^k}{k!},$$

and hence

$$\operatorname{Erf} z = z\Phi(\tfrac{1}{2}, \tfrac{3}{2}; -z^2). \tag{9.13.1}$$

Similarly, the complementary error function (2.1.6) can be written in the form

$$\operatorname{Erfc} z = \int_z^{\infty} e^{-t^2}\, dt = \tfrac{1}{2} z e^{-z^2} \int_0^{\infty} \frac{e^{-z^2 s}}{\sqrt{1+s}}\, ds,$$

if we set $t = z\sqrt{1+s}$. Then, according to the integral representation (9.11.6),[39]

$$\operatorname{Erfc} z = \tfrac{1}{2} z e^{-z^2} \Psi(1, \tfrac{3}{2}; z^2),$$

or

$$\operatorname{Erfc} z = \tfrac{1}{2} e^{-z^2} \Psi(\tfrac{1}{2}, \tfrac{1}{2}; z^2), \qquad |\arg z| < \frac{\pi}{2}, \tag{9.13.2}$$

where we have used (9.10.8).

3. *The function $F(z)$.* Next we consider the function $F(z)$, related to the probability integral of imaginary argument (see Sec. 2.3). It follows from (2.3.4) that

$$F(z) = \sum_{k=0}^{\infty} \frac{(-1)^k 2^k z^{2k+1}}{1\cdot 3 \cdots (2k+1)} = z \sum_{k=0}^{\infty} \frac{(1)_k (-z^2)^k}{k!(\frac{3}{2})_k},$$

and hence

$$F(z) = z\Phi(1, \tfrac{3}{2}; -z^2). \tag{9.13.3}$$

4. *Fresnel integrals.* Combining (2.4.6), (2.1.5) and (9.13.1), we find that

$$\begin{aligned} C(z) &= \frac{z}{2}\left[\Phi\left(\frac{1}{2}, \frac{3}{2}; \frac{\pi i z^2}{2}\right) + \Phi\left(\frac{1}{2}, \frac{3}{2}; -\frac{\pi i z^2}{2}\right)\right], \\ S(z) &= \frac{z}{2i}\left[\Phi\left(\frac{1}{2}, \frac{3}{2}; \frac{\pi i z^2}{2}\right) - \Phi\left(\frac{1}{2}, \frac{3}{2}; -\frac{\pi i z^2}{2}\right)\right]. \end{aligned} \tag{9.13.4}$$

5. *The exponential integral.* By definition,

$$\operatorname{Ei}(-z) = -\int_z^{\infty} \frac{e^{-t}}{t}\, dt, \qquad |\arg z| < \pi$$

---

[39] In the derivation we assume that $z > 0$, and then use analytic continuation to extend (9.13.2) into the domain $|\arg z| < \pi/2$.

[cf. (3.1.2)], and hence, setting $t = z(1 + s)$ and using the integral representation (9.11.6), we have

$$\mathrm{Ei}(-z) = -e^{-z}\int_0^\infty \frac{e^{-zs}}{1+s}\,ds = -e^{-z}\Psi(1, 1; z),$$

or

$$\mathrm{Ei}(z) = -e^{z}\Psi(1, 1; -z), \qquad |\arg(-z)| < \pi. \tag{9.13.5}$$

6. *The sine and cosine integrals.* Combining (3.3.6) and (9.13.5), we find that

$$\mathrm{Ci}(z) = -\frac{1}{2}e^{-iz}\Psi(1, 1; ze^{\pi i/2}) - \frac{1}{2}e^{iz}\Psi(1, 1; ze^{-\pi i/2}), \qquad |\arg z| < \frac{\pi}{2},$$

$$\mathrm{Si}(z) = \frac{\pi}{2} + \frac{1}{2i}e^{-iz}\Psi(1, 1; ze^{\pi i/2}) - \frac{1}{2i}e^{iz}\Psi(1, 1; ze^{-\pi i/2}), \ |\arg z| < \frac{\pi}{2}. \tag{9.13.6}$$

7. *The logarithmic integral.* It is an immediate consequence of (3.4.3) and (9.13.5) that

$$\mathrm{li}(z) = -z\Psi(1, 1; -\log z), \qquad |\arg z| < \pi, \qquad |\arg(1 - z)| < \pi. \tag{9.13.7}$$

8. *Hermite polynomials.* According to (4.9.2), the even Hermite polynomials can be written in the form

$$H_{2n}(z) = \sum_{k=0}^{n}(-1)^k\frac{(2n)!}{k!(2n-2k)!}(2z)^{2n-2k} = (-1)^n(2n)!\sum_{k=0}^{n}\frac{(-1)^k(2z)^{2k}}{(n-k)!(2k)!}$$

$$= (-1)^n\frac{(2n)!}{n!}\sum_{k=0}^{n}\frac{(-n)_k(2z)^{2k}}{(2k)!} = (-1)^n\frac{(2n)!}{n!}\sum_{k=0}^{n}\frac{(-n)_k(z^2)^k}{(\frac{1}{2})_k k!},$$

since

$$(2k)! = 2^{2k}(\tfrac{1}{2})_k k!,$$

and therefore

$$H_{2n}(z) = (-1)^n\frac{(2n)!}{n!}\Phi(-n, \tfrac{1}{2}; z^2). \tag{9.13.8}$$

For the odd Hermite polynomials we have the analogous formula

$$H_{2n+1}(z) = (-1)^n\frac{(2n+1)!}{n!}2z\Phi(-n, \tfrac{3}{2}; z^2). \tag{9.13.9}$$

9. *Laguerre polynomials.* It follows from (4.17.2) that

$$L_n^\alpha(z) = \sum_{k=0}^{n}\frac{\Gamma(n+\alpha+1)}{\Gamma(k+\alpha+1)}\frac{(-z)^k}{k!(n-k)!} = \frac{(\alpha+1)_n}{n!}\sum_{k=0}^{n}\frac{(-n)_k z^k}{(\alpha+1)_k k!},$$

and hence

$$L_n^\alpha(z) = \frac{(\alpha+1)_n}{n!}\Phi(-n, \alpha+1; z). \tag{9.13.10}$$

10. *Cylinder functions.* Assuming temporarily that $\operatorname{Re} \nu > -\frac{1}{2}$, we set $s = \frac{1}{2}(1 + t)$ in the integral representation (5.10.3), obtaining

$$J_\nu(z) = \frac{2^{2\nu}(z/2)^\nu e^{-iz}}{\Gamma(\frac{1}{2})\Gamma(\nu + \frac{1}{2})} \int_0^1 e^{2izs} s^{\nu - \frac{1}{2}}(1 - s)^{\nu - \frac{1}{2}}\, ds.$$

Therefore, according to (9.11.1),

$$J_\nu(z) = \frac{2^{2\nu}(z/2)^\nu e^{-iz}\Gamma(\nu + \frac{1}{2})}{\Gamma(\frac{1}{2})\Gamma(2\nu + 1)} \Phi(\nu + \tfrac{1}{2}, 2\nu + 1; 2iz),$$

or

$$J_\nu(z) = \frac{(z/2)^\nu}{\Gamma(\nu + 1)} e^{-iz}\Phi(\nu + \tfrac{1}{2}, 2\nu + 1; 2iz), \qquad |\arg z| < \pi, \quad (9.13.11)$$

where we have used the duplication formula (1.2.3) for the gamma function. Then we use the principle of analytic continuation to show that (9.13.11) holds for arbitrary ν.

Similar representations can be obtained for the other cylinder functions. For example, it follows from (5.6.4), (9.13.11) and (9.10.3) that[40]

$$H_\nu^{(1)}(z) = -\frac{2i}{\sqrt{\pi}} e^{i(z - \nu\pi)}(2z)^\nu \Psi(\nu + \tfrac{1}{2}, 2\nu + 1; 2ze^{-\pi i/2}),$$
$$-\frac{\pi}{2} < \arg z < \pi, \quad (9.13.12)$$

$$H_\nu^{(2)}(z) = \frac{2i}{\sqrt{\pi}} e^{-i(z - \nu\pi)}(2z)^\nu \Psi(\nu + \tfrac{1}{2}, 2\nu + 1; 2ze^{\pi i/2}),$$
$$-\pi < \arg z < \frac{\pi}{2}. \quad (9.13.13)$$

Then, using (5.7.6), we obtain the following representations of the Bessel functions of imaginary argument:

$$I_\nu(z) = \frac{(z/2)^\nu}{\Gamma(\nu + 1)} e^{-z}\Phi(\nu + \tfrac{1}{2}, 2\nu + 1; 2z), \qquad |\arg z| < \pi, \quad (9.13.14)$$

$$K_\nu(z) = \sqrt{\pi}(2z)^\nu e^{-z}\Psi(\nu + \tfrac{1}{2}, 2\nu + 1; 2z), \qquad |\arg z| < \pi. \quad (9.13.15)$$

11. *Whittaker functions.* A class of functions related to the confluent hypergeometric functions, and often encountered in the applications, consists of the *Whittaker functions*, defined by the formulas[41]

$$\begin{aligned} M_{k,\mu}(z) &= z^{\mu + \frac{1}{2}} e^{-z/2}\Phi(\tfrac{1}{2} - k + \mu, 2\mu + 1; z), \quad |\arg z| < \pi, \\ W_{k,\mu}(z) &= z^{\mu + \frac{1}{2}} e^{-z/2}\Psi(\tfrac{1}{2} - k + \mu, 2\mu + 1; z), \quad |\arg z| < \pi. \end{aligned} \quad (9.13.16)$$

[40] We also use formulas (9.11.2) and (1.2.2–3).

[41] E. T. Whittaker and G. N. Watson, *op. cit.*, Chap. 16.

9.14. Generalized Hypergeometric Functions

Consider the power series

$$\sum_{k=0}^{\infty} \frac{\prod_{r=1}^{p} (\alpha_r)_k}{\prod_{s=1}^{q} (\gamma_s)_k} \frac{z^k}{k!} = \sum_{k=0}^{\infty} \frac{(\alpha_1)_k \cdots (\alpha_p)_k}{(\gamma_1)_k \cdots (\gamma_q)_k} \frac{z^k}{k!}, \tag{9.14.1}$$

where p and q are nonnegative integers ($p,q = 0, 1, 2, \ldots$) satisfying the condition $p \leqslant q + 1$, z is a complex variable, α_r and γ_s are arbitrary parameters (except that $\gamma_s \neq 0, -1, -2, \ldots$), and $(\lambda)_k = \Gamma(\lambda + k)/\Gamma(\lambda)$.[42] Using the ratio test, we see at once that the radius of convergence of the series (9.14.1) equals ∞ if $p \leqslant q$ and 1 if $p = q + 1$. The sum of the series (9.14.1) is called the *generalized hypergeometric function*, and is denoted by the symbol

$${}_pF_q\left(\begin{matrix}\alpha_1, \ldots, \alpha_p; z \\ \gamma_1, \ldots, \gamma_q\end{matrix}\right),$$

or more concisely, by ${}_pF_q(\alpha_r; \gamma_s; z)$, i.e.,

$${}_pF_q(\alpha_r; \gamma_s; z) = \sum_{k=0}^{\infty} \frac{\prod_{r=1}^{p} (\alpha_r)_k}{\prod_{s=1}^{q} (\gamma_s)_k} \frac{z^k}{k!}. \tag{9.14.2}$$

Clearly, ${}_pF_q(\alpha_r; \gamma_s; z)$ is an entire function of z if $p \leqslant q$. The function ${}_{q+1}F_q(\alpha_r; \gamma_s; z)$ is originally defined only in the disk $|z| < 1$, but can be extended outside this disk by using analytic continuation.

The following are the simplest generalized hypergeometric functions:

$${}_0F_0(\alpha_r; \gamma_s; z) = \sum_{k=0}^{\infty} \frac{z^k}{k!} = e^z,$$

$${}_1F_0(\alpha_r; \gamma_s; z) = \sum_{k=0}^{\infty} \frac{(\alpha_1)_k}{k!} z^k = (1 - z)^{-\alpha_1},$$

$${}_0F_1(\alpha_r; \gamma_s; z) = \sum_{k=0}^{\infty} \frac{z^k}{(\gamma_1)_k k!} = \Gamma(\gamma_1) z^{-(\gamma_1 - 1)/2} I_{\gamma_1 - 1}(2z^{1/2}),$$

$${}_1F_1(\alpha_r; \gamma_s; z) = \sum_{k=0}^{\infty} \frac{(\alpha_1)_k}{(\gamma_1)_k} \frac{z^k}{k!} = \Phi(\alpha_1, \gamma_1; z),$$

$${}_2F_1(\alpha_r; \gamma_s; z) = \sum_{k=0}^{\infty} \frac{(\alpha_1)_k (\alpha_2)_k}{(\gamma_1)_k k!} = F(\alpha_1, \alpha_2; \gamma_1; z).$$

---

[42] As usual, the meaningless products

$$\prod_{r=1}^{0} \cdots, \quad \prod_{s=1}^{0} \cdots,$$

which appear when $p = 0$ or $q = 0$, are set equal to 1.

The last two examples show that the hypergeometric functions considered in this chapter are special cases of the more general function (9.14.2).

Some features of the theory of ordinary hypergeometric functions can be carried over to the case of generalized hypergeometric functions. For example, it is easily seen that the function $u = {}_pF_q(\alpha_r; \gamma_s; z)$ is a particular solution of the linear differential equation

$$\left[\delta \prod_{s=1}^{q} (\delta + \gamma_s - 1) - z \prod_{r=1}^{p} (\delta + \alpha_r)\right] u = 0 \tag{9.14.3}$$

of order $q + 1$, where δ denotes the operator $z(d/dz)$.[43] This equation reduces to (9.10.1) if $p = q = 1$, and to the hypergeometric equation (9.2.16) of $p = 2, q = 1$. There is a well-developed theory of generalized hypergeometric functions, with appropriate recurrence relations, integral representations, etc.[44]

PROBLEMS

1. Starting from the integral representation (9.1.6), prove that

$$F(\alpha, \beta; \gamma; x + i0) - F(\alpha, \beta; \gamma; x - i0)$$
$$= \frac{2\pi i \Gamma(\gamma)}{\Gamma(\alpha)\Gamma(\beta)\Gamma(1 + \gamma - \alpha - \beta)} (x - 1)^{\gamma - \alpha - \beta} F(\gamma - \alpha, \gamma - \beta; 1 + \gamma - \alpha - \beta; 1 - x),$$
$$x > 1, \quad \gamma \neq 0, -1, -2, \ldots$$

Hint. During the proof, assume that $\operatorname{Re} \alpha < 1$, $\operatorname{Re} \gamma > \operatorname{Re} \beta > 0$, and then use analytic continuation.

Comment. This formula shows why the cut $[1, \infty]$ is necessary in defining $F(\alpha, \beta; \gamma; z)$ for $\alpha, \beta \neq 0, -1, -2, \ldots$

2. Derive the formulas

$$\frac{d}{dz}(z^\alpha F) = \alpha z^{\alpha-1} F(\alpha + 1), \quad \frac{d}{dz}(z^{\gamma-1} F) = (\gamma - 1) z^{\gamma-2} F(\gamma - 1),$$

where the notation is the same as in Sec. 9.2.

3. Prove the following identities:

$$F(2\alpha, 2\beta; \alpha + \beta + \tfrac{1}{2}; \tfrac{1}{2}) = \frac{\Gamma(\alpha + \beta + \frac{1}{2})\Gamma(\frac{1}{2})}{\Gamma(\alpha + \frac{1}{2})\Gamma(\beta + \frac{1}{2})}, \quad \alpha + \beta + \tfrac{1}{2} \neq 0, -1, -2, \ldots,$$

$$F(\alpha, \beta; 1 + \alpha - \beta; -1) = 2^{-\alpha} \frac{\Gamma(1 + \alpha - \beta)\Gamma(\frac{1}{2})}{\Gamma\left(1 - \beta + \frac{\alpha}{2}\right)\Gamma\left(\frac{1}{2} + \frac{\alpha}{2}\right)},$$
$$1 + \alpha - \beta \neq 0, -1, -2, \ldots$$

---

[43] Note that applying δ to u corresponds to multiplying u by k.

[44] For a summary of the theory and references for further reading, see the Bateman Manuscript Project, *Higher Transcendental Functions, Vol. 1*, Chap. 4. Some new results are given by N. E. Nörlund, *Sur les fonctions hypergéométriques d'ordre supérieur*, Mat.-Fys. Skr. Danske Vid. Selsk., **1**, no. 2 (1956)

4. Show that the *hypergeometric polynomials* $F(-n, \beta; \gamma; z)$ $(n = 0, 1, 2, \ldots, \gamma \neq 0, -1, -2, \ldots)$ can be defined as the expansion coefficients of the generating function

$$w(z, t) = (1 - t)^{\beta-\gamma}(1 - t + zt)^{-\beta} = \sum_{n=0}^{\infty} \frac{(\gamma)_n}{n!} F(-n, \beta; \gamma; z)t^n,$$

$$|t| < \min\{1, |z - 1|^{-1}\}.$$

5. Derive the integral representation

$$\frac{\Gamma(\alpha)\Gamma(\beta)}{\Gamma(\gamma)} F(\alpha, \beta; \gamma; z) = \frac{1}{2\pi i}\int_{c-i\infty}^{c+i\infty} \frac{\Gamma(\alpha + s)\Gamma(\beta + s)\Gamma(-s)}{\Gamma(\gamma + s)}(-z)^s\, ds,$$

$$\operatorname{Re}\alpha > 0, \quad \operatorname{Re}\beta > 0, \quad |\arg(-z)| < \pi, \quad \gamma \neq 0, -1, -2, \ldots,$$

where $\min\{\operatorname{Re}\alpha, \operatorname{Re}\beta\} < c < 0$.

Hint. Complete the contour of integration on the right with the arc of a circle of radius $R_n = n + \frac{1}{2}$ $(n \to \infty)$, and then use residue theory.

Comment. The restrictions imposed on the parameters can be eliminated by suitably deforming the contour of integration.[45]

6. Using term-by-term integration, verify the following formulas:

$$F(\alpha, \beta; \gamma; z) = \frac{\Gamma(\gamma)}{\Gamma(c)\Gamma(\gamma - c)}\int_0^1 t^{c-1}(1 - t)^{\gamma-c-1}F(\alpha, \beta; c; zt)\, dt,$$

$$\operatorname{Re}\gamma > \operatorname{Re}c > 0, \quad |\arg(1 - z)| < \pi,$$

$$F(\alpha, \beta; \gamma + 1; z) = \gamma\int_0^1 F(\alpha, \beta; \gamma; zt)t^{\gamma-1}\, dt, \quad \operatorname{Re}\gamma > 0, \quad |\arg(1 - z)| < \pi.$$

7. By analogy with Sec. 9.10, the *hypergeometric function of the second kind* $G(\alpha, \beta; \gamma; z)$ can be defined as

$$G(\alpha, \beta; \gamma; z) = \frac{\Gamma(1 - \gamma)}{\Gamma(\alpha - \gamma + 1)\Gamma(\beta - \gamma + 1)} F(\alpha, \beta; \gamma; z)$$

$$+ \frac{\Gamma(\gamma - 1)}{\Gamma(\alpha)\Gamma(\beta)} z^{1-\gamma}F(1 + \alpha - \gamma, 1 + \beta - \gamma; 2 - \gamma; z),$$

$$|\arg z| < \pi, \quad |\arg(1 - z)| < \pi, \quad \gamma \neq 0, \pm 1, \pm 2, \ldots$$

Prove that $G(\alpha, \beta; \gamma; z)$ satisfies the relation

$$G(\alpha, \beta; \gamma; z) = z^{1-\gamma}G(\alpha - \gamma + 1, \beta - \gamma + 1; 2 - \gamma; z).$$

8. Repeating the considerations of Sec. 9.10, show that $G(\alpha, \beta; \gamma; z)$ is an entire function of α, β, γ, and derive the formula

$$G(\alpha, \beta; n + 1; z) = \frac{(-1)^{n+1}}{\Gamma(\alpha - n)\Gamma(\beta - n)}\sum_{k=0}^{\infty} \frac{(\alpha)_k(\beta)_k}{(n + k)!k!} z^k$$

$$\times [\psi(\alpha + k) + \psi(\beta + k) - \psi(1 + k) - \psi(n + 1 + k) + \log z]$$

$$+ \frac{1}{\Gamma(\alpha)\Gamma(\beta)}\sum_{k=0}^{n-1} \frac{(-1)^k(n - k - 1)!(\alpha - n)_k(\beta - n)_k}{k!} z^{k-n},$$

$$|\arg z| < \pi, \quad |z| < 1, \quad n = 0, 1, 2, \ldots, \quad \alpha, \beta \neq 0, -1, -2, \ldots$$

---

[45] E. T. Whittaker and G. N. Watson, *op. cit.*, p. 286.

9. Prove that the functions $F(\alpha, \beta; \gamma; z)$ and $G(\alpha, \beta; \gamma; z)$ are a pair of solutions of the hypergeometric equation (9.2.16) with Wronskian

$$W\{F(\alpha, \beta; \gamma; z), G(\alpha, \beta; \gamma; z)\} = -\frac{\Gamma(\gamma)}{\Gamma(\alpha)\Gamma(\beta)} z^{-\gamma}(1 - z)^{\gamma-\alpha-\beta-1},$$

$$|\arg(1 - z)| < \pi, \quad |\arg z| < \pi, \quad \gamma \neq 0, -1, -2, \ldots$$

Comment. It follows that the two solutions are linearly independent if $\alpha, \beta \neq 0, -1, -2, \ldots$

10. Find differentiation formulas and recurrence relations for the function $G(\alpha, \beta; \gamma; z)$.

Hint. Use the corresponding relations for the function $F(\alpha, \beta; \gamma; z)$.

11. Derive the integral representation

$$\frac{\Gamma(\alpha)}{\Gamma(\gamma)} \Phi(\alpha, \gamma; z) = \frac{1}{2\pi i} \int_{c-i\infty}^{c+i\infty} \frac{\Gamma(\alpha + s)\Gamma(-s)}{\Gamma(\gamma + s)} (-z)^s \, ds,$$

$$\operatorname{Re} \alpha > 0, \quad -\operatorname{Re} \alpha < c < 0, \quad \gamma \neq 0, -1, -2, \ldots \quad |\arg(-z)| < \frac{\pi}{2}.$$

Hint. Use residue theory.

12. Derive the integral representation

$$\Phi(\alpha, \gamma; z) = \frac{\Gamma(\gamma)}{\Gamma(\gamma - \alpha)} e^z z^{(1-\gamma)/2} \int_0^\infty e^{-t} t^{\frac{1}{2}(\gamma-1)-\alpha} J_{\gamma-1}(2\sqrt{zt}) \, dt.$$

$$\operatorname{Re}(\gamma - \alpha) > 0, \quad |\arg z| < \pi, \quad \gamma \neq 0, -1, -2, \ldots$$

Hint. Expand the Bessel function in power series, and then integrate term by term.

13. Derive the integral representation

$$\Psi(\alpha, \gamma; z) = \frac{2z^{(1-\gamma)/2}}{\Gamma(\alpha)\Gamma(\alpha - \gamma + 1)} \int_0^\infty e^{-t} t^{\alpha - \frac{1}{2}(1+\gamma)} K_{\gamma-1}(2\sqrt{zt}) \, dt,$$

$$\operatorname{Re} \alpha > 0, \quad \operatorname{Re}(\alpha - \gamma) > -1, \quad |\arg z| < \pi,$$

where $K_\nu(z)$ is Macdonald's function.

14. Prove the formulas

$$\Phi(\alpha, \gamma; z) = \frac{\Gamma(\gamma)}{\Gamma(c)\Gamma(\gamma - c)} \int_0^1 t^{c-1}(1 - t)^{\gamma-c-1} \Phi(\alpha, c; zt) \, dt,$$

$$\operatorname{Re} \gamma > \operatorname{Re} c > 0,$$

$$\Phi(\alpha, \gamma + 1; z) = \gamma \int_0^1 \Phi(\alpha, \gamma; zt) t^{\gamma-1} \, dt, \qquad \operatorname{Re} \gamma > 0.$$

15. Show that the Laplace transform of $\Phi(\alpha, \gamma; x)$ is

$$\overline{\Phi}(\alpha, \gamma; x) = \frac{1}{p} F\left(\alpha, 1; \gamma; \frac{1}{p}\right).$$

16. Verify that the Whittaker functions $M_{k,\mu}(z)$ and $W_{k,\mu}(z)$ are a pair of solutions of *Whittaker's equation*

$$u'' + \left(-\frac{1}{4} + \frac{k}{z} + \frac{\frac{1}{4} - \mu^2}{z^2}\right)u = 0,$$

with Wronskian

$$W\{M_{k,\mu}(z), W_{k,\mu}(z)\} = -\frac{\Gamma(2\mu + 1)}{\Gamma(\frac{1}{2} - k + \mu)}, \qquad 2\mu + 1 \neq 0, -1, -2, \ldots$$

Hint. Use the definitions (9.13.16).

17. Derive the integral representation[46]

$$W_{k,\mu}(z) = \frac{z^k e^{-z/2}}{\Gamma(\mu - k + \frac{1}{2})} \int_0^\infty e^{-t} t^{\mu - k - \frac{1}{2}} \left(1 + \frac{t}{z}\right)^{\mu + k - \frac{1}{2}} dt,$$
$$\operatorname{Re}(\mu - k + \tfrac{1}{2}) > 0, \qquad |\arg z| < \pi.$$

18. Using the result of the preceding problem, prove the asymptotic formula

$$W_{k,\mu}(z) \approx e^{-z/2} z^k, \qquad |z| \to \infty, \quad |\arg z| \leqslant \pi - \delta.$$

19. Using the results of Sec. 9.13, derive the following representations of various special functions in terms of $W_{k,\mu}(z)$:

$$\operatorname{Erfc} z = \frac{1}{2\sqrt{z}} e^{-z^2/2} W_{-\frac{1}{4},\frac{1}{4}}(z^2), \qquad |\arg z| < \frac{\pi}{2},$$

$$\operatorname{Ei}(z) = -\frac{1}{\sqrt{-z}} e^{z/2} W_{-\frac{1}{2},0}(-z), \qquad |\arg(-z)| < \pi,$$

$$\operatorname{li}(z) = -\sqrt{\frac{z}{-\log z}}\, W_{-\frac{1}{2},0}(-\log z), \qquad |\arg z| < \pi, \quad |\arg(1 - z)| < \pi,$$

$$K_\nu(z) = \sqrt{\frac{\pi}{2z}}\, W_{0,\nu}(2z), \qquad |\arg z| < \pi.$$

20. Prove that

$$\frac{d}{dz}\, {}_pF_q(\alpha_r; \gamma_s; z) = \frac{\prod_{r=1}^{p} \alpha_r}{\prod_{s=1}^{q} \gamma_s}\, {}_pF_q(\alpha_r + 1; \gamma_s + 1; z).$$

21. Prove that

$${}_{p+1}F_{q+1}(\alpha_r; \gamma_s; z)$$
$$= \frac{\Gamma(\gamma_{q+1})}{\Gamma(\alpha_{p+1})\Gamma(\gamma_{q+1} - \alpha_{p+1})} \int_0^1 t^{\alpha_{p+1} - 1}(1 - t)^{\gamma_{q+1} - \alpha_{p+1} - 1}\, {}_pF_q(\alpha_r; \gamma_s; zt)\, dt,$$
$$\operatorname{Re}\gamma_{q+1} > \operatorname{Re}\alpha_{p+1} > 0,$$

where $|\arg(1 - z)| < \pi$ if $p = q + 1$.

22. Derive the formula

$$[F(\alpha, \beta; \alpha + \beta + \tfrac{1}{2}; z)]^2 = {}_3F_2\left(\begin{matrix} 2\alpha, 2\beta, \alpha + \beta; z \\ \alpha + \beta + \frac{1}{2}, 2\alpha + 2\beta \end{matrix}\right).$$

---

[46] E. T. Whittaker and G. N. Watson, *op. cit.*, p. 340.

Hint. Find a third-order linear differential equation satisfied by the square of the function $F(\alpha, \beta; \alpha + \beta + \frac{1}{2}; z)$,[47] and show that the function

$$ {}_3F_2\left(\begin{matrix} 2\alpha, 2\beta; \alpha + \beta; z \\ \alpha + \beta + \frac{1}{2}; 2\alpha + 2\beta \end{matrix}\right) $$

is the solution of this equation which is analytic in a neighborhood of the point $z = 0$.

---

[47] E. T. Whittaker and G. N. Watson, *op. cit.*, Problems 10–11, p. 298.

10

PARABOLIC CYLINDER FUNCTIONS

10.1. Separation of Variables in Laplace's Equation in Parabolic Coordinates

To solve the boundary value problems of potential theory for a domain whose surface is an infinite parabolic cylinder, it is appropriate to use a coordinate system such that the cylinder corresponds to a constant value of one of the coordinates. Thus, let x, y and z be a system of rectangular coordinates with the z-axis parallel to the generator of the cylinder and the x-axis along the axis of symmetry of any one of the parabolas in which the planes perpendicular to the z-axis intersect the cylinder. Choosing the origin at the focus of this parabola, we introduce a three-dimensional system of *parabolic coordinates* α, β, z, related to the rectangular coordinates x, y, z by the formulas

$$x = \frac{c}{2}(\alpha^2 - \beta^2), \qquad y = c\alpha\beta, \quad z = z, \tag{10.1.1}$$

where

$$-\infty < \alpha < \infty, \quad 0 \leqslant \beta < \infty, \quad -\infty < z < \infty,$$

and $c > 0$ is a scale factor. The corresponding triply orthogonal system of surfaces consists of the parabolic cylinders $\alpha = \text{const}$ with foci at the origin,[1] described by the equation

$$y^2 = -2c\alpha^2\left(x - \frac{c\alpha^2}{2}\right), \tag{10.1.2}$$

[1] The surface $\alpha = \text{const} > 0$, is the half of the parabolic cylinder (10.1.2) with $y > 0$, and the surface $\alpha = -\text{const}$ is the other half, as indicated in Figure 38.

the parabolic cylinders β = const with foci at the origin, described by the equation

$$y^2 = 2c\beta^2\left(x + \frac{c\beta^2}{2}\right), \tag{10.1.3}$$

and the planes z = const (see Figure 38). In particular, given a parabolic cylinder with equation

$$y^2 = 2p\left(x + \frac{p}{2}\right) \tag{10.1.4}$$

in standard form,[2] suppose we choose the product $c\beta_0^2$ equal to p. Then the cylinder (10.1.4) has equation $\beta = \beta_0$ in the coordinates α, β, z, and the domain inside the cylinder to the values $0 \leqslant \beta < \beta_0$, while the domain outside the cylinder corresponds to the values $\beta_0 < \beta < \infty$.

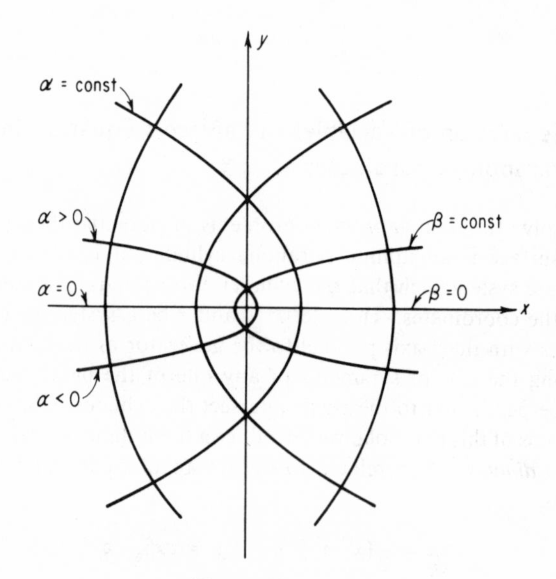

FIGURE 38

It is an immediate consequence of (10.1.1) that the square of the element of arc length in the coordinates α, β, z is

$$ds^2 = c^2(\alpha^2 + \beta^2)(d\alpha^2 + d\beta^2) + dz^2. \tag{10.1.5}$$

Therefore the metric coefficients are

$$h_\alpha = h_\beta = c\sqrt{\alpha^2 + \beta^2}, \qquad h_z = 1,$$

[2] Here p is the distance from the focus (at the origin) to the directrix.

and Laplace's equation takes the form [cf. (8.1.3)]

$$\nabla^2 u = \frac{1}{c^2(a^2+\beta^2)}\left[\frac{\partial^2 u}{\partial\alpha^2}+\frac{\partial^2 u}{\partial\beta^2}+c^2(\alpha^2+\beta^2)\frac{\partial^2 u}{\partial z^2}\right]=0. \qquad (10.1.6)$$

Now suppose we look for solutions of (10.1.6) of the form

$$u = \mathrm{A}(\alpha)\mathrm{B}(\beta)\mathrm{Z}(z). \qquad (10.1.7)$$

Then the variables separate, and we obtain

$$\frac{1}{c^2(\alpha^2+\beta^2)}\left[\frac{1}{\mathrm{A}}\frac{d^2\mathrm{A}}{d\alpha^2}+\frac{1}{\mathrm{B}}\frac{d^2\mathrm{B}}{d\beta^2}\right]=-\frac{1}{Z}\frac{d^2Z}{dz^2}=\lambda^2,$$

where λ is an arbitrary constant. It follows that

$$\frac{d^2\mathrm{Z}}{dz^2}+\lambda^2\mathrm{Z}=0,$$

$$\frac{1}{\mathrm{A}}\frac{d^2\mathrm{A}}{d\alpha^2}+\frac{1}{\mathrm{B}}\frac{d^2\mathrm{B}}{d\beta^2}-\lambda^2c^2(\alpha^2+\beta^2)=0. \qquad (10.1.8)$$

The last equation, in turn, can hold only if

$$\frac{d^2\mathrm{A}}{d\alpha^2}+(\mu-\lambda^2c^2\alpha^2)\mathrm{A}=0, \qquad (10.1.9)$$

$$\frac{d^2\mathrm{B}}{d\beta^2}-(\mu+\lambda^2c^2\beta^2)\mathrm{B}=0, \qquad (10.1.10)$$

where μ is again a constant. Thus Laplace's equation has infinitely many solutions of the form (10.1.7), depending on two arbitrary parameters λ and μ.

In most physical problems, the parameter λ is a positive real number (cf. Sec. 9.10). Then, introducing new variables

$$\xi=\sqrt{\lambda c}\,\alpha,\quad \eta=\sqrt{\lambda c}\,\beta,\qquad -\infty<\xi<\infty,\quad 0\leqslant\eta<\infty,$$

and a new parameter ν related to μ by the formula

$$\mu=\lambda c(2\nu+1),$$

we reduce equations (10.1.9–10) to the form

$$\frac{d^2\mathrm{A}}{d\xi^2}+(2\nu+1-\xi^2)\mathrm{A}=0, \qquad (10.1.11)$$

$$\frac{d^2\mathrm{B}}{d\eta^2}-(2\nu+1+\eta^2)\mathrm{B}=0. \qquad (10.1.12)$$

10.2. Hermite Functions

We now investigate equations (1.10.11–12), which, as just shown, arise when separating Laplace's equation in parabolic coordinates. Clearly, the problem reduces to studying the linear differential equation

$$u''+(2\nu+1-z^2)u=0 \qquad (10.2.1)$$

for arbitrary real or complex z and ν. If we make the substitution

$$u = e^{-z^2/2}v, \tag{10.2.2}$$

(10.2.1) goes into the equation

$$v'' - 2zv' + 2\nu v = 0, \tag{10.2.3}$$

which for nonnegative integral $\nu = n$ $(n = 0, 1, 2, \ldots)$ is just the differential equation (4.10.4) for the Hermite polynomials studied in Chapter 4. Therefore, in the case where the parameter ν is arbitrary, it is natural to call the solutions of (10.2.3) *Hermite functions*, while the corresponding solutions of (10.2.1) are called *parabolic cylinder functions.*[3]

The Hermite functions can be expressed in terms of the confluent hypergeometric function $\Phi(\alpha, \gamma; z)$. In fact, if we choose $t = z^2$ as a new independent variable, equation (10.2.3) goes into

$$t\frac{d^2v}{dt^2} + \left(\frac{1}{2} - t\right)\frac{dv}{dt} + \frac{\nu}{2}v = 0, \tag{10.2.4}$$

which is the special case of equation (9.10.1) corresponding to the parameter values

$$\alpha = -\frac{\nu}{2}, \qquad \gamma = \frac{1}{2}.$$

Therefore, according to (9.10.2), the general solution of the differential equation (10.2.4) is

$$v = A\Phi\left(-\frac{\nu}{2}, \frac{1}{2}; t\right) + B\sqrt{t}\,\Phi\left(\frac{1-\nu}{2}, \frac{3}{2}; t\right), \tag{10.2.5}$$

or

$$v = A\Phi\left(-\frac{\nu}{2}, \frac{1}{2}; z^2\right) + Bz\Phi\left(\frac{1-\nu}{2}, \frac{3}{2}; z^2\right), \tag{10.2.6}$$

after returning to the original variable z. In particular, choosing the constants A and B to be

$$A = \frac{2^\nu\Gamma(\frac{1}{2})}{\Gamma\left(\frac{1-\nu}{2}\right)}, \qquad B = \frac{2^\nu\Gamma(-\frac{1}{2})}{\Gamma\left(-\frac{\nu}{2}\right)}, \tag{10.2.7}$$

---

[3] The definition given here differs somewhat from that prevalent in the literature (see the Bateman Manuscript Project, *Higher Transcendental Functions, Vol. 2*, Chap. 8), where the term *parabolic cylinder function* refers to a solution of the equation

$$u'' + \left(\nu + \frac{1}{2} - \frac{z^2}{4}\right)u = 0,$$

which reduces to (10.2.1) if we make the substitution $z = \sqrt{2}t$. One of the solutions of this equation is the function $D_\nu(z)$, related to our function $H_\nu(z)$ [see (10.2.8)] by the formula

$$D_\nu(z) = 2^{-\nu/2}e^{-z^2/4}H_\nu\left(\frac{z}{\sqrt{2}}\right).$$

we arrive at the solution

$$v = H_\nu(z) = \frac{2^\nu \Gamma(\frac{1}{2})}{\Gamma\left(\frac{1-\nu}{2}\right)} \Phi\left(-\frac{\nu}{2}, \frac{1}{2}; z^2\right) + \frac{2^\nu \Gamma(-\frac{1}{2})}{\Gamma\left(-\frac{\nu}{2}\right)} z\Phi\left(\frac{1-\nu}{2}, \frac{3}{2}; z^2\right), \quad (10.2.8)$$

which we call the *Hermite function* (*of degree* ν).[4] It follows from (10.2.8) and the known properties of the gamma function and the confluent hypergeometric function that $H_\nu(z)$ is an entire function both of the variable z and the parameter ν.

If $\nu = n$ ($n = 0, 1, 2, \ldots$), one of the terms in (10.2.8) vanishes and the other reduces to a polynomial in z. Using formulas (1.2.1–3) from the theory of the gamma function, we find after some simple calculations that

$$\begin{aligned} H_{2m}(z) &= (-1)^m \frac{(2m)!}{m!} \Phi(-m, \tfrac{1}{2}; z^2), \\ H_{2m+1}(z) &= (-1)^m \frac{(2m+1)!}{m!} 2z\Phi(-m, \tfrac{3}{2}; z^2). \end{aligned} \quad (10.2.9)$$

Comparing these formulas with (9.13.8–9), we see that if $\nu = n$, the function $H_\nu(z)$ reduces to the Hermite polynomial of degree n.

If $\nu \neq 0, 1, 2, \ldots$, the general solution of equation (10.2.3) can be expressed in terms of Hermite functions. In fact, since equation (10.2.3) does not change if we replace z by $-z$, the function $v_2 = H_\nu(-z)$, as well as the function $v_1 = H_\nu(z)$, is a solution of (10.2.3). By the usual method (cf. Sec. 5.9), it is easily shown that the pair of solutions v_1, v_2 has a Wronskian of the form

$$W\{v_1, v_2\} = Ce^{z^2},$$

where C is a constant. Setting $z = 0$ and taking account of the formulas

$$H_\nu(0) = \frac{2^\nu \Gamma(\frac{1}{2})}{\Gamma\left(\frac{1-\nu}{2}\right)}, \qquad H'_\nu(0) = \frac{2^\nu \Gamma(-\frac{1}{2})}{\Gamma\left(-\frac{\nu}{2}\right)}, \quad (10.2.10)$$

which are immediate consequences of (10.2.8), we find that

$$C = W\{v_1, v_2\}_{z=0} = -\frac{2^{2\nu+1}\Gamma(\frac{1}{2})\Gamma(-\frac{1}{2})}{\Gamma\left(\frac{1-\nu}{2}\right)\Gamma\left(-\frac{\nu}{2}\right)} = \frac{2^{\nu+1}\sqrt{\pi}}{\Gamma(-\nu)},$$

[4] It should be noted that according to (9.10.3), the Hermite function $H_\nu(z)$ bears the following simple relation to the confluent hypergeometric function of the second kind:

$$H_\nu(z) = 2^\nu \Psi\left(-\frac{\nu}{2}, \frac{1}{2}; z^2\right).$$

where in the last step we have used formulas (1.2.2-3) from the theory of the gamma function. It follows that

$$W\{H_\nu(z), H_\nu(-z)\} = \frac{2^{\nu+1}\sqrt{\pi}}{\Gamma(-\nu)} e^{z^2}. \tag{10.2.11}$$

Therefore, if $\nu \neq 0, 1, 2, \ldots$, the solutions $H_\nu(z)$ and $H_\nu(-z)$ are linearly independent and the general solution of (10.2.3) can be written in the form

$$\nu = MH_\nu(z) + NH_\nu(-z). \tag{10.2.12}$$

However, suppose $\nu = n$ $(n = 0, 1, 2, \ldots)$, so that $W \equiv 0$. Then $H_\nu(z)$ and $H_\nu(-z)$ are linearly dependent, and in fact,

$$H_n(-z) = (-1)^n H_n(z). \tag{10.2.13}$$

Therefore the right-hand side of (10.2.12) is no longer the general solution of (10.2.3).

To obtain an expression for the general solution of (10.2.3) which is suitable for arbitrary values of the parameter ν, we first observe that the substitution

$$v = e^{z^2}w, \qquad \zeta = iz$$

transforms (10.2.3) into the equation

$$w'' - 2\zeta w' - 2(\nu + 1)w = 0, \tag{10.2.14}$$

which is the same as (10.2.3) except that ν has been replaced by $-\nu - 1$. It follows that the functions

$$v_3 = e^{z^2}H_{-\nu-1}(iz), \qquad v_4 = e^{z^2}H_{-\nu-1}(-iz) \tag{10.2.15}$$

are also solutions of equation (10.2.3). Calculating the Wronskians

$$\begin{aligned} W\{H_\nu(z), e^{z^2}H_{-\nu-1}(iz)\} &= e^{z^2 - \frac{1}{2}(\nu+1)\pi i}, \\ W\{H_\nu(z), e^{z^2}H_{-\nu-1}(-iz)\} &= e^{z^2 + \frac{1}{2}(\nu+1)\pi i}, \end{aligned} \tag{10.2.16}$$

we find that each of the solutions (10.2.15) is linearly independent of $H_\nu(z)$. Therefore, for arbitrary ν, the general solution of (10.2.3) can be written in either of the following equivalent forms:

$$v = MH_\nu(z) + Ne^{z^2}H_{-\nu-1}(iz) = PH_\nu(z) + Qe^{z^2}H_{-\nu-1}(-iz). \tag{10.2.17}$$

Finally, comparing (10.2.17) and (10.2.2), we find the following expressions for the general parabolic cylinder function:

$$\begin{aligned} u &= Me^{-z^2/2}H_\nu(z) + Ne^{z^2/2}H_{-\nu-1}(iz) \\ &= Pe^{-z^2/2}H_\nu(z) + Qe^{z^2/2}H_{-\nu-1}(-iz). \end{aligned} \tag{10.2.18}$$

10.3. Some Relations Satisfied by the Hermite Functions

In the preceding section, it was shown that each of the functions

$$\begin{aligned} v_1 &= H_\nu(z), & v_3 &= e^{z^2}H_{-\nu-1}(iz), \\ v_2 &= H_\nu(-z), & v_4 &= e^{z^2}H_{-\nu-1}(-iz) \end{aligned} \tag{10.3.1}$$

is a solution of equation (10.2.3). Since a second-order linear differential equation cannot have three linearly independent solutions, it must be possible to write each of the functions (10.3.1) as a linear combination of any two others. In particular, if $\nu \neq -1, -2, \ldots,$[5] there must exist a relation of the form

$$H_\nu(z) = Me^{z^2}H_{-\nu-1}(iz) + Ne^{z^2}H_{-\nu-1}(-iz). \tag{10.3.2}$$

To determine the constants M and N, we use the conditions (10.2.10), obtaining the system of equations

$$M + N = \frac{2^{2\nu+1}\Gamma\left(1 + \dfrac{\nu}{2}\right)}{\Gamma\left(\dfrac{1-\nu}{2}\right)}, \qquad M - N = \frac{2^{2\nu+1}\Gamma\left(\dfrac{1+\nu}{2}\right)}{i\Gamma\left(-\dfrac{\nu}{2}\right)}.$$

Transforming the right-hand sides of these equations by using formulas (1.2.2–3) from the theory of the gamma function, we find that

$$M + N = \frac{2^{\nu+1}\Gamma(\nu+1)}{\sqrt{\pi}}\cos\frac{\nu\pi}{2}, \qquad M - N = \frac{2^{\nu+1}\Gamma(\nu+1)}{\sqrt{\pi}}\, i\sin\frac{\nu\pi}{2}. \tag{10.3.3}$$

Solving the system (10.3.3) and substituting the resulting values of M and N into (10.3.2), we arrive at the relation

$$H_\nu(z) = \frac{2^\nu\Gamma(\nu+1)}{\sqrt{\pi}}e^{z^2}[e^{\nu\pi i/2}H_{-\nu-1}(iz) + e^{-\nu\pi i/2}H_{-\nu-1}(-iz)]. \tag{10.3.4}$$

Formula (10.3.4) remains valid for negative integral ν if we take the right-hand side to mean its limit as $\nu \to -n$ $(n = 1, 2, \ldots)$. Replacing z by $-z$ in (10.3.4), we obtain the relation

$$H_\nu(-z) = \frac{2^\nu\Gamma(\nu+1)}{\sqrt{\pi}}e^{z^2}[e^{\nu\pi i/2}H_{-\nu-1}(-iz) + e^{-\nu\pi i/2}H_{-\nu-1}(iz)]. \tag{10.3.5}$$

---

[5] If $\nu \neq -1, -2, \ldots,$ then

$$W\{e^{z^2}H_{-\nu-1}(iz), e^{z^2}H_{-\nu-1}(-iz)\} = \frac{i\sqrt{\pi}}{2^\nu\Gamma(\nu+1)}e^{z^2} \neq 0.$$

Further relations can be deduced from (10.3.4–5) by purely algebraic operations. For example, we have

$$H_\nu(z) = e^{\nu\pi i} H_\nu(-z) + \frac{2^{\nu+1}\sqrt{\pi}}{\Gamma(-\nu)} e^{z^2 + \frac{1}{2}(\nu+1)\pi i} H_{-\nu-1}(-iz), \tag{10.3.6}$$

$$H_\nu(z) = e^{-\nu\pi i} H_\nu(-z) + \frac{2^{\nu+1}\sqrt{\pi}}{\Gamma(-\nu)} e^{z^2 - \frac{1}{2}(\nu+1)\pi i} H_{-\nu-1}(iz), \tag{10.3.7}$$

and so on.

10.4. Recurrence Relations for the Hermite Functions

The Hermite function $H_\nu(z)$ satisfies simple recurrence relations which generalize the corresponding formulas for Hermite polynomials (see Sec. 4.10) to the case where the degree ν is an arbitrary complex number. To derive these recurrence relations, we first make a preliminary transformation of (10.2.8), which leads to a simple power series representation of $H_\nu(z)$. Replacing the hypergeometric functions in (10.2.8) by their explicit series representations [cf. (9.9.1)], and using the formulas

$$\Gamma\left(-\frac{\nu}{2}\right)\Gamma\left(\frac{1-\nu}{2}\right) = 2^{\nu+1}\sqrt{\pi}\,\Gamma(-\nu), \tag{10.4.1}$$

$$\Gamma^2\left(\frac{1}{2}\right) = -\Gamma\left(-\frac{1}{2}\right)\Gamma\left(\frac{3}{2}\right) = \pi,$$

implied by (1.2.2–3), we have

$$H_\nu(z) = \frac{\sqrt{\pi}}{2\Gamma(-\nu)}\left[\sum_{k=0}^{\infty} \frac{\Gamma\left(k - \frac{\nu}{2}\right)}{k!\,\Gamma\left(k + \frac{1}{2}\right)} z^{2k} - \sum_{k=0}^{\infty} \frac{\Gamma\left(k + \frac{1-\nu}{2}\right)}{k!\,\Gamma\left(k + \frac{3}{2}\right)} z^{2k+1}\right]$$

$$= \frac{\sqrt{\pi}}{2\Gamma(-\nu)}\left[\sum_{k=0}^{\infty} \frac{\Gamma\left(\frac{2k-\nu}{2}\right)}{\Gamma\left(\frac{2k+1}{2}\right)\Gamma\left(\frac{2k+2}{2}\right)} z^{2k} - \sum_{k=0}^{\infty} \frac{\Gamma\left(\frac{2k+1-\nu}{2}\right)}{\Gamma\left(\frac{2k+2}{2}\right)\Gamma\left(\frac{2k+3}{2}\right)} z^{2k+1}\right]$$

$$= \frac{\sqrt{\pi}}{2\Gamma(-\nu)} \sum_{m=0}^{\infty} \frac{(-1)^m \Gamma\left(\frac{m-\nu}{2}\right)}{\Gamma\left(\frac{m+1}{2}\right)\Gamma\left(\frac{m+2}{2}\right)} z^m. \tag{10.4.2}$$

Since, according to (1.2.3),

$$2^m \Gamma\left(\frac{m+1}{2}\right)\Gamma\left(\frac{m+2}{2}\right) = \sqrt{\pi}\,\Gamma(m+1) = \sqrt{\pi}\,m!,$$

formula (10.4.2) can be simplified to[6]

$$H_\nu(z) = \frac{1}{2\Gamma(-\nu)} \sum_{m=0}^{\infty} \frac{(-1)^m \Gamma\left(\frac{m-\nu}{2}\right)}{m!} (2z)^m, \qquad |z| < \infty. \quad (10.4.3)$$

This expansion, which is of independent interest, allows us to give a very simple derivation of the required recurrence relations.

Differentiating the series (10.4.3) and introducing the new summation index $n = m - 1$, we find that

$$\begin{aligned} H_\nu'(z) &= \frac{1}{2\Gamma(-\nu)} \sum_{m=1}^{\infty} \frac{2(-1)^m \Gamma\left(\frac{m-\nu}{2}\right)}{(m-1)!} (2z)^{m-1} \\ &= -\frac{1}{2\Gamma(-\nu)} \sum_{n=0}^{\infty} \frac{2(-1)^n \Gamma\left(\frac{n+1-\nu}{2}\right)}{n!} (2z)^n \\ &= -\frac{2\Gamma(1-\nu)}{\Gamma(-\nu)} H_{\nu-1}(z) = 2\nu H_{\nu-1}(z). \end{aligned}$$

Thus the Hermite function $H_\nu(z)$ satisfies the recurrence relation

$$H_\nu'(z) = 2\nu H_{\nu-1}(z), \quad (10.4.4)$$

which generalizes formula (4.10.2). Next we differentiate (10.4.4), obtaining

$$H_\nu''(z) = 2\nu H_{\nu-1}'(z),$$

which, together with the differential equation (10.2.3) written in the form

$$H_\nu''(z) - 2zH_\nu'(z) + 2\nu H_\nu(z) = 0,$$

implies

$$2\nu H_{\nu-1}'(z) = 2zH_\nu'(z) - 2\nu H_\nu(z). \quad (10.4.5)$$

Using (10.4.4) to eliminate $H_{\nu-1}'(z)$ and $H_\nu'(z)$ from (10.4.5), we obtain

$$H_\nu(z) - 2zH_{\nu-1}(z) + 2(\nu - 1)H_{\nu-2}(z) = 0. \quad (10.4.6)$$

Finally, replacing ν by $\nu + 1$ in (10.4.6) leads to another recurrence relation

$$H_{\nu+1}(z) - 2zH_\nu(z) + 2\nu H_{\nu-1}(z) = 0, \quad (10.4.7)$$

which agrees with our previous formula (4.10.1) when ν is a positive integer.

---

[6] Because of the intervention of the duplication formula (10.4.1), the series (10.4.3) can be used for nonnegative integral $\nu = n$ only if we agree that the indeterminate ratio

$$\frac{\Gamma(-1)}{\Gamma(-2)}$$

is formally equal to -4 [the value consistent with (10.4.1)], and all other indeterminate expressions are evaluated with this in mind.

10.5. Integral Representations of the Hermite Functions

Various integral representations of the Hermite functions $H_\nu(z)$ involving contour integrals or definite integrals can be derived by summing the series defining $H_\nu(z)$. The simplest such representation is obtained from (10.4.3) by assuming that $\operatorname{Re}\nu < 0$ and replacing $\Gamma[\frac{1}{2}(m-\nu)]$ by an integral of the type (1.1.1). This gives

$$\begin{aligned} H_\nu(z) &= \frac{1}{2\Gamma(-\nu)} \sum_{m=0}^{\infty} \frac{(-1)^m (2z)^m}{m!} \int_0^\infty e^{-s} s^{\frac{1}{2}(m-\nu)-1}\, ds \\ &= \frac{1}{2\Gamma(-\nu)} \int_0^\infty e^{-s} s^{-\frac{1}{2}\nu-1}\, ds \sum_{m=0}^{\infty} \frac{(-1)^m (2z\sqrt{s})^m}{m!} \\ &= \frac{1}{2\Gamma(-\nu)} \int_0^\infty e^{-s-2z\sqrt{s}} s^{-\frac{1}{2}\nu-1}\, ds, \end{aligned} \tag{10.5.1}$$

where reversing the order of summation and integration is justified by an absolute convergence argument. Introducing the new variable of integration $t = \sqrt{s}$, we can write (10.5.1) in the form

$$H_\nu(z) = \frac{1}{\Gamma(-\nu)} \int_0^\infty e^{-t^2-2tz} t^{-\nu-1}\, dt, \qquad \operatorname{Re}\nu < 0. \tag{10.5.2}$$

This formula resembles the integral representations of Sec. 4.11, derived earlier for the Hermite polynomials. In particular, it follows from (10.5.2) that the Hermite functions of negative integral degree can be expressed in closed form in terms of the complementary error function (2.1.6). In fact, setting $\nu = -1$ in (10.5.2), we obtain

$$H_{-1}(z) = \int_0^\infty e^{-t^2-2tz}\, dt = e^{z^2} \int_0^\infty e^{-(t+z)^2}\, dt = e^{z^2} \int_z^\infty e^{-s^2}\, ds,$$

i.e.,

$$H_{-1}(z) = e^{z^2} \operatorname{Erfc} z, \tag{10.5.3}$$

and in general

$$H_{-n-1}(z) = \frac{(-1)^n}{2^n n!} \frac{d^n}{dz^n} (e^{z^2} \operatorname{Erfc} z), \qquad n = 0, 1, 2, \ldots \tag{10.5.4}$$

Another important integral representation of $H_\nu(z)$ can be deduced from (10.3.4) by replacing the Hermite functions in the right-hand side by integrals of the form (10.5.2). Under the assumption that $\operatorname{Re}\nu > -1$, this gives

$$H_\nu(z) = \frac{2^\nu e^{z^2}}{\sqrt{\pi}} \left[e^{\nu\pi i/2} \int_0^\infty e^{-t^2-2izt} t^\nu\, dt + e^{-\nu\pi i/2} \int_0^\infty e^{-t^2+2izt} t^\nu\, dt \right],$$

or

$$H_\nu(z) = \frac{2^{\nu+1} e^{z^2}}{\sqrt{\pi}} \int_0^\infty e^{-t^2} t^\nu \cos\left(2zt - \frac{\nu\pi}{2}\right) dt, \qquad \operatorname{Re}\nu > -1. \tag{10.5.5}$$

Formula (10.5.5) is the generalization of the integral representations (4.11.2–3) of the Hermite polynomials, to which it reduces when $\nu = n$ ($n = 0, 1, 2, \ldots$).

Some other integral representations of the Hermite functions are given in Problems 1–4 at the end of this chapter.

10.6. Asymptotic Representations of the Hermite Functions for Large $|z|$

To derive asymptotic representations of the Hermite functions $H_\nu(z)$ for large $|z|$ and fixed $|\nu|$, we first assume that $\operatorname{Re} \nu < 0$, $|\arg z| < \pi/2$. Then, using (10.5.2) to represent $H_\nu(z)$, we replace e^{-t^2} by its Taylor series expansion with remainder, i.e.,

$$e^{-t^2} = \sum_{k=0}^{n} \frac{(-1)^k t^{2k}}{k!} + \omega_n(t), \tag{10.6.1}$$

where

$$|\omega_n(t)| \leqslant \frac{t^{2n+2}}{(n+1)!}.$$

Integrating term by term and noting that

$$\int_0^\infty e^{-2tz} t^{2k-\nu-1}\, dt = \frac{\Gamma(2k-\nu)}{(2z)^{2k-\nu}}, \qquad k = 0, 1, 2, \ldots \tag{10.6.2}$$

if $\operatorname{Re} z > 0$, $\operatorname{Re} \nu < 0$ [cf. (1.5.1)], we find that

$$H_\nu(z) = (2z)^\nu \left[\sum_{k=0}^{n} \frac{(-1)^k(-\nu)_{2k}}{k!} (2z)^{-2k} + r_n(z)\right], \tag{10.6.3}$$

where

$$r_n(z) = \frac{(2z)^{-\nu}}{\Gamma(-\nu)} \int_0^\infty \omega_n(t) e^{-2tz} t^{-\nu-1}\, dt$$

and

$$(-\nu)_0 = 1, \quad (-\nu)_{2k} = \frac{\Gamma(-\nu+2k)}{\Gamma(-\nu)} = (-\nu)(-\nu+1)\cdots(-\nu+2k-1)$$

($k = 1, 2, \ldots$). Now suppose that

$$|\arg z| \leqslant \frac{\pi}{2} - \delta,$$

where $\delta > 0$ is arbitrarily small. Then it is easily seen that

$$|r_n(z)| \leqslant \frac{(2|z|)^{-\operatorname{Re}\nu} e^{\frac{1}{2}\pi|\operatorname{Im}\nu|}}{|\Gamma(-\nu)|(n+1)!} \int_0^\infty e^{-2t|z|\sin\delta} t^{2n+1-\operatorname{Re}\nu}\, dt = O(|z|^{-2n-2})$$

(cf. footnote 37, p. 269), and hence (10.6.3) can be written in the form

$$H_\nu(z) = (2z)^\nu\left[\sum_{k=0}^{n} \frac{(-1)^k(-\nu)_{2k}}{k!}(2z)^{-2k} + O(|z|^{-2n-2})\right]. \quad (10.6.4)$$

Next we show that (10.6.5) remains valid for arbitrary ν. In fact, let the condition $\operatorname{Re}\nu < 0$ be replaced by the weaker condition $\operatorname{Re}\nu < 1$. Then, using the recurrence relation (10.4.7), we represent $H_\nu(z)$ in the form

$$H_\nu(z) = 2zH_{\nu-1}(z) - 2(\nu - 1)H_{\nu-2}(z), \quad (10.6.5)$$

where the real part of the degree of each Hermite function on the right is negative. Applying (10.6.4) to each of these functions, and making some simple calculations, we obtain an expansion of the same form as (10.6.4), thereby extending (10.6.4) to the case $\operatorname{Re}\nu < 1$. Repeating this argument as often as necessary, we find that (10.6.4) is valid for any value of ν. Moreover, by slightly modifying the method used to prove (10.6.4),[7] we can extend the result to the larger sector

$$|\arg z| \leqslant \frac{3\pi}{4} - \delta.$$

Thus, finally, we arrive at the following asymptotic representation of $H_\nu(z)$ for large z and fixed ν:

$$H_\nu(z) = (2z)^\nu\left[\sum_{k=0}^{n} \frac{(-1)^k}{k!}(-\nu)_{2k}(2z)^{-2k} + O(|z|^{-2n-2})\right], \quad |\arg z| \leqslant \frac{3\pi}{4} - \delta. \quad (10.6.6)$$

Asymptotic representations of $H_\nu(z)$ which are valid in other sectors of the complex plane can be derived from (10.6.6) by using the relations (10.3.6–7). For example, if

$$\frac{\pi}{4} < \arg z < \frac{5\pi}{4},$$

then

$$|\arg(-z)| = |\arg z - \pi| < \frac{3\pi}{4}, \qquad |\arg(-iz)| = \left|\arg z - \frac{\pi}{2}\right| < \frac{3\pi}{4}.$$

Therefore, applying (10.6.6) to each Hermite function in the right-hand side of (10.3.6), we find that

$$H_\nu(z) = (2z)^\nu\left[\sum_{k=0}^{n} \frac{(-1)^k}{k!}(-\nu)_{2k}(2z)^{-2k} + O(|z|^{-2n-2})\right]$$
$$- \frac{\sqrt{\pi}e^{\nu\pi i}}{\Gamma(-\nu)} e^{z^2}z^{-\nu-1}\left[\sum_{k=0}^{n} \frac{(\nu+1)_{2k}}{k!}(2z)^{-2k} + O(|z|^{-2n-2})\right],$$
$$\frac{\pi}{4} + \delta \leqslant \arg z \leqslant \frac{5\pi}{4} - \delta. \quad (10.6.7)$$

[7] Instead of (10.5.2), use the integral representation

$$H_\nu(z) = \frac{1}{\Gamma(-\nu)}\int_0^{\infty \cdot e^{i\theta}} e^{-t^2-2tz}t^{-\nu-1}\,dt,$$

where $|\theta| < \pi/4$ and the integration is along the ray $\arg t = \theta$.

Similarly, it follows from (10.3.7) and (10.6.6) that

$$H_\nu(z) = (2z)^\nu\Big[\sum_{k=0}^{n} \frac{(-1)^k}{k!}(-\nu)_{2k}(2z)^{-2k} + O(|z|^{-2n-2})\Big]$$

$$- \frac{\sqrt{\pi}e^{-\nu\pi i}}{\Gamma(-\nu)} e^{z^2} z^{-\nu-1}\Big[\sum_{k=0}^{n} \frac{(\nu+1)_{2k}}{k!}(2z)^{-2k} + O(|z|^{-2n-2})\Big],$$

$$-\Big(\frac{5\pi}{4} - \delta\Big) \leqslant \arg z \leqslant -\Big(\frac{\pi}{4} + \delta\Big). \qquad (10.6.8)$$

Together, formulas (10.6.6–8) give a complete description of the behavior of the function $H_\nu(z)$ for large $|z|$. These formulas do not contradict each other in their common regions of applicability, since the second terms of (10.6.7–8) are small compared to the first terms if

$$-\frac{3\pi}{4} < \arg z < -\frac{\pi}{4}, \qquad \frac{\pi}{4} < \arg z < \frac{3\pi}{4},$$

and can therefore be included in the term $O(|z|^{-2n-2})$.

Finally, we note that (10.6.4) is an immediate consequence of the asymptotic representation (9.12.3) for the confluent hypergeometric function of the second kind and the fact that

$$H_\nu(z) = 2^\nu\, \Psi\Big(-\frac{\nu}{2}, \frac{1}{2}; z^2\Big)$$

(cf. footnote 4, p. 285).

10.7. The Dirichlet Problem for a Parabolic Cylinder

The special functions studied in this chapter allow us to solve the boundary value problems of potential theory for the case of a domain bounded by a parabolic cylinder. To find the appropriate set of solutions of Laplace's equation, we introduce the parabolic coordinates (10.1.1) and look for solutions in the form of the product (10.1.7), thereby arriving at equations (10.1.8–10). If we require that the solutions be bounded in the whole domain, in particular at infinity, it must be assumed that the parameter λ is real.[8] Then the corresponding solution of (10.1.8) is

$$Z = C\cos\lambda z + D\sin\lambda z, \qquad \lambda \geqslant 0, \qquad (10.7.1)$$

which is bounded for $-\infty < z < \infty$.

Introducing the new parameter ν related to μ by the formula

$$\mu = \lambda c(2\nu + 1),$$

[8] Without loss of generality, we can assume that λ is nonnegative, since changing the sign of λ does not affect the separation constant λ^2.

and using (10.2.18), we find that the general solution of (10.1.9) can be written in the form

$$\mathrm{A} = Me^{-\lambda c\alpha^2/2}H_\nu(\sqrt{\lambda c}\,\alpha) + Ne^{\lambda c\alpha^2/2}H_{-\nu-1}(i\sqrt{\lambda c}\,\alpha). \tag{10.7.2}$$

According to the asymptotic formulas of Sec. 10.6,

$$H_\nu(\sqrt{\lambda c}\,\alpha) \approx (2\sqrt{\lambda c}\,\alpha)^\nu, \qquad \alpha \to \infty,$$

$$H_{-\nu-1}(i\sqrt{\lambda c}\,\alpha) \approx e^{-\frac{1}{2}(\nu+1)\pi i}(2\sqrt{\lambda c}\,\alpha)^{-\nu-1}, \qquad \alpha \to \infty,$$

and hence we must set $N = 0$ if the solutions are to be bounded. Moreover, for $\nu \neq 0, 1, 2, \ldots$, we have

$$H_\nu(\sqrt{\lambda c}\,\alpha) \approx \frac{\sqrt{\pi}}{\Gamma(-\nu)}\, e^{\lambda c\alpha^2}(\sqrt{\lambda c}\,|\alpha|)^{-\nu-1}, \qquad \alpha \to -\infty$$

and therefore we must also set $M = 0$. It follows that unless ν is a non-negative integer, there are no solutions which are bounded as $\alpha \to \pm\infty$ (except the trivial solution identically equal to zero).

For integral $\nu = n$ $(n = 0, 1, 2, \ldots)$, the Hermite functions reduce to Hermite polynomials, and the solution of equation (10.1.9) bounded in the interval $(-\infty, \infty)$ is

$$\mathrm{A} = Me^{-\lambda c\alpha^2/2}H_n(\sqrt{\lambda c}\,\alpha), \qquad n = 0, 1, 2, \ldots \tag{10.7.3}$$

Substituting the corresponding value $\mu = \lambda c(2n + 1)$ into (10.1.10), we can write the general solution of this equation as

$$\mathrm{B} = Pe^{\lambda c\beta^2/2}H_n(i\sqrt{\lambda c}\,\beta) + Qe^{-\lambda c\beta^2/2}H_{-n-1}(\sqrt{\lambda c}\,\beta) \tag{10.7.4}$$

[cf. (10.2.18)]. Combining (10.7.1) and (10.7.3, 4), we see that Laplace's equation has infinitely many solutions of the form

$$u = u_{\lambda,n} = e^{-\lambda c\alpha^2/2}H_n(\sqrt{\lambda c}\,\alpha)\Big[P_{\lambda,n}e^{\lambda c\beta^2/2}H_n(i\sqrt{\lambda c}\,\beta) + Q_{\lambda,n}e^{-\lambda c\beta^2/2}H_{-n-1}(\sqrt{\lambda c}\,\beta)\Big]\begin{matrix}\cos \lambda z\\ \sin \lambda z\end{matrix},$$

$$\lambda \geqslant 0, \qquad n = 0, 1, 2, \ldots, \tag{10.7.5}$$

which are bounded for $-\infty < \alpha < \infty$, $-\infty < z < \infty$. For the exterior problem, β varies over the interval $\beta_0 < \beta < \infty$, where the surface of the parabolic cylinder corresponds to $\beta = \beta_0$, and hence we have to set $P_{\lambda,n} = 0$, in view of the asymptotic formulas

$$H_n(i\sqrt{\lambda c}\,\beta) \approx i^n(2\sqrt{\lambda c}\,\beta)^n, \qquad \beta \to \infty,$$

$$H_{-n-1}(\sqrt{\lambda c}\,\beta) \approx (2\sqrt{\lambda c}\,\beta)^{-n-1}, \qquad \beta \to \infty.$$

We now show that $Q_{\lambda,n}$ must be set equal to 0 if the solutions (10.7.5) are to be harmonic in the case of the interior problem, where $0 \leqslant \beta < \beta_0$. Here the decisive consideration is the behavior of grad u near the singular curve of the transformation (10.1.1), i.e., the line $\alpha = \beta = 0$ on which the Jacobian $\partial(x, y, z)/\partial(\alpha, \beta, z)$ vanishes. It is an immediate consequence of (10.1.5) that

$$(\text{grad } u)^2 = \frac{1}{c^2(\alpha^2 + \beta^2)} \left[\left(\frac{\partial u}{\partial \alpha}\right)^2 + \left(\frac{\partial u}{\partial \beta}\right)^2\right] + \left(\frac{\partial u}{\partial z}\right)^2.$$

Since the denominator in the right-hand side vanishes on the curve $\alpha = \beta = 0$, a necessary condition for grad u to be finite is that the expression in brackets should also vanish for $\alpha = \beta = 0$, i.e., that $Q_{\lambda,n} = 0$, since (10.7.5) implies[9]

$$\left[\left(\frac{\partial u}{\partial \alpha}\right)^2 + \left(\frac{\partial u}{\partial \beta}\right)^2\right]_{\alpha=\beta=0} = \left(\sqrt{\lambda c}\, Q_{\lambda,n} \begin{matrix}\cos \lambda z\\ \sin \lambda z\end{matrix}\right)^2 = 0.$$

Moreover, this condition is also sufficient, since it is easily verified that if $Q_{\lambda,n} = 0$, then the expression

$$\left[\left(\frac{\partial u}{\partial \alpha}\right)^2 + \left(\frac{\partial u}{\partial \beta}\right)^2\right]$$

is divisible by $\alpha^2 + \beta^2$, so that grad u is well-behaved on the line $\alpha = \beta = 0$.[10] Thus the appropriate particular solutions of Laplace's equation are

$$u = u_{\lambda,n} = P_{\lambda,n} e^{-(\lambda c/2)(\alpha^2 - \beta^2)} H_n(\sqrt{\lambda c}\,\alpha) H_n(i\sqrt{\lambda c}\,\beta) \begin{matrix}\cos \lambda z\\ \sin \lambda z\end{matrix},$$

$$\lambda \geqslant 0, \quad n = 0, 1, 2, \ldots \qquad (10.7.6)$$

for the interior problem, and

$$u = u_{\lambda,n} = Q_{\lambda,n} e^{-(\lambda c/2)(\alpha^2 + \beta^2)} H_n(\sqrt{\lambda c}\,\alpha) H_{-n-1}(\sqrt{\lambda c}\,\beta) \begin{matrix}\cos \lambda z\\ \sin \lambda z\end{matrix},$$

$$\lambda \geqslant 0, \quad n = 0, 1, 2, \ldots \qquad (10.7.7)$$

for the exterior problem.

Boundary value problems involving parabolic cylinders are solved by superposition of the particular solutions (10.7.6–7). For example, consider the interior Dirichlet problem, assuming, for simplicity, that the function $f = f(\alpha, z)$ appearing in the boundary condition

$$u|_{\beta=\beta_0} = f \qquad (10.7.8)$$

---

[9] In the course of the calculations, we use the formulas

$$H_n(0)H_n'(0) = 0, \qquad H_n(0)H'_{-n-1}(0) = -\cos\frac{n\pi}{2}, \qquad H_n'(0)H_{-n-1}(0) = \sin\frac{n\pi}{2},$$

$$n = 0, 1, 2, \ldots,$$

which follow from (10.2.10).

[10] Cf. the analogous treatment for an oblate spheroid on p. 217.

is an even function of z, which implies that the same is true of the solution $u = u(\alpha, \beta, z)$.[11] Suppose that f can be expanded in a Fourier integral

$$f = \int_0^\infty f_\lambda(\alpha) \cos \lambda z \, d\lambda, \qquad -\infty < z < \infty, \tag{10.7.9}$$

where

$$f_\lambda = \frac{2}{\pi} \int_0^\infty f \cos \lambda z \, dz, \tag{10.7.10}$$

and moreover suppose that the solution u can also be represented as a Fourier integral

$$u = \int_0^\infty u_\lambda(\alpha, \beta) \cos \lambda z \, d\lambda, \qquad -\infty < z < \infty. \tag{10.7.11}$$

Then, according to (10.7.6), we can look for $u_\lambda(\alpha, \beta)$ in the form of a series

$$u_\lambda(\alpha, \beta) = \sum_{n=0}^\infty P_{\lambda,n} e^{-(\lambda c/2)(\alpha^2 - \beta^2)} H_n(\sqrt{\lambda c}\,\alpha) H_n(i\sqrt{\lambda c}\,\beta),$$
$$-\infty < \alpha < \infty, \quad 0 \leqslant \beta < \beta_0, \tag{10.7.12}$$

and we have the condition

$$f_\lambda(\alpha) = \sum_{n=0}^\infty P_{\lambda,n} e^{-(\lambda c/2)(\alpha^2 - \beta_0^2)} H_n(\alpha\sqrt{\lambda c})\, H_n(i\sqrt{\lambda c}\,\beta_0),$$
$$-\infty < \alpha < \infty \tag{10.7.13}$$

for determining the coefficients $P_{\lambda,n}$. Assuming that $f(\alpha)$ satisfies the conditions of Theorem 2, p. 71, we find that

$$P_{\lambda,n} e^{\lambda c \beta_0^2/2} H_n(i\sqrt{\lambda c}\,\beta_0) = \frac{\sqrt{\lambda c}}{2^n n! \sqrt{\pi}} \int_{-\infty}^\infty e^{-\lambda c \alpha^2/2} f_\lambda(\alpha) H_n(\sqrt{\lambda c}\,\alpha) \, d\alpha, \tag{10.7.14}$$

and hence the expansion coefficient $u_\lambda(\alpha, \beta)$ is given by the sum

$$u_\lambda(\alpha, \beta) = \sum_{n=0}^\infty e^{-(\lambda c/2)(\alpha^2 - \beta^2 + \beta_0^2)} \frac{H_n(i\sqrt{\lambda c}\,\beta)}{H_n(i\sqrt{\lambda c}\,\beta_0)} H_n(\sqrt{\lambda c}\,\alpha)$$
$$\times \frac{\sqrt{\lambda c}}{2^n n! \sqrt{\pi}} \int_{-\infty}^\infty e^{-\lambda c \alpha^2/2} f_\lambda(\alpha) H_n(\sqrt{\lambda c}\,\alpha) \, d\alpha. \tag{10.7.15}$$

Substituting (10.7.15) into (10.7.11), we obtain the formal solution of our problem.

---

[11] The case where f is an odd function of z is handled in the same way. Then the solution in the general case is represented as the sum of the solutions of the two simpler problems with the following even and odd boundary conditions:

$$f_1 = \tfrac{1}{2}[f(\alpha, z) + f(\alpha, -z)], \qquad f_2 = \tfrac{1}{2}[f(\alpha, z) - f(\alpha, -z)].$$

10.8. Application to Quantum Mechanics

The Schrödinger equation for a linear harmonic oscillator of mass m, angular frequency ω_0 and total energy E has the form

$$\frac{d^2\psi}{dx^2} + \left(\frac{2mE}{\hbar^2} - \frac{m^2\omega_0^2}{\hbar^2}\right)\psi = 0, \tag{10.8.1}$$

where ψ is the wave function and $\hbar$ is Planck's constant.[12] In quantum mechanics, it is required to find the values of E for which (10.8.1) has bounded solutions in the interval $-\infty < x < \infty$. If we set

$$\mu = \frac{2mE}{\hbar^2}, \qquad \lambda c = \frac{m\omega_0}{\hbar},$$

equation (10.8.1) coincides with equation (10.1.9). It follows from the results of Sec. 10.7 that the solutions of (10.8.1) are bounded in $(-\infty, \infty)$ only if

$$\mu = \lambda c(2n + 1), \qquad n = 0, 1, 2, \ldots,$$

i.e., only if

$$\frac{2mE}{\hbar^2} = (2n + 1)\frac{m\omega_0}{\hbar}, \qquad n = 0, 1, 2, \ldots,$$

which implies

$$E = E_n = (n + \tfrac{1}{2})\hbar\omega_0, \qquad n = 0, 1, 2, \ldots \tag{10.8.2}$$

The corresponding wave functions can be expressed in terms of Hermite polynomials.

PROBLEMS

1. Derive the following integral representations of the Hermite functions:

$$H_\nu(z) = \frac{2^{\nu+1}}{\Gamma\left(\frac{1-\nu}{2}\right)} \int_0^\infty e^{-t^2} t^{-\nu}(t^2 + z^2)^{\nu/2}\, dt, \qquad \operatorname{Re} \nu < 1, \quad |\arg z| < \frac{\pi}{2},$$

$$H_\nu(z) = \frac{2^{\nu+1}}{\Gamma\left(-\frac{\nu}{2}\right)} z \int_0^\infty e^{-t^2} t^{-\nu-1}(t^2 + z^2)^{(\nu-1)/2}\, dt, \qquad \operatorname{Re} \nu < 0, \quad |\arg z| < \frac{\pi}{2}.$$

Hint. Use formulas (9.10.3) and (9.11.6), and the representation of $H_\nu(z)$ in terms of $\Psi(\alpha, \gamma; z)$, the confluent hypergeometric function of the second kind (see footnote 4, p. 285).

---

[12] See D. Bohm, *Quantum Theory*, Prentice-Hall, Inc., Englewood Cliffs, N.J. (1963), p. 296.

2. Derive the following integral representation of the product of two Hermite functions:

$$H_\mu(z)H_\nu(z) = \frac{\Gamma(-\mu-\nu)}{\Gamma(-\mu)\Gamma(-\nu)}\int_0^{\pi/2} H_{\mu+\nu}[z(\cos\varphi+\sin\varphi)]\cos^{-\mu-1}\varphi\,\sin^{-\nu-1}\varphi\,d\varphi,$$

$$\operatorname{Re}\mu < 0, \quad \operatorname{Re}\nu < 0.$$

Hint. Use (10.5.1) and transform to polar coordinates in the double integral.

3. Prove the integral representation

$$H_\mu(z)H_\nu(z) = \frac{1}{\Gamma(-\mu-\nu)}\int_0^\infty e^{-t^2-2zt}t^{-\mu-\nu-1}{}_2F_2\left(\begin{matrix}-\mu, -\nu; \tfrac{1}{2}t^2\\ -\dfrac{\mu+\nu}{2}, \dfrac{1-\mu-\nu}{2}\end{matrix}\right)dt,$$

where ${}_2F_2$ is a generalized hypergeometric function (see Sec. 9.14).

Hint. Use (10.5.1) to represent the left-hand side as a double integral over the square $0 \leqslant s < \infty$, $0 \leqslant t < \infty$, and then transform to the new variables $u = s + t$, $v = t/s$.

4. Prove the formulas

$$[H_\nu(z)]^2 = \frac{1}{\Gamma(-2\nu)}\int_0^\infty e^{-t^2-2zt}t^{-2\nu-1}\Phi\left(-\nu, -\nu+\frac{1}{2}; \frac{t^2}{2}\right)dt, \qquad \operatorname{Re}\nu < 0,$$

$$H_\nu(z)H_{\nu+1}(z) = \frac{1}{\Gamma(-2\nu-1)}\int_0^\infty e^{-t^2-2zt}t^{-2\nu-2}\,\Phi\left(-\nu-1, -\nu-\frac{1}{2}; \frac{t^2}{2}\right)dt,$$

$$\operatorname{Re}\nu < -\tfrac{1}{2}.$$

5. Show that the Hermite functions satisfy the integral equation

$$x^{-(\nu+1)/2}H_\nu(x) = 2\int_0^\infty (xy)^{1/2}J_{-\nu/2}(2xy)y^{-(\nu+1)/2}H_\nu(y)\,dy,$$

$$0 < x < \infty, \quad \operatorname{Re}\nu < 1.$$

6. Show that the Hermite functions of half-integral degree can be expressed in terms of the cylinder functions of imaginary argument. In particular, prove the relation

$$H_{-1/2}(z) = \left(\frac{z}{2\pi}\right)^{1/2}e^{z^2/2}K_{1/4}\left(\frac{z^2}{4}\right), \qquad |\arg z| < \frac{\pi}{2}.$$

Hint. Use the integral representation (10.5.1), and make the change of variable $t = 2z\sinh^2(\theta/4)$.

7. Prove the formula

$$K_0\left(\frac{x^2+y^2}{2}\right) = 2\sum_{n=0}^\infty \frac{(-1)^n\Gamma(n+\frac{1}{2})}{n!}e^{-(x^2+y^2)/2}H_{2n}(x)H_{-2n-1}(y),$$

$$-\infty < x < \infty, \quad 0 \leqslant y < \infty,$$

where $K_0(z)$ is Macdonald's function.

Hint. Apply Theorem 2, p. 71 and the result of Problem 1.

8. Consider the system of *paraboloidal coordinates* α, β, φ related to the rectangular coordinates x, y, z by the formulas

$$x = c\alpha\beta \cos \varphi, \quad y = c\alpha\beta \sin \varphi, \quad z = \frac{c}{2}(\alpha^2 - \beta^2),$$

where $0 \leqslant \alpha < \infty$, $0 \leqslant \beta < \infty$, $-\pi < \varphi \leqslant \pi$, and $c > 0$ is a scale factor. In this coordinate system, the surfaces $\alpha = \text{const}$, $\beta = \text{const}$ are paraboloids of revolution instead of parabolic cylinders, as in (10.1.1). Find the square of the element of arc length, the metric coefficients and Laplace's equation in the system α, β, φ. Show that separation of variables is possible in Laplace's equation written in the coordinates α, β, φ, and find the appropriate particular solutions, both for the interior and the exterior problem.

BIBLIOGRAPHY

Bailey, W. N., *Generalized Hypergeometric Series*, Cambridge Tracts in Mathematics and Mathematical Physics, No. 32, Cambridge University Press, London (1935).

Bateman, H., *Partial Differential Equations of Mathematical Physics*, Cambridge University Press, London (1959).

Bateman, H., *The Mathematical Analysis of Electrical and Optical Wave-Motion on the Basis of Maxwell's Equations*, Dover Publications, Inc., New York (1955).

Bateman Manuscript Project, see works by A. Erdélyi *et al.* cited below.

Bowman, F., *Introduction to Bessel Functions*, Dover Publications, Inc., New York (1958).

Buchholz, H., *Die Konfluente Hypergeometrische Funktion mit Besonderer Berücksichtigung Ihrer Anwendungen*, Springer-Verlag, Berlin (1953).

Buronova, N. M., *A Guide to Mathematical Tables, Supplement No. 1 to A Guide to Mathematical Tables by A. V. Lebedev and R. M. Fedorova* (translated by D. G. Fry), Pergamon Press, Inc., New York (1960).

Copson, E. T., *An Introduction to the Theory of Functions of a Complex Variable*, Oxford University Press, London (1935).

Courant, R. and D. Hilbert, *Methods of Mathematical Physics*, Interscience Publishers, New York, *Volume I* (1953), *Volume II* (1962).

De Bruijn, N. G., *Asymptotic Methods in Analysis*, Interscience Publishers, Inc., New York (1958).

Erdélyi, A., *Asymptotic Expansions*, Dover Publications, Inc., New York (1956).

Erdélyi, A., W. Magnus, F. Oberhettinger and F. G. Tricomi, *Higher Transcendental Functions* (in three volumes), based, in part, on notes left by Harry Bateman, McGraw-Hill Book Co., New York (1953).

Erdélyi, A., W. Magnus, F. Oberhettinger and F. G. Tricomi, *Tables of Integral Transforms* (in two volumes), based, in part, on notes left by Harry Bateman, McGraw-Hill Book Co., New York (1954).

Fletcher, A., J. C. P. Miller, L. Rosenhead and L. J. Comrie, *An Index of Mathematical Tables* (in two volumes), second edition, Addison-Wesley Publishing Co., Inc., Reading, Mass. (1962).

Frank, P. and R. von Mises, *Die Differential- und Integralgleichungen der Mechanik und Physik* (in two volumes), second enlarged edition, Dover Publications, Inc., New York (1961).

Gray, A. and G. B. Mathews, *A Treatise on Bessel Functions and Their Applications to Physics*, second edition, prepared by A. Gray and T. M. MacRobert, Macmillan and Co., Ltd., London (1952). (Dover Reprint)

Heatley, A. H., *Some Integrals, Differential Equations, and Series Related to the Modified Bessel Function of the First Kind*, University of Toronto Studies, Mathematical Series, No. 7, University of Toronto Press, Toronto (1939).

Hobson, E. W., *The Theory of Spherical and Ellipsoidal Harmonics*, Cambridge University Press, London (1931).

Hochstadt, H., *Special Functions of Mathematical Physics*, Holt, Rinehart and Winston, Inc., New York (1961).

Jackson, D., *Fourier Series and Orthogonal Polynomials*, Carus Mathematical Monograph No. 6, Mathematical Association of America, State University of New York, Buffalo, N.Y. (1941).

Jahnke, E. and F. Emde, *Tables of Higher Functions*, sixth edition, revised by F. Lösch, McGraw-Hill Book Co., New York (1960).

Jeffreys, H., *Asymptotic Approximations*, Oxford University Press, London (1962).

Jeffreys, H. and B. S. Jeffreys, *Methods of Mathematical Physics*, third edition, Cambridge University Press, London (1956).

Kampé de Fériet, J., *La Fonction Hypergéométrique*, Mémorial des Sciences Mathématiques, Fascicule 85, Gauthier-Villars, Paris (1937).

Klein, F., *Vorlesungen über die Hypergeometrische Funktion*, Springer-Verlag, Berlin (1933).

Lebedev, A. V. and R. M. Fedorova, *A Guide to Mathematical Tables* (translated by D. G. Fry), Pergamon Press, Inc., New York (1960). See supplement by N. M. Buronova cited above.

Lense, J., *Reihenentwicklungen in der Mathematischen Physik*, third edition, Walter de Gruyter & Co., Berlin (1953).

Lense, J. *Kugelfunktionen*, second edition, Akademische Verlagsgesellschaft, Geest & Portig K.-G., Leipzig (1954).

Lösch, F. and F. Schoblik, *Die Fakultät (Gammafunktion) und Verwandte Funktionen mit Besonderer Berücksichtigung Ihrer Anwendungen*, B. G. Teubner, Leipzig (1951).

Luke, Y. L., *Integrals of Bessel Functions*, McGraw-Hill Book Co., New York (1962).

MacRobert, T. M., *Spherical Harmonics, An Elementary Treatise on Harmonic Functions with Applications*, second edition, Methuen and Co., Ltd., London (1947).

MacRobert, T. M., *Functions of a Complex Variable*, fifth edition, Macmillan and Co., Ltd., London (1962).

McLachlan, N. S., *Bessel Functions for Engineers*, second edition, Oxford University Press, London (1955).

Magnus, W. and F. Oberhettinger, *Formulas and Theorems for the Functions of Mathematical Physics* (translated by J. Wermer), Chelsea Publishing Co., New York (1954).

Morse, P. M. and H. Feshbach, *Methods of Theoretical Physics* (in two volumes), McGraw-Hill Book Co., New York (1953).

Nielsen, N., *Handbuch der Theorie der Zylinderfunktionen*, B. G. Teubner, Leipzig (1904).

Nielsen, N., *Handbuch der Theorie der Gammafunktion*, B. G. Teubner, Leipzig (1906).

Nielsen, N., *Theorie des Integrallogarithmus und Verwandter Transzendenten*, B. G. Teubner, Leipzig (1906).

Petiau, G., *La Théorie des Fonctions de Bessel*, Centre National de la Recherche Scientifique, Paris (1955).

Rainville, E. D., *Special Functions*, The Macmillan Co., New York (1960).

Relton, F. E., *Applied Bessel Functions*, Blackie and Son, Ltd., London (1946).

Rey Pastor, J. and A. De Castro Brzezicki, *Funciones de Bessel, Teoria Matematica y Aplicaciones a la Ciencia y a la Técnica*, Editorial Dossat, S. A., Madrid (1958).

Robin, L., *Fonctions Sphériques de Legendre et Fonctions Sphéroïdales*, Gauthier-Villars, Paris, *Volume I* (1957), *Volume II* (1958), *Volume III* (1959).

Rosser, J. B., *Theory and Application of* $\int_0^z e^{-x^2}\,dx$ *and* $\int_0^z e^{-p^2y^2}\,dy \int_0^y e^{-x^2}\,dx$, Mapleton House, Brooklyn, N.Y. (1948).

Ryshik, I. M. and I. S. Gradstein, *Tables of Series, Products, and Integrals*, VEB Deutscher Verlag der Wissenschaften, Berlin (1957).

Sansone, G., *Orthogonal Functions* (translated by A. H. Diamond), Interscience Publishers, New York (1959).

Shohat, J. A., E. Hille and J. L. Walsh, *A Bibliography on Orthogonal Polynomials*, Bulletin National Research Council, No. 103, National Research Council of the National Academy of Sciences, Washington, D.C. (1940).

Slater, L. J., *Confluent Hypergeometric Functions*, Cambridge University Press, London (1960).

Smirnov, V. I., *Lehrgang der Höheren Matematik*, VEB Deutscher Verlag der Wissenschaften, Berlin, *Volume III, Part 2* (1955), *Volume IV* (1958).

Sneddon, I. N., *Fourier Transforms*, McGraw-Hill Book Co., New York (1951).

Sneddon, I. N., *Special Functions of Mathematical Physics and Chemistry*, second edition, Oliver and Boyd, London (1961).

Snow, C., *The Hypergeometric and Legendre Functions with Applications to Integral Equations of Potential Theory*, National Bureau of Standards Applied Mathematics Series, No. 19, U.S. Government Printing Office, Washington, D.C. (1952).

Sommerfeld, A., *Partial Differential Equations in Physics* (translated by E. G. Straus), Academic Press Inc., New York (1949).

Sternberg, W. J. and T. L. Smith, *The Theory of Potential and Spherical Harmonics*, University of Toronto Press, Toronto (1952).

Szegö, G., *Orthogonal Polynomials*, revised edition, American Mathematical Society, New York (1959).

Tikhonov, A. N. and A. A. Samarski, *Differentialgleichungen der Mathematischen Physik*, VEB Deutscher Verlag der Wissenschaften, Berlin (1959).

Tolstov, G. P., *Fourier Series* (translated by R. A. Silverman), Prentice-Hall, Inc., Englewood Cliffs, N.J. (1962).

Tricomi, F. G., *Funzioni Ipergeometriche Confluenti*, Edizioni Cremonese, Rome (1954).

Tricomi, F. G., *Vorlesungen über Orthogonalreihen*, Springer-Verlag, Berlin (1955).

Tricomi, F. G., *Fonctions Hypergéométriques Confluentes*, Mémorial des Sciences Mathématiques, Fascicule 140, Gauthier-Villars, Paris (1960).

Truesdell, C., *An Essay toward a Unified Theory of Special Functions Based upon the Functional Equation* $\frac{\partial}{\partial z} F(z, \alpha) = F(z, \alpha + 1)$, Princeton University Press, Princeton, N.J. (1948).

Watson, G. N., *A Treatise on the Theory of Bessel Functions*, second edition, Cambridge University Press, London (1962).

Weyrich, R., *Die Zylinderfunktionen und Ihre Anwendungen*, B. G. Teubner, Leipzig (1937).

Whittaker, E. T. and G. N. Watson, *A Course of Modern Analysis*, fourth edition, Cambridge University Press, London (1963).

INDEX

A CATALOGUE OF SELECTED DOVER BOOKS IN ALL FIELDS OF INTEREST

A CATALOGUE OF SELECTED DOVER BOOKS IN ALL FIELDS OF INTEREST

CELESTIAL OBJECTS FOR COMMON TELESCOPES, T. W. Webb. The most used book in amateur astronomy: inestimable aid for locating and identifying nearly 4,000 celestial objects. Edited, updated by Margaret W. Mayall. 77 illustrations. Total of 645pp. 5⅜ x 8½.
20917-2, 20918-0 Pa., Two-vol. set $10.00

HISTORICAL STUDIES IN THE LANGUAGE OF CHEMISTRY, M. P. Crosland. The important part language has played in the development of chemistry from the symbolism of alchemy to the adoption of systematic nomenclature in 1892. ". . . wholeheartedly recommended,"—Science. 15 illustrations. 416pp. of text. 5⅝ x 8¼. 63702-6 Pa. $7.50

BURNHAM'S CELESTIAL HANDBOOK, Robert Burnham, Jr. Thorough, readable guide to the stars beyond our solar system. Exhaustive treatment, fully illustrated. Breakdown is alphabetical by constellation: Andromeda to Cetus in Vol. 1; Chamaeleon to Orion in Vol. 2; and Pavo to Vulpecula in Vol. 3. Hundreds of illustrations. Total of about 2000pp. 6⅛ x 9¼.
23567-X, 23568-8, 23673-0 Pa., Three-vol. set $32.85

THEORY OF WING SECTIONS: INCLUDING A SUMMARY OF AIR-FOIL DATA, Ira H. Abbott and A. E. von Doenhoff. Concise compilation of subatomic aerodynamic characteristics of modern NASA wing sections, plus description of theory. 350pp. of tables. 693pp. 5⅜ x 8½.
60586-8 Pa. $9.95

DE RE METALLICA, Georgius Agricola. Translated by Herbert C. Hoover and Lou H. Hoover. The famous Hoover translation of greatest treatise on technological chemistry, engineering, geology, mining of early modern times (1556). All 289 original woodcuts. 638pp. 6¾ x 11.
60006-8 Clothbd. $19.95

THE ORIGIN OF CONTINENTS AND OCEANS, Alfred Wegener. One of the most influential, most controversial books in science, the classic statement for continental drift. Full 1966 translation of Wegener's final (1929) version. 64 illustrations. 246pp. 5⅜ x 8½. (EBE) 61708-4 Pa. $5.00

THE PRINCIPLES OF PSYCHOLOGY, William James. Famous long course complete, unabridged. Stream of thought, time perception, memory, experimental methods; great work decades ahead of its time. Still valid, useful; read in many classes. 94 figures. Total of 1391pp. 5⅜ x 8½.
20381-6, 20382-4 Pa., Two-vol. set $17.90

YUCATAN BEFORE AND AFTER THE CONQUEST, Diego de Landa. First English translation of basic book in Maya studies, the only significant account of Yucatan written in the early post-Conquest era. Translated by distinguished Maya scholar William Gates. Appendices, introduction, 4 maps and over 120 illustrations added by translator. 162pp. 5⅜ x 8½.
23622-6 Pa. $3.00

THE MALAY ARCHIPELAGO, Alfred R. Wallace. Spirited travel account by one of founders of modern biology. Touches on zoology, botany, ethnography, geography, and geology. 62 illustrations, maps. 515pp. 5⅜ x 8½.
20187-2 Pa. $6.95

THE DISCOVERY OF THE TOMB OF TUTANKHAMEN, Howard Carter, A. C. Mace. Accompany Carter in the thrill of discovery, as ruined passage suddenly reveals unique, untouched, fabulously rich tomb. Fascinating account, with 106 illustrations. New introduction by J. M. White. Total of 382pp. 5⅜ x 8½. (Available in U.S. only) 23500-9 Pa. $5.50

THE WORLD'S GREATEST SPEECHES, edited by Lewis Copeland and Lawrence W. Lamm. Vast collection of 278 speeches from Greeks up to present. Powerful and effective models; unique look at history. Revised to 1970. Indices. 842pp. 5⅜ x 8½. 20468-5 Pa. $9.95

THE 100 GREATEST ADVERTISEMENTS, Julian Watkins. The priceless ingredient; His master's voice; 99 44/100% pure; over 100 others. How they were written, their impact, etc. Remarkable record. 130 illustrations. 233pp. 7⅞ x 10 3/5. 20540-1 Pa. $6.95

CRUICKSHANK PRINTS FOR HAND COLORING, George Cruickshank. 18 illustrations, one side of a page, on fine-quality paper suitable for watercolors. Caricatures of people in society (c. 1820) full of trenchant wit. Very large format. 32pp. 11 x 16. 23684-6 Pa. $6.00

THIRTY-TWO COLOR POSTCARDS OF TWENTIETH-CENTURY AMERICAN ART, Whitney Museum of American Art. Reproduced in full color in postcard form are 31 art works and one shot of the museum. Calder, Hopper, Rauschenberg, others. Detachable. 16pp. 8¼ x 11.
23629-3 Pa. $3.50

MUSIC OF THE SPHERES: THE MATERIAL UNIVERSE FROM ATOM TO QUASAR SIMPLY EXPLAINED, Guy Murchie. Planets, stars, geology, atoms, radiation, relativity, quantum theory, light, antimatter, similar topics. 319 figures. 664pp. 5⅜ x 8½.
21809-0, 21810-4 Pa., Two-vol. set $11.00

EINSTEIN'S THEORY OF RELATIVITY, Max Born. Finest semi-technical account; covers Einstein, Lorentz, Minkowski, and others, with much detail, much explanation of ideas and math not readily available elsewhere on this level. For student, non-specialist. 376pp. 5⅜ x 8½.
60769-0 Pa. $5.00

THE SENSE OF BEAUTY, George Santayana. Masterfully written discussion of nature of beauty, materials of beauty, form, expression; art, literature, social sciences all involved. 168pp. 5⅜ x 8½. 20238-0 Pa. **$3.50**

ON THE IMPROVEMENT OF THE UNDERSTANDING, Benedict Spinoza. Also contains *Ethics, Correspondence,* all in excellent R. Elwes translation. Basic works on entry to philosophy, pantheism, exchange of ideas with great contemporaries. 402pp. 5⅜ x 8½. 20250-X Pa. $5.95

THE TRAGIC SENSE OF LIFE, Miguel de Unamuno. Acknowledged masterpiece of existential literature, one of most important books of 20th century. Introduction by Madariaga. 367pp. 5⅜ x 8½.
20257-7 Pa. $6.00

THE GUIDE FOR THE PERPLEXED, Moses Maimonides. Great classic of medieval Judaism attempts to reconcile revealed religion (Pentateuch, commentaries) with Aristotelian philosophy. Important historically, still relevant in problems. Unabridged Friedlander translation. Total of 473pp. 5⅜ x 8½. 20351-4 Pa. $6.95

THE I CHING (THE BOOK OF CHANGES), translated by James Legge. Complete translation of basic text plus appendices by Confucius, and Chinese commentary of most penetrating divination manual ever prepared. Indispensable to study of early Oriental civilizations, to modern inquiring reader. 448pp. 5⅜ x 8½. 21062-6 Pa. $6.00

THE EGYPTIAN BOOK OF THE DEAD, E. A. Wallis Budge. Complete reproduction of Ani's papyrus, finest ever found. Full hieroglyphic text, interlinear transliteration, word for word translation, smooth translation. Basic work, for Egyptology, for modern study of psychic matters. Total of 533pp. 6½ x 9¼. (USCO) 21866-X Pa. $8.50

THE GODS OF THE EGYPTIANS, E. A. Wallis Budge. Never excelled for richness, fullness: all gods, goddesses, demons, mythical figures of Ancient Egypt; their legends, rites, incarnations, variations, powers, etc. Many hieroglyphic texts cited. Over 225 illustrations, plus 6 color plates. Total of 988pp. 6⅛ x 9¼. (EBE)
22055-9, 22056-7 Pa., Two-vol. set $20.00

THE STANDARD BOOK OF QUILT MAKING AND COLLECTING, Marguerite Ickis. Full information, full-sized patterns for making 46 traditional quilts, also 150 other patterns. Quilted cloths, lame, satin quilts, etc. 483 illustrations. 273pp. 6⅞ x 9⅝. 20582-7 Pa. $5.95

CORAL GARDENS AND THEIR MAGIC, Bronsilaw Malinowski. Classic study of the methods of tilling the soil and of agricultural rites in the Trobriand Islands of Melanesia. Author is one of the most important figures in the field of modern social anthropology. 143 illustrations. Indexes. Total of 911pp. of text. 5⅝ x 8¼. (Available in U.S. only)
23597-1 Pa. $12.95

THE PHILOSOPHY OF HISTORY, Georg W. Hegel. Great classic of Western thought develops concept that history is not chance but a rational process, the evolution of freedom. 457pp. 5⅜ x 8½. 20112-0 Pa. $6.00

LANGUAGE, TRUTH AND LOGIC, Alfred J. Ayer. Famous, clear introduction to Vienna, Cambridge schools of Logical Positivism. Role of philosophy, elimination of metaphysics, nature of analysis, etc. 160pp. 5⅜ x 8½. (USCO) 20010-8 Pa. $2.50

A PREFACE TO LOGIC, Morris R. Cohen. Great City College teacher in renowned, easily followed exposition of formal logic, probability, values, logic and world order and similar topics; no previous background needed. 209pp. 5⅜ x 8½. 23517-3 Pa. $4.95

REASON AND NATURE, Morris R. Cohen. Brilliant analysis of reason and its multitudinous ramifications by charismatic teacher. Interdisciplinary, synthesizing work widely praised when it first appeared in 1931. Second (1953) edition. Indexes. 496pp. 5⅜ x 8½. 23633-1 Pa. $7.50

AN ESSAY CONCERNING HUMAN UNDERSTANDING, John Locke. The only complete edition of enormously important classic, with authoritative editorial material by A. C. Fraser. Total of 1176pp. 5⅜ x 8½. 20530-4, 20531-2 Pa., Two-vol. set $16.00

HANDBOOK OF MATHEMATICAL FUNCTIONS WITH FORMULAS, GRAPHS, AND MATHEMATICAL TABLES, edited by Milton Abramowitz and Irene A. Stegun. Vast compendium: 29 sets of tables, some to as high as 20 places. 1,046pp. 8 x 10½. 61272-4 Pa. $17.95

MATHEMATICS FOR THE PHYSICAL SCIENCES, Herbert S. Wilf. Highly acclaimed work offers clear presentations of vector spaces and matrices, orthogonal functions, roots of polynomial equations, conformal mapping, calculus of variations, etc. Knowledge of theory of functions of real and complex variables is assumed. Exercises and solutions. Index. 284pp. 5⅝ x 8¼. 63635-6 Pa. $5.00

THE PRINCIPLE OF RELATIVITY, Albert Einstein et al. Eleven most important original papers on special and general theories. Seven by Einstein, two by Lorentz, one each by Minkowski and Weyl. All translated, unabridged. 216pp. 5⅜ x 8½. 60081-5 Pa. $3.50

THERMODYNAMICS, Enrico Fermi. A classic of modern science. Clear, organized treatment of systems, first and second laws, entropy, thermodynamic potentials, gaseous reactions, dilute solutions, entropy constant. No math beyond calculus required. Problems. 160pp. 5⅜ x 8½. 60361-X Pa. $4.00

ELEMENTARY MECHANICS OF FLUIDS, Hunter Rouse. Classic undergraduate text widely considered to be far better than many later books. Ranges from fluid velocity and acceleration to role of compressibility in fluid motion. Numerous examples, questions, problems. 224 illustrations. 376pp. 5⅝ x 8¼. 63699-2 Pa. $7.00

THE AMERICAN SENATOR, Anthony Trollope. Little known, long unavailable Trollope novel on a grand scale. Here are humorous comment on American vs. English culture, and stunning portrayal of a heroine/villainess. Superb evocation of Victorian village life. 561pp. 5⅜ x 8½.
23801-6 Pa. $7.95

WAS IT MURDER? James Hilton. The author of *Lost Horizon* and *Goodbye, Mr. Chips* wrote one detective novel (under a pen-name) which was quickly forgotten and virtually lost, even at the height of Hilton's fame. This edition brings it back—a finely crafted public school puzzle resplendent with Hilton's stylish atmosphere. A thoroughly English thriller by the creator of Shangri-la. 252pp. 5⅜ x 8. (Available in U.S. only)
23774-5 Pa. $3.00

CENTRAL PARK: A PHOTOGRAPHIC GUIDE, Victor Laredo and Henry Hope Reed. 121 superb photographs show dramatic views of Central Park: Bethesda Fountain, Cleopatra's Needle, Sheep Meadow, the Blockhouse, plus people engaged in many park activities: ice skating, bike riding, etc. Captions by former Curator of Central Park, Henry Hope Reed, provide historical view, changes, etc. Also photos of N.Y. landmarks on park's periphery. 96pp. 8½ x 11. 23750-8 Pa. $4.50

NANTUCKET IN THE NINETEENTH CENTURY, Clay Lancaster. 180 rare photographs, stereographs, maps, drawings and floor plans recreate unique American island society. Authentic scenes of shipwreck, lighthouses, streets, homes are arranged in geographic sequence to provide walking-tour guide to old Nantucket existing today. Introduction, captions. 160pp. 8⅞ x 11¾. 23747-8 Pa. $7.95

STONE AND MAN: A PHOTOGRAPHIC EXPLORATION, Andreas Feininger. 106 photographs by *Life* photographer Feininger portray man's deep passion for stone through the ages. Stonehenge-like megaliths, fortified towns, sculpted marble and crumbling tenements show textures, beauties, fascination. 128pp. 9¼ x 10¾. 23756-7 Pa. $5.95

CIRCLES, A MATHEMATICAL VIEW, D. Pedoe. Fundamental aspects of college geometry, non-Euclidean geometry, and other branches of mathematics: representing circle by point. Poincare model, isoperimetric property, etc. Stimulating recreational reading. 66 figures. 96pp. 5⅝ x 8¼.
63698-4 Pa. $3.50

THE DISCOVERY OF NEPTUNE, Morton Grosser. Dramatic scientific history of the investigations leading up to the actual discovery of the eighth planet of our solar system. Lucid, well-researched book by well-known historian of science. 172pp. 5⅜ x 8½. 23726-5 Pa. $3.50

THE DEVIL'S DICTIONARY. Ambrose Bierce. Barbed, bitter, brilliant witticisms in the form of a dictionary. Best, most ferocious satire America has produced. 145pp. 5⅜ x 8½. 20487-1 Pa. $2.50

THE COMPLETE BOOK OF DOLL MAKING AND COLLECTING, Catherine Christopher. Instructions, patterns for dozens of dolls, from rag doll on up to elaborate, historically accurate figures. Mould faces, sew clothing, make doll houses, etc. Also collecting information. Many illustrations. 288pp. 6 x 9. 22066-4 Pa. $4.95

THE DAGUERREOTYPE IN AMERICA, Beaumont Newhall. Wonderful portraits, 1850's townscapes, landscapes; full text plus 104 photographs. The basic book. Enlarged 1976 edition. 272pp. 8¼ x 11¼. 23322-7 Pa. $7.95

CRAFTSMAN HOMES, Gustav Stickley. 296 architectural drawings, floor plans, and photographs illustrate 40 different kinds of "Mission-style" homes from *The Craftsman* (1901-16), voice of American style of simplicity and organic harmony. Thorough coverage of Craftsman idea in text and picture, now collector's item. 224pp. 8⅛ x 11. 23791-5 Pa. $6.50

PEWTER-WORKING: INSTRUCTIONS AND PROJECTS, Burl N. Osborn. & Gordon O. Wilber. Introduction to pewter-working for amateur craftsman. History and characteristics of pewter; tools, materials, step-by-step instructions. Photos, line drawings, diagrams. Total of 160pp. 7⅞ x 10¾. 23786-9 Pa. $3.50

THE GREAT CHICAGO FIRE, edited by David Lowe. 10 dramatic, eye-witness accounts of the 1871 disaster, including one of the aftermath and rebuilding, plus 70 contemporary photographs and illustrations of the ruins—courthouse, Palmer House, Great Central Depot, etc. Introduction by David Lowe. 87pp. 8¼ x 11. 23771-0 Pa. $4.00

SILHOUETTES: A PICTORIAL ARCHIVE OF VARIED ILLUSTRATIONS, edited by Carol Belanger Grafton. Over 600 silhouettes from the 18th to 20th centuries include profiles and full figures of men and women, children, birds and animals, groups and scenes, nature, ships, an alphabet. Dozens of uses for commercial artists and craftspeople. 144pp. 8⅜ x 11¼. 23781-8 Pa. $4.50

ANIMALS: 1,419 COPYRIGHT-FREE ILLUSTRATIONS OF MAMMALS, BIRDS, FISH, INSECTS, ETC., edited by Jim Harter. Clear wood engravings present, in extremely lifelike poses, over 1,000 species of animals. One of the most extensive copyright-free pictorial sourcebooks of its kind. Captions. Index. 284pp. 9 x 12. 23766-4 Pa. $8.95

INDIAN DESIGNS FROM ANCIENT ECUADOR, Frederick W. Shaffer. 282 original designs by pre-Columbian Indians of Ecuador (500-1500 A.D.). Designs include people, mammals, birds, reptiles, fish, plants, heads, geometric designs. Use as is or alter for advertising, textiles, leathercraft, etc. Introduction. 95pp. 8¾ x 11¼. 23764-8 Pa. $4.50

SZIGETI ON THE VIOLIN, Joseph Szigeti. Genial, loosely structured tour by premier violinist, featuring a pleasant mixture of reminiscenes, insights into great music and musicians, innumerable tips for practicing violinists. 385 musical passages. 256pp. 5⅝ x 8¼. 23763-X Pa. $4.00

TONE POEMS, SERIES II: TILL EULENSPIEGELS LUSTIGE STREICHE, ALSO SPRACH ZARATHUSTRA, AND EIN HELDENLEBEN, Richard Strauss. Three important orchestral works, including very popular *Till Eulenspiegel's Marry Pranks,* reproduced in full score from original editions. Study score. 315pp. 9⅜ x 12¼. (Available in U.S. only)
23755-9 Pa. $8.95

TONE POEMS, SERIES I: DON JUAN, TOD UND VERKLARUNG AND DON QUIXOTE, Richard Strauss. Three of the most often performed and recorded works in entire orchestral repertoire, reproduced in full score from original editions. Study score. 286pp. 9⅜ x 12¼. (Available in U.S. only)
23754-0 Pa. $8.95

11 LATE STRING QUARTETS, Franz Joseph Haydn. The form which Haydn defined and "brought to perfection." *(Grove's).* 11 string quartets in complete score, his last and his best. The first in a projected series of the complete Haydn string quartets. Reliable modern Eulenberg edition, otherwise difficult to obtain. 320pp. 8⅜ x 11¼. (Available in U.S. only)
23753-2 Pa. $8.95

FOURTH, FIFTH AND SIXTH SYMPHONIES IN FULL SCORE, Peter Ilyitch Tchaikovsky. Complete orchestral scores of Symphony No. 4 in F Minor, Op. 36; Symphony No. 5 in E Minor, Op. 64; Symphony No. 6 in B Minor, "Pathetique," Op. 74. Bretikopf & Hartel eds. Study score. 480pp. 9⅜ x 12¼.
23861-X Pa. $10.95

THE MARRIAGE OF FIGARO: COMPLETE SCORE, Wolfgang A. Mozart. Finest comic opera ever written. Full score, not to be confused with piano renderings. Peters edition. Study score. 448pp. 9⅜ x 12¼. (Available in U.S. only)
23751-6 Pa. $12.95

"IMAGE" ON THE ART AND EVOLUTION OF THE FILM, edited by Marshall Deutelbaum. Pioneering book brings together for first time 38 groundbreaking articles on early silent films from *Image* and 263 illustrations newly shot from rare prints in the collection of the International Museum of Photography. A landmark work. Index. 256pp. 8¼ x 11.
23777-X Pa. $8.95

AROUND-THE-WORLD COOKY BOOK, Lois Lintner Sumption and Marguerite Lintner Ashbrook. 373 cooky and frosting recipes from 28 countries (America, Austria, China, Russia, Italy, etc.) include Viennese kisses, rice wafers, London strips, lady fingers, hony, sugar spice, maple cookies, etc. Clear instructions. All tested. 38 drawings. 182pp. 5⅜ x 8.
23802-4 Pa. $2.75

THE ART NOUVEAU STYLE, edited by Roberta Waddell. 579 rare photographs, not available elsewhere, of works in jewelry, metalwork, glass, ceramics, textiles, architecture and furniture by 175 artists—Mucha, Seguy, Lalique, Tiffany, Gaudin, Hohlwein, Saarinen, and many others. 288pp. 8⅜ x 11¼.
23515-7 Pa. $8.95

THE CURVES OF LIFE, Theodore A. Cook. Examination of shells, leaves, horns, human body, art, etc., in "*the* classic reference on how the golden ratio applies to spirals and helices in nature . . . . "—Martin Gardner. 426 illustrations. Total of 512pp. 5⅜ x 8½. 23701-X Pa. **$6.95**

AN ILLUSTRATED FLORA OF THE NORTHERN UNITED STATES AND CANADA, Nathaniel L. Britton, Addison Brown. Encyclopedic work covers 4666 species, ferns on up. Everything. Full botanical information, illustration for each. This earlier edition is preferred by many to more recent revisions. 1913 edition. Over 4000 illustrations, total of 2087pp. 6⅛ x 9¼. 22642-5, 22643-3, 22644-1 Pa., Three-vol. set **$28.50**

MANUAL OF THE GRASSES OF THE UNITED STATES, A. S. Hitchcock, U.S. Dept. of Agriculture. The basic study of American grasses, both indigenous and escapes, cultivated and wild. Over 1400 species. Full descriptions, information. Over 1100 maps, illustrations. Total of 1051pp. 5⅜ x 8½. 22717-0, 22718-9 Pa., Two-vol. set **$17.00**

THE CACTACEAE,, Nathaniel L. Britton, John N. Rose. Exhaustive, definitive. Every cactus in the world. Full botanical descriptions. Thorough statement of nomenclatures, habitat, detailed finding keys. The one book needed by every cactus enthusiast. Over 1275 illustrations. Total of 1080pp. 8 x 10¼. 21191-6, 21192-4 Clothbd., Two-vol. set **$50.00**

AMERICAN MEDICINAL PLANTS, Charles F. Millspaugh. Full descriptions, 180 plants covered: history; physical description; methods of preparation with all chemical constituents extracted; all claimed curative or adverse effects. 180 full-page plates. Classification table. 804pp. 6½ x 9¼. 23034-1 Pa. **$13.95**

A MODERN HERBAL, Margaret Grieve. Much the fullest, most exact, most useful compilation of herbal material. Gigantic alphabetical encyclopedia, from aconite to zedoary, gives botanical information, medical properties, folklore, economic uses, and much else. Indispensable to serious reader. 161 illustrations. 888pp. 6½ x 9¼. (Available in U.S. only) 22798-7, 22799-5 Pa., Two-vol. set **$15.00**

THE HERBAL or GENERAL HISTORY OF PLANTS, John Gerard. The 1633 edition revised and enlarged by Thomas Johnson. Containing almost 2850 plant descriptions and 2705 superb illustrations, Gerard's *Herbal* is a monumental work, the book all modern English herbals are derived from, the one herbal every serious enthusiast should have in its entirety. Original editions are worth perhaps $750. 1678pp. 8½ x 12¼. 23147-X Clothbd. **$75.00**

MANUAL OF THE TREES OF NORTH AMERICA, Charles S. Sargent. The basic survey of every native tree and tree-like shrub, 717 species in all. Extremely full descriptions, information on habitat, growth, locales, economics, etc. Necessary to every serious tree lover. Over 100 finding keys. 783 illustrations. Total of 986pp. 5⅜ x 8½. 20277-1, 20278-X Pa., Two-vol. set **$12.00**

GREAT NEWS PHOTOS AND THE STORIES BEHIND THEM, John Faber. Dramatic volume of 140 great news photos, 1855 through 1976, and revealing stories behind them, with both historical and technical information. Hindenburg disaster, shooting of Oswald, nomination of Jimmy Carter, etc. 160pp. 8¼ x 11. 23667-6 Pa. **$6.00**

CRUICKSHANK'S PHOTOGRAPHS OF BIRDS OF AMERICA, Allan D. Cruickshank. Great ornithologist, photographer presents 177 closeups, groupings, panoramas, flightings, etc., of about 150 different birds. Expanded *Wings in the Wilderness*. Introduction by Helen G. Cruickshank. 191pp. 8¼ x 11. 23497-5 Pa. $7.95

AMERICAN WILDLIFE AND PLANTS, A. C. Martin, et al. Describes food habits of more than 1000 species of mammals, birds, fish. Special treatment of important food plants. Over 300 illustrations. 500pp. 5⅜ x 8½. 20793-5 Pa. **$6.50**

THE PEOPLE CALLED SHAKERS, Edward D. Andrews. Lifetime of research, definitive study of Shakers: origins, beliefs, practices, dances, social organization, furniture and crafts, impact on 19th-century USA, present heritage. Indispensable to student of American history, collector. 33 illustrations. 351pp. 5⅜ x 8½. 21081-2 Pa. $4.50

OLD NEW YORK IN EARLY PHOTOGRAPHS, Mary Black. New York City as it was in 1853-1901, through 196 wonderful photographs from N.-Y. Historical Society. Great Blizzard, Lincoln's funeral procession, great buildings. 228pp. 9 x 12. 22907-6 Pa. $8.95

MR. LINCOLN'S CAMERA MAN: MATHEW BRADY, Roy Meredith. Over 300 Brady photos reproduced directly from original negatives, photos. Jackson, Webster, Grant, Lee, Carnegie, Barnum; Lincoln; Battle Smoke, Death of Rebel Sniper, Atlanta Just After Capture. Lively commentary. 368pp. 8⅜ x 11¼. 23021-X Pa. $11.95

TRAVELS OF WILLIAM BARTRAM, William Bartram. From 1773-8, Bartram explored Northern Florida, Georgia, Carolinas, and reported on wild life, plants, Indians, early settlers. Basic account for period, entertaining reading. Edited by Mark Van Doren. 13 illustrations. 141pp. 5⅜ x 8½. 20013-2 Pa. $6.00

THE GENTLEMAN AND CABINET MAKER'S DIRECTOR, Thomas Chippendale. Full reprint, 1762 style book, most influential of all time; chairs, tables, sofas, mirrors, cabinets, etc. 200 plates, plus 24 photographs of surviving pieces. 249pp. 9⅞ x 12¾. 21601-2 Pa. $8.95

AMERICAN CARRIAGES, SLEIGHS, SULKIES AND CARTS, edited by Don H. Berkebile. 168 Victorian illustrations from catalogues, trade journals, fully captioned. Useful for artists. Author is Assoc. Curator, Div. of Transportation of Smithsonian Institution. 168pp. 8½ x 9½. 23328-6 Pa. $5.00

ILLUSTRATED GUIDE TO SHAKER FURNITURE, Robert Meader. Director, Shaker Museum, Old Chatham, presents up-to-date coverage of all furniture and appurtenances, with much on local styles not available elsewhere. 235 photos. 146pp. 9 x 12. 22819-3 Pa. $6.95

COOKING WITH BEER, Carole Fahy. Beer has as superb an effect on food as wine, and at fraction of cost. Over 250 recipes for appetizers, soups, main dishes, desserts, breads, etc. Index. 144pp. 5⅜ x 8½. (Available in U.S. only) 23661-7 Pa. $3.00

STEWS AND RAGOUTS, Kay Shaw Nelson. This international cookbook offers wide range of 108 recipes perfect for everyday, special occasions, meals-in-themselves, main dishes. Economical, nutritious, easy-to-prepare: goulash, Irish stew, boeuf bourguignon, etc. Index. 134pp. 5⅜ x 8½. 23662-5 Pa. $3.95

DELICIOUS MAIN COURSE DISHES, Marian Tracy. Main courses are the most important part of any meal. These 200 nutritious, economical recipes from around the world make every meal a delight. "I . . . have found it so useful in my own household,"—*N.Y. Times*. Index. 219pp. 5⅜ x 8½. 23664-1 Pa. $3.95

FIVE ACRES AND INDEPENDENCE, Maurice G. Kains. Great back-to-the-land classic explains basics of self-sufficient farming: economics, plants, crops, animals, orchards, soils, land selection, host of other necessary things. Do not confuse with skimpy faddist literature; Kains was one of America's greatest agriculturalists. 95 illustrations. 397pp. 5⅜ x 8½. 20974-1 Pa. $4.95

A PRACTICAL GUIDE FOR THE BEGINNING FARMER, Herbert Jacobs. Basic, extremely useful first book for anyone thinking about moving to the country and starting a farm. Simpler than Kains, with greater emphasis on country living in general. 246pp. 5⅜ x 8½. 23675-7 Pa. $3.95

PAPERMAKING, Dard Hunter. Definitive book on the subject by the foremost authority in the field. Chapters dealing with every aspect of history of craft in every part of the world. Over 320 illustrations. 2nd, revised and enlarged (1947) edition. 672pp. 5⅜ x 8½. 23619-6 Pa. $8.95

THE ART DECO STYLE, edited by Theodore Menten. Furniture, jewelry, metalwork, ceramics, fabrics, lighting fixtures, interior decors, exteriors, graphics from pure French sources. Best sampling around. Over 400 photographs. 183pp. 8⅜ x 11¼. 22824-X Pa. $6.95

ACKERMANN'S COSTUME PLATES, Rudolph Ackermann. Selection of 96 plates from the *Repository of Arts,* best published source of costume for English fashion during the early 19th century. 12 plates also in color. Captions, glossary and introduction by editor Stella Blum. Total of 120pp. 8⅜ x 11¼. 23690-0 Pa. $5.00

THE ANATOMY OF THE HORSE, George Stubbs. Often considered the great masterpiece of animal anatomy. Full reproduction of 1766 edition, plus prospectus; original text and modernized text. 36 plates. Introduction by Eleanor Garvey. 121pp. 11 x 14¾. 23402-9 Pa. **$8.95**

BRIDGMAN'S LIFE DRAWING, George B. Bridgman. More than 500 illustrative drawings and text teach you to abstract the body into its major masses, use light and shade, proportion; as well as specific areas of anatomy, of which Bridgman is master. 192pp. 6½ x 9¼. (Available in U.S. only) 22710-3 Pa. **$4.50**

ART NOUVEAU DESIGNS IN COLOR, Alphonse Mucha, Maurice Verneuil, Georges Auriol. Full-color reproduction of *Combinaisons ornementales* (c. 1900) by Art Nouveau masters. Floral, animal, geometric, interlacings, swashes—borders, frames, spots—all incredibly beautiful. 60 plates, hundreds of designs. 9⅜ x 8-1/16. 22885-1 Pa. **$4.50**

FULL-COLOR FLORAL DESIGNS IN THE ART NOUVEAU STYLE, E. A. Seguy. 166 motifs, on 40 plates, from *Les fleurs et leurs applications decoratives* (1902): borders, circular designs, repeats, allovers, "spots." All in authentic Art Nouveau colors. 48pp. 9⅜ x 12¼. 23439-8 Pa. **$6.00**

A DIDEROT PICTORIAL ENCYCLOPEDIA OF TRADES AND INDUSTRY, edited by Charles C. Gillispie. 485 most interesting plates from the great French Encyclopedia of the 18th century show hundreds of working figures, artifacts, process, land and cityscapes; glassmaking, papermaking, metal extraction, construction, weaving, making furniture, clothing, wigs, dozens of other activities. Plates fully explained. 920pp. 9 x 12. 22284-5, 22285-3 Clothbd., Two-vol. set **$50.00**

HANDBOOK OF EARLY ADVERTISING ART, Clarence P. Hornung. Largest collection of copyright-free early and antique advertising art ever compiled. Over 6,000 illustrations, from Franklin's time to the 1890's for special effects, novelty. Valuable source, almost inexhaustible.
Pictorial Volume. Agriculture, the zodiac, animals, autos, birds, Christmas, fire engines, flowers, trees, musical instruments, ships, games and sports, much more. Arranged by subject matter and use. 237 plates. 288pp. 9 x 12. 20122-8 Clothbd. **$15.00**

Typographical Volume. Roman and Gothic faces ranging from 10 point to 300 point, "Barnum," German and Old English faces, script, logotypes, scrolls and flourishes, 1115 ornamental initials, 67 complete alphabets, more. 310 plates. 320pp. 9 x 12. 20123-6 Clothbd. $15.00

CALLIGRAPHY (CALLIGRAPHIA LATINA), J. G. Schwandner. High point of 18th-century ornamental calligraphy. Very ornate initials, scrolls, borders, cherubs, birds, lettered examples. 172pp. 9 x 13. 20475-8 Pa. **$7.95**